Entrepreneurship and Skill Development in Horticultural Processing

About the Editors

Prof. (Dr.) K.P. Sudheer, obtained his B Tech Agricultural Engineering degree from Kerala Agricultural University (KAU), M Tech Agricultural Processing from Tamil Nadu Agricultural University(TNAU), PhD Agricultural Engineering from IARI New Delhi and Post doctorate from KU Leuven Belgium. Prof. Sudheer started his career as scientist in Agro Processing Division, CIAE (ICAR), Bhopal. He is presently working as ICAR National Fellow and Project Coordinator of Centre of Excellence in Post-harvest Technology, College of Horticulture, Vellanikkara, KAU.

Professor Sudheer is actively involved in teaching various subjects for B Tech Food Engineering, B Tech Agricultural Engineering, M. Tech Agricultural Processing and Food Engineering, Ph D Agricultural Processing and Food Engineering, and in research programmes under AICRP on Post-harvest Engineering & Technology. He is also guiding Doctoral and Masters research scholars and has many external aided research projects funded by Ministry of Food Processing Industries, Kerala State Council for Science, Technology & Environment, Ministry of Rural Development, NABARD, Food Corporation of India, Government of Kerala. Prof. Sudheer is also acting as the Project Coordinator of Food and Agricultural Process Engineering research group in KAU. He has 150 research papers in reputed national & international journals, and proceedings of seminar/workshops. He authored five text books and many bulletins in the field of Post-harvest Technology. Prof. Sudheer is the recipient of prestigious Normal E Borlaug Fellowship by USDA. He has also received many international and national fellowships for his research programmes including NUFFIC Fellowship from Netherlands, VLIR-UDC fellowship from Belgium, CINADCO fellowship from Israel, ERASMUS MUNDUS Fellowship from Sweden. Prof. Sudheer has received the Krishi Vigyan Award - 2015 for the best Agricultural Scientist, from the Government of Kerala.

Prof. (Dr.) V. Indira, took her Master's degree in Foods and Nutrition from Sri. Avinashilingam Home Science College for Women, Coimbatore in 1978. She joined KAU as Junior Assistant Professor of Nutrition in the Department of Processing Technology, College of Horticulture in 1979. In 1993, she took her Ph.D. in Foods and Nutrition from KAU. She was former Professor & Head, Department of Home Science, College of Horticulture, Vellanikkara.

Dr. Indira, had extensive teaching, research and extension experience in KAU. She handled UG courses in food and nutrition for Agriculture, Dairy Science and Technology, and Veterinary and Animal Science students, and M. Sc. and Ph.D. courses for the students of Department of Home Science, College of Horticulture, KAU, Vellanikkara. She implemented two externally aided projects and guided four Ph.D and 24 M. Sc. students in the Department of Home Science as major advisor. She organized various training programmes in fruit and vegetable preservation, mushroom cultivation, food and nutrition. She has in her credit about 50 research publications and many bulletins and books.

Entrepreneurship and Skill Development in Horticultural Processing

K.P. Sudheer *(ICAR National Fellow)*

Professor & Head, Department of Agricultural Engineering
Project Coordinator, Centre of Excellence in Post-harvest Technology
College of Horticulture, Vellanikkara
Kerala Agricultural University
Thrissur, Kerala- 680 656

V. Indira

Former Professor & Head
Department of Home Science
College of Horticulture, Vellanikkara
Kerala Agricultural University
Thrissur, Kerala- 680 656

CRC Press
Taylor & Francis Group
Boca Raton London New York

CRC Press is an imprint of the
Taylor & Francis Group, an **informa** business

NEW INDIA PUBLISHING AGENCY
New Delhi – 110 034

First published 2022
by CRC Press
4 Park Square, Milton Park, Abingdon, Oxon, OX14 4RN

and by CRC Press
6000 Broken Sound Parkway NW, Suite 300, Boca Raton, FL 33487-2742

© 2022 selection and editorial matter, NIPA; individual chapters, the contributors

CRC Press is an imprint of Taylor & Francis Group, an Informa business

Print edition not for sale in South Asia (India, Sri Lanka, Nepal, Bangladesh, Pakistan or Bhutan).

British Library Cataloguing-in-Publication Data
A catalogue record for this book is available from the British Library

Library of Congress Cataloging-in-Publication Data
A catalog record has been requested

ISBN: 978-1-032-15893-8 (hbk)
ISBN: 978-1-003-24613-8 (ebk)

DOI: 10.1201/9781003246138

Dedicated to
Budding Entrepreneurs

Prof. (Dr) P. Rajendran
Vice Chancellor
Kerala Agricultural University

Foreword

Horticultural Processing is one of the world's largest and important industries feeding the world population. Though a global leader in horticultural produces, India is yet to tap the true potential of this sector and convert it into an engine of growth. Horticultural processing is a sunrise sector in view of its large potential for growth and likely socio-economic impact, specifically on employment and income generation. It is heartening to observe that more and more entrepreneurs are taking up horticulture processing as a commercial venture. At the same time, though the production is high, the proportion that is subjected to processing is meagre and post-harvest losses are also too huge. Innovations in processing methods and processing ventures on horticultural crops are the need of the hour.

Concurrently there is also an urgent need to extend government/financial support to set up economically viable processing units in the production catchments. These model units should be equipped with latest technologies and skills by frequently organizing different entrepreneurial/motivational and skill oriented workshops/trainings.

The publication entitled, "Entrepreneurship and Skill Development in Horticultural Processing" provides valuable information in the area of processing and value addition of horticultural crops. The nineteen chapters in post-harvest engineering and technology of various horticultural produces penned by the academicians and researchers in respective fields are comprehensive and educative. The publication also dwells on the latest innovations in the horticultural processing sector which will aid entrepreneurship.

I appreciate the authors for bringing out such a useful publication at a time when the Indian food industry and academia are looking forward for information and inspiration. This book will be a valuable document for entrepreneurs, students, researchers, teachers and all stake holders in the area of food processing.

I compliment Prof (Dr.) K.P. Sudheer, ICAR National Fellow, Professor & Head, Department of Agricultural Engineering and Dr.V. Indira, Former Professor & Head, Department of Home Science, College of Horticulture, Vellanikkara, for their professional leadership and efforts to bring out this publication.

Thrissur

Prof. (Dr) P. Rajendran
Vice Chancellor
Kerala Agricultural University

Preface

India has become the second largest producer of fruits and vegetables in the world. Considering the availability of wide raw material base that the country offers, along with the consumers of over one billion people, the processing industry holds tremendous opportunities for large investment. It is imperative to process and preserve the available fruits and vegetables in seasons of plenty not only to avoid wastage but also to confirm that it is available during off seasons and at reasonable price. New business ventures for fruit and vegetable processing sector must be established to reduce the wastage and ensure the availability during off-season.

The fruit and vegetable processing industry in the country is in a nascent and primitive stage. Number of establishments in the organized sector is far too few compared with several developed and developing nations. The technologies adopted for processing, preservation and value addition in medium, small and micro industries are primitive and outdated. A prerequisite for adapting to newer technologies is lack of awareness about recent developments in the areas of post-harvest technology and food engineering. This book investigates the present status and future trends of some existing, new and emerging food preservation technologies and their potential application in processing ventures.

The book mainly comprises of novel food processing techniques and the equipment requirement for installation of such system in place. The book also provides the scope and opportunities of entrepreneurship in the major horticultural crops like banana, mango, pine-apple, cashew, cocoa, coconut, spices, tuber crops, and some under-utilized fruits and vegetables. The book also enlightens the readers about the marketing strategies, business plan preparation, safety and quality issues etc., which are essential for starting a food enterprise. It covers almost all important aspects of entrepreneurship development in food processing sector. Its special feature is the treatment given to the horticultural processing and entrepreneurship in an integrated manner. The authorship of various chapters comes from professors, scientists and research scholars who have compiled the scattered information from different sources at one place.

The readership of the book would vest in the academicians, researchers, students, industrialists and those engaged in the profession. We are sure that this book will lead to more scientific and technological approach to entrepreneurship in food processing sector. This approach can considerably reduce the post-harvest losses, add value to these commodities and increase the profit of farmers.

We acknowledge from the core of our heart to all the contributors of this book who have put-in a lot of scholarly efforts in bringing out a state-of-art technology. Their sincere effort made towards collection of information from literature consulted during the preparation of the manuscript is greatly appreciated. With deep sense of pride and dignity, authors express heartfelt sense of gratitude and regards to Dr. K. Alagusundaram, Deputy Director General (Engg), ICAR, New Delhi, Prof.(Dr.)P. Rajendran Vice-Chancellor, Kerala Agricultural University, Dr. P. Indira Devi, Director of Research, KAU, Dr. Jiju P Alex, Director of Extension, KAU, Dr. George Thomas C, Associate Dean, College of Horticulture, and Dr. K.P. Visalakshi, Former Head, Department of Agricultural Engineering, College of Horticulture, KAU with whose guidance, scientific knowledge, constructive criticism and constant encouragement, we have been able to publish this manuscript.

The purpose of writing this book would be achieved if it could present the essence of entrepreneurship in the right perspective to the concerned stake holders. There is no limit to perfection, hence constructive criticism is welcome for its improvement in future.

Prof. (Dr.). K. P. Sudheer
Prof. (Dr.). V. Indira

Contents

15. Hand Holding Entrepreneurs in Honey Processing and Value Addition 323

Mani Chellappan

16. Value Addition in Mushroom Processing- A Potential Sector for Budding Entrepreneurs 345

Lakshmy P S, Arun Prasath V, & Suman K T

17. Importance of Quality and Safety of Processed Products 367

Sudheer K P & Indira V

18. Marketing Strategies and Supply Chain of Horticultural
Products 395

Ranjit Kumar E G

List of Contributors

1. **Anjineyulu Kothakota,** Department of Food & Agricultural Process Engineering Kelappaji, College of Agricultural Engineering & Technology, Tavanur, Kerala Agricultural University, Kerala-679 573

2. **Arun Prasath V,** Department of Food & Agricultural Process Engineering, Agricultural Engineering College & Research Institute, Tamil Nadu Agricultural University Coimbatore-641 003

3. **Binoo P Bonny,** Professor & Head, Communication Centre, Kerala Agricultural University, Mannuthy, Thrissur, Kerala-680 651

4. **Geethalekshmi P R**, Assistant Professor, Department of Post-harvest Technology, College of Agriculture, Kerala Agricultural University, Vellayani, Thiruvananthapuram Kerala -695 522

5. **Hebbar K B,** Head & Principal Scientist, Physiology, Biochemistry & Post-harvest Technology, Central Plantation Crops Research Institute, Kasaragode, Kerala-671 124

6. **Indira V.,** Former Professor & Head, Department of Home Science, College of Horticulture Vellanikkara, Kerala Agricultural University, Thrissur, Kerala- 680 656

7. **Jayashree E,** Principal Scientist, Indian Institute of Spices Research, Moozhikkal Calicut, Kerala -673 012

8. **Jiju P Alex,** Director of Extension, Kerala Agricultural University, Mannuthy, Thrissur, Kerala 680 651

9. **Jyothi A N,** Principal Scientist, Crop Utilization Division, Central Tuber Crops Research Institute, Sreekariyam, Thiruvananthapuram, Kerala-695 017

10. **Kalpana Rayaguru,** Associate Professor, Department of Agril Processing and Food Engg., College of Agricultural Engineering, Orissa University of Agriculture and Technology, Bhubaneswar- 751 003, Odisha

11. **K P, Sudheer,** Professor & Head, ICAR National Fellow, Project Coordinator, Centre of Excellence in Post-harvest Technology, Department of Agricultural Engineering, College of Horticulture, Vellanikkara, Kerala Agricultural University, Thrissur, Kerala- 680 656

12. **Lakshmy P S,** Consultant (Food and Nutrition), Sreelakshmy, Kuttath Lane, Koorkanchery, Thrissur, Kerala- 680 007

13. **Mani Chellappan,** Professor & Head, Department of Entomology, College of Horticulture, Vellanikkara, Thrissur, Kerala Agricultural University, Thrissur-680 656

14. **Manikantan M R,** Senior Scientist, Physiology, Biochemistry & Post-harvest Technology, Central Plantation Crops Research Institute, Kudlu (PO), Kasaragode, Kerala- 671 124

15. **Manjunath Shetty**, Research Scholar, Department of Post-harvest Technology, College of Agriculture, Kerala Agricultural University, Vellayani, Thiruvananthapuram, Kerala-695522.

16. **Mathew A C,** Principal Scientist, Physiology, Biochemistry & Post -harvest Technology, Central Plantation Crops Research Institute, Kasaragode, Kerala-671124.

17. **Mini C,** Professor & Head, Department of Post-harvest Technology, College of Agriculture, Kerala Agricultural University, Vellayani, Thiruvananthapuram, Kerala -695 522

18. **Minimol J S,** Assistant Professor, Cocoa Research Station, Kerala Agricultural University, Vellanikkara,Thrissur, Kerala -680656

19. **Padmaja G,** Principal Scientist & Head (Rtd), Crop Utilization Division, Central Tuber Crops Research Institute, Sreekariyam, Thiruvananthapuram, Kerala-695 017

20. **Pandiselvam R,** Scientist, Physiology, Biochemistry & Post-harvest Technology, Central Plantation Crops Research Institute, Kasaragode, Kerala-671 124

21. **Preethi Manniledam,** Manager, Technology Business Incubator, National Institute of Technology, Calicut, Kerala-673 601

22. **Radha Ramanan T,** Assistant Professor, Mechanical Engineering, School of Management Studies, National Institute of Technology, Calicut, Kerala-673 601

23. **Ranjit Kumar E G,** Professor & Director, MBA (AGM), CCBM, Vellanikkara, Kerala Agricultural University, Kerala- 680 656

24. **Ravindra Naik,** Principal Scientist, Central Institute of Agricultural Engineering, Regional Centre, Coimbatore, Tamil Nadu-641 007

25. **Sagarika N,** PhD Scholar, Department of Food Process Engineering, College of Food Processing Technology & Bio Energy, Anand Agricultural University, Anand-388110, Gujarat

26. **Sajeev M S,** Principal Scientist, Crop Utilization Division, Central Tuber Crops Research Institute, Sreekariyam, Thiruvananthapuram, Kerala-695017

27. **Sangeetha K Prathap,** Assistant Professor, School of Management Studies, Cochin University of Science & Technology, Cochin, Kerala -682022

28. **Sanjaya K Dash,** Dean, College of Agricultural Engineering, Orissa University of Agriculture and Technology, Bhubaneswar-751 003, Odisha

29. **Shameena Beegum,** Scientist, Physiology, Biochemistry and Post-harvest Technology Division, Central Plantation Crops Research Institute, Kudlu (PO), Kasaragode, Kerala-671 124

30. **Sheriff J T,** Principal Scientist & Head, Crop Utilization Division, Central Tuber Crops Research Institute, Sreekariyam, Thiruvananthapuram, Kerala-695017

31. **Sobhana A,** Professor & Head, Cashew Research Station, Madakkathara, Kerala Agricultural University, Vellanikkara, Thrissur, Kerala-680656

32. **Suma B,** Professor & Head, Cocoa Research Station, Kerala Agricultural University, Vellanikkara, Thrissur, Kerala-680 656

33. **Suman K T,** Assistant Professor, Krishi Vigyan Kendra, Vellanikkara, Thrissur, Kerala-680 656

1

An Introduction to Entrepreneurial Opportunities in Horticultural Processing Sector

Binoo P Bonny & Sudheer K P

Introduction

Indian agriculture is in a phase of transition from subsistence to high-tech agribusiness. This has ushered in an era where entrepreneurship is viewed as a strategic intervention in rural development. As a result, promotion of rural enterprises has gained great significance as an instrument for improved farm earnings, employment and women empowerment. It implies development of agriculture through farm diversification, value addition and development of agro-processing industries that provides autonomy, independence and reduced need for social support to the farmers. Improved farm-industry linkages along with great export potential will be added advantages. In this development scenario, agriculture, specifically high value horticultural crops like fruits, vegetables, spices, flowers, plantation crops, medicinal and aromatic plants among others that ensure maximum returns to the growers with multiple scopes for value addition hold great promise. In fact, horticultural crops become a necessary condition for improving the living standards of farmers and ensuring nutritional security of the nation. The concept of nutritional security goes beyond food security to include adequate availability of micronutrients such as vitamins and minerals in addition to calorie and protein needs of the people supplied through food grains. Therefore, the spectacular increase in food grain production and a reasonable overall agricultural performance has failed to eradicate malnutrition as can be inferred from Table 1. It indicates that the Global Hunger Index (GHI), a more comprehensive tool used to measure and track hunger, ranks India (97 out of 118 developing countries) in the category of serious hunger behind Bangladesh (90/118). More disturbing are the rates of growing malnutrition among Indian households. As such, nutritional security becomes a more relevant indicator of food security as we move up the development ladder

Table 1: Global Hunger Index (GHI) and its three components of India in comparison with neighboring countries

Country	Proportion of undernourished in the population (%)		Prevalence of under weight children below five years (%)		U₅ Mortality Rate (per 1000 live birth)		GHI score	
	2010	2015	2003-08	2014	2010	2015	2008	2016
India	16	15	43.5	43.5	60	48	36	28.5 (97)
Bangladesh	17	16	41.3	36.8	50	38	32.4	27.1 (90)
China	13	9	6.0	3.4	16	11	11.5	7.7 (29)
Pakistan	22	22	25.3	30.9	92	81	35.1	33.4 (107)

Figure in parenthesis is GHI rank out of 118 countries

and increased availability of fruits and vegetables which are rich sources of micronutrients and health-giving phytochemicals like antioxidants and detoxifying agents become more critical. In line with this, a major shift in consumption pattern to fresh and processed fruits and vegetables is also evident that holds great promise for huge influx of investment in the horticulture sector with diversification and value addition holding the premium.

Competitive advantage of India

As food grain production saturated markets, a trend towards diversification set in that lead to the emergence of Horticulture as a core sector with a crop coverage of 21 million ha and an annual production of 240.5 million tones. Potential to produce a wide variety of horticultural crops under varied agro-climatic conditions is often expounded as the strength of Indian agriculture. Diversity of Indian horticultural crops is evident from the fact that it produces about 70 varieties of vegetables and equally diverse varieties of fruits. It is home to 70% of tropical and sub-tropical and 30% of temperate types of vegetables and fruits. The importance of horticultural sector in Indian economy can be visualized from the fact that it accounts for 30.4% of India's agricultural GDP that too from 11% of cropped area. Changes in pattern of domestic demand and, to some extent, export demand in the wake of trade liberalisation, resulted in changes in resource use and increasing diversification of enterprises.

However, it is often recognized that a strong link between farmers and consumers is essential to materialise the full benefits of agriculture reforms. National Horticulture Mission has made some progress in respect of this by increasing the production along with integration of technologies to contain post harvest losses and marketing. It provides opportunities for the educated farmers and unemployed rural youth to harness the business potential of the sector. But, there is a need to remove the distortions in the present supply chain and create better integration between different links of the supply chain so that horticulture can be developed into an important source of gainful employment. Entrepreneurship stands as a vehicle to improve this process through efficient assembling and harnessing of various inputs, bearing the risks, innovating and adopting good agricultural practices (GAP) of production to reduce the cost and increase the quality and quantity of produce, expanding the horizons of the market and co-ordinating and managing the manufacturing unit at various levels.

Increased opportunities and World Trade Organization (WTO) agreements have necessitated the need to bring agricultural development to the global competitive level that opened new vistas for agri-business in India. It has given rise to a New Agricultural Policy of India that accorded agriculture the status of an industry. By now, the signs of agricultural development are perceptible in marked shift from staples to cash crops. Strong agricultural production base and accelerated economic growth holds a significant potential for the food processing industry that provides a strong link between agriculture and consumers. But, the industry in India remains in a nascent and primitive stage with only a few numbers in the organized sector compared with developed and developing nations. The technologies adopted by these medium, small and micro industries for processing, preservation and value addition are also primitive and outdated. Lack of awareness about recent developments in the areas of post harvest technology and food engineering hinder the process of adopting newer technologies. Hence, concerted efforts were involved at all levels to boost entrepreneurship development in the sector.

History of entrepreneurship development in India

Small and medium scale enterprises has been recognized for its employment generation potential from early sixties and promotional packages in the form of financial help and incentives, infrastructural facilities, and technical and managerial guidance provided through various supporting organizations of the Central, State and local levels. But, experience indicated that these were not sufficient to solicit adequate response from the entrepreneurs and it was realized that emphasis on human development is a necessary condition for entrepreneurship development. Concerted efforts on entrepreneurship development in India actually started with the establishment of Small Industry Extension and Training Institute (SIET) in 1962, now called the National Institute of Small Industry Extension Training (NISIET), at Hyderabad. SIET pioneered the entrepreneurship development research of India in collaboration with Prof. David C. McClelland of Harvard University. His work popular as the *Kakinada Experiment* proved that, through proper education and training, vital quality of an entrepreneur, which McClelland called as the 'need for achievement' (n' ach) can be developed. It made people appreciate the need for and importance of entrepreneurial training, now popularly known as EDPs (Entrepreneurship Development Programmes), to induce motivation and competence among the young prospective entrepreneurs.

Gujarat Industrial Investment Corporation (GIIC) was the first to start a three-month training programme on entrepreneurship development in 1970. This was followed by the establishment of North Eastern Council (NEC) in 1972, a major initiative to foster economic development in North East India. North Eastern Industrial and Technical Consultancy Organization, (NEITCO) and Entrepreneurial Motivation Training Centres (EMTCs) in District headquarters of Assam were also established to impart training on entrepreneurship development.

Encouraging results of the EDPs organised under SIET, Small Industry Development Organisation (SIDO) through Small Industry Services Institute (SISI), Industrial Development Bank of India (IDBI) and Technical Consultancy Organisations (TCOs) culminated to the establishment of Centre for Entrepreneurship Development (CED), Ahmedabad in 1979. CED, Ahmedabad was the first centre of its kind wholly committed to the cause of entrepreneurship development. CED, Ahmedabad was successful in inspiring the national level financial institutions such as IDBI, IFCI, ICICI and SBI along with the Gujarat Government to sponsor a National Resource Organisation, called Entrepreneurship Development Institute of India (EDI), Ahmedabad, in 1983. Ever since its inception EDI has successfully discharged the entrusted responsibilities of extension and institutionalization of entrepreneurship development activities in the country.

Government of India also established the National Institute for Entrepreneurship and Small Business Development (NIESBUD) in 1983 for coordinating entrepreneurship development activities in the country. Subsequently, State Governments of Bihar, Goa, Gujarat, Himachal Pradesh, Jammu & Kashmir, Karnataka, Kerala, Madhya Pradesh, Maharashtra, Odisha, Tamil Nadu, and Uttar Pradesh established state-level Centre for Entrepreneurship Development (CED) or Institute of Entrepreneurship Development (IED) with the support from national level financial institutions. According to the study of NIESBUD, there are some 686 organisations involved in conducting EDPs in the country.

Policy support and promotion of innovations

India is in a phase of economic transition where entrepreneurship development holds the centre stage with flagship government programmes like *Start up India* and *Make in India* among others. This phase of economy is best explained in terms of Nee's (1989) market transition theory. It places entrepreneurship and private economy at the centre of transforming economies as illustrated by the case of China. He conjectures that economic reform is a process in which markets replace political authority in the distribution of resources and predicts that human capital factors gradually replace political factors as determinants of socioeconomic success. Accordingly, an entrepreneur has been conceptualized as one who is proactive through his visionary objectives and begins with the outcome in mind backed by a positive and unyielding determination following sequential structure of activities in a systematic way from the very start of the business enterprise.

Based on this conceptualisation, EDPs were designed to unleash the talent of potential entrepreneurs and establishment of profitable small-micro enterprises, and its management. It is in this context, agricultural enterprises especially food processing sector gained phenomenal role in the future of our country. It became imperative to understand and enhance the process of agricultural innovations and technologies that can influence the present and future agricultural development in India. Inclusive Market-Oriented Development (IMOD) approach was stressed so that transformation of impoverished subsistence farming to prosperous market orientation can be ensured. IMOD is a cycle in which value-adding innovations (technical, policy, institutional and others) enable the poor to capture larger rewards from markets, while managing their risks. Harnessing markets specifically to benefit the poor, designing innovations to enable the farmers to become a part of the value chain through multi-level stakeholder partnerships and managing the risks form the essential elements of IMOD. Farmers become part of this link by engaging them as entrepreneurs through small- and medium-term enterprises (SMEs). Strategies and Support Mechanisms developed to support the prospective entrepreneurs include business incubation, post harvest technologies and product development, food safety, and funding.

Government of India has also established food parks and cold chains in different parts of the country to help small and medium entrepreneurs who find it difficult to invest in capital intensive facilities. Keeping this in mind, Government has initiated technology business incubation units to promote entrepreneurship and agro- industry which will open the vistas of incubation landscape to the micro segment of the vast rural economy.

Mega food parks

The Scheme of mega food park aims to bring together farmers, processors and retailers so as to ensure maximizing value addition, minimizing wastage, increasing farmers' income and creating employment opportunities by providing a mechanism to link agricultural production to the market particularly in rural sector. The Scheme is based on cluster approach and envisages a well-defined agri/horticultural-processing zone containing state-of-the art processing facilities with support infrastructure and well-established supply chain. Mega food parks functional under the scheme till mid 2017 are Patanjali Food and Herbal Park, Haridwar; Srini Food Park, Chittoor; North East Mega Food Park, Nalbari; International Mega Food Park, Fazilka; Integrated Food Park,Tumkur; Jharkhand Mega Food Park, Ranchi; Indus Mega Food Park, Khargoan; Jangipur Bengal Mega Food Park, Murshidabad and MITS Mega Food Park Pvt Ltd, Rayagada. It is expected that on an average, each project will have around 30-35 food processing units with a collective investment of Rs.250 crores that would eventually lead to an annual turnover of about Rs. 450-500 crores and creation of direct and indirect employment to the extent of about 30,000 persons.

Project components

The scheme aims to facilitate the establishment of a strong food processing industry backed by an efficient supply chain, which includes Collection Centres, Primary Processing Centres (PPC), Central Processing Centres (CPC) and Cold Chain Infrastructure. Illustration of the Mega Food Park scheme is depicted as Fig.1.

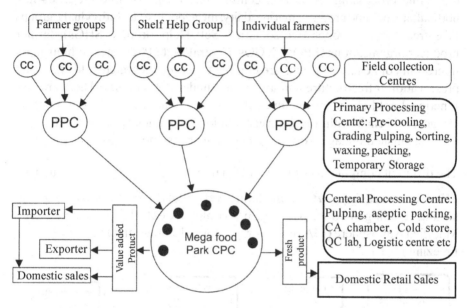

Fig. 1: Components of the Mega Food Park scheme (*Source*: mofpi.nic.in)

Collection Centres and Primary Processing Centres (*CC &PPC*): These components have cleaning, grading, sorting and packing facilities, dry warehouses, specialized cold stores including pre-cooling chambers, ripening chambers, reefer vans, mobile pre-coolers, mobile collection vans etc.

Central Processing Centres (*CPC*): Includes common facilities like testing laboratory, cleaning, grading, sorting and packing facilities, dry warehouses, specialized storage facilities including controlled atmosphere chambers, pressure ventilators, variable humidity stores, pre-cooling chambers, ripening chambers, cold chain infrastructure including reefer vans, packaging unit, irradiation facilities, steam sterilization units, steam generating units, food incubation cum development centres etc.

Land requirement: Around 50- 100 acres of land is required for establishing the CPC, though the actual requirement of land would depend upon the business plan and varies from region to region. Land required for setting up of PPCs and CCs would be in addition to this.

Cold chain, value addition and preservation infrastructure scheme

Objective of the scheme is to provide integrated cold chain and preservation infrastructure facilities without any break from the farm gate to the consumer. It include pre-cooling facilities at production sites, reefer vans, mobile cooling units as well as value addition centres with infrastructural facilities like processing/ multi-line processing/ collection centres, etc. for perishable produces like horticulture, organic produce, marine, dairy, meat, poultry etc. Individuals, groups of entrepreneurs, cooperative societies, Self Help Groups (SHGs), farmer producer organisations (FPOs), NGOs, Central/State PSUs etc. with business interest in cold chain solutions are eligible to set up integrated cold chain and preservation infrastructure and avail grant under the scheme. There are 236 functional cold chains as on June 2017 in the country. An illustration of an efficient cold chain with storage, transportation and minimal processing facilities with established linkage from farm gate to the consumer, is given as Fig.2.

The different components of the Cold Chain projects are discussed as under:

Minimal processing centre at the farm level will have facilities for weighing, sorting, grading waxing, packing, pre-cooling, Control Atmosphere (CA)/ Modified Atmosphere (MA) cold storage, normal storage and Individual Quick Freezing (IQF).

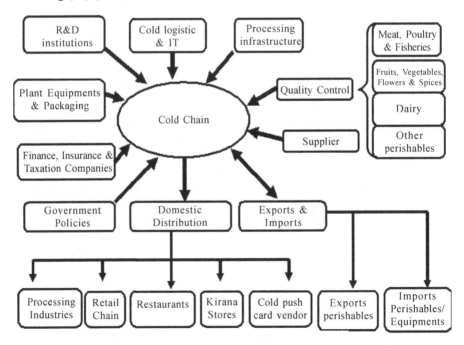

Fig. 2: Components of the Cold Supply Chain Scheme (*Source*: mofpi.nic.in)

Mobile pre-cooling vans and reefer trucks: These facilities enable smooth transportation with minimal loss and wastes.

Distribution hubs: These are equipped with multi products and multi Control Atmosphere (CA)/ Modified Atmosphere (MA) chambers/ cold storage/ variable humidity chambers, packing facility, Cleaning in Process (CIP) fog treatment, Individual Quick Freezing (IQF) and blast freezing.

Irradiation facility: Quarantine and other phytosanitary requirements can be assured using the facility.

Agri Business Incubator

Agri business incubators (ABI) are powerful economic development tool to promote entrepreneurship development in food processing sector. They promote growth through innovation and application of technology, support economic development strategies for small business development, and encourage growth from within local economies, while also providing a mechanism for technology transfer. The ABI's would primarily focus on those technologies which needs support for commercialization and further proliferation. These can act as a growth driver in the low end spectrum of the incubation eco-system. The components under ABI include mentoring support in business and technology plans, entrepreneurship cum skill development, identification of appropriate technology, hands on experience on processing machineries, product development (process protocols for various value added products), project report preparation, marketing assistance, professional assistance to make the enterprise successful and achieve higher growth.

The first Agri-Business Incubation (ABI) Program in India was launched at ICRISAT, in 2003. It was envisaged as an initiative of the International Crops Research Institute for the Semi-Arid Tropics (ICRISAT) in partnership with India's Department of Science and Technology (DST). ABI aimed at promoting agricultural technologies developed by ICRISAT, other R&D centres of excellence, universities, and other institutions, separately and jointly. Its approach features a dual service and outreach strategy. The service strategy focuses business development on five strategic areas, building on the expertise of ICRISAT and its partners viz. seed, biofuels, ventures to develop particular innovations (products or services), farming (high-value crops), and agricultural biotechnology. The outreach strategy involves collaborative business incubation to bring a wider range of expertise and resources to bear on business development to foster agricultural development in other regions.

A recent survey found that their numbers had grown from 10 in 2000 to 30 business incubators and science and technology parks involved in the commercialization of software and other engineering technologies in 2009 (NSTEDB and ISBA 2009). Of the 495 ventures that graduated from the business incubators in India, 387 remained in business. More than 10,000 jobs were created through these ventures.

Objectives of Agri Business Incubator

The ABI facility envisages design in agri-market-oriented development plan that seeks to improve farmers' livelihoods. The potential of rural food processing industry to tackle this challenge is yet to be fully exploited. Employment is much higher in the food sector than any other sector. Therefore, role of ABI becomes vital for a rapid transformation of the rural economy in India and aims to set a benchmark in the field of food processing, to check post harvest losses. ABI has effective and cost-attractive processing systems for agricultural commodities, particularly fruits and vegetables, spices, coconut and rice. The main objectives of the Agri business Incubator are as follows:

- To facilitate creation of agri-business enterprises through technology development and commercialisation

- Development and establishment of standardization protocols for commercial production of food products *viz.* microencapsulated spice and fruit powders, ready to cook/eat food products, extruded snacks, vacuum fried fruit and vegetable products

- Development and establishment of business plan models for food processing industries based on developed technologies/products

- Design and development of gender friendly processing equipment to cater the needs of micro- small and medium scale processing industries

- Conducting demonstrations and trainings to potential food processing entrepreneurs

KAU Agri Business Incubator

The state funded centre established at Kerala Agricultural University has developed many innovative process protocols and food processing machineries to cater the needs of emerging food processing sector. The KAU ABI has provided entrepreneur support to several food processing industries. The centre has also contributed towards the design and development of women friendly/ gender friendly small scale processing tools. The ABI has played significant role in promoting entrepreneurship development in jackfruit processing, banana processing and rice processing sector in the state. The ABI also conducts regular workshops on entrepreneurship developments in food processing sector to potential food entrepreneurs.

- The centre has provided entrepreneur support to nine processing industries (two rice mills, banana based ethnic mix, dried fruits and vegetables, spice powders, thermal processed tender jackfruit, intermediate moisture ripe banana and jackfruit, passion fruit processing) and two are in progress. These processing industries provides a regular income to the rural youth specially women group.

- The centre has developed gender friendly processing equipment viz. jackfruit slicer, combo drier suitable for blanching and drying of fruits and vegetables, banana slicer, edible wax applicator, chapathi maker, coconut scraper, etc for micro- small and medium scale processing industries.

- Process protocol for vacuum fried fruits & vegetables with enhanced nutritive value and consumer acceptance (better colour and 50% reduced oil uptake). The frying oil could be reused more than 40 times compared to the conventional frying system.

- Process protocol for retort pouch packaging of tender jackfruit which could extend the shelf life about six months.

- Process protocol for the production of microencapsulated whey-melon juice, and banana pseudostem juice-horse gram extract powder.

- Developed hot extruded ready to eat snack food from healthy ingredients viz; medicinal rice, nendran banana, ragi and corn

- Retort pouch packaging of ethnic food products viz. Jackfruit varatty and Ramessery idly.

- Development of cold extruded nutraceutical pasta/noodles with grains (specialty rice like Rakthasali, Kumkumasali, Njavara, etc), corn, and vegetables as ingredients.

- Developed a UV assisted ohmic heating system for preservation of liquid juice.

- Process protocol for microencapsulation of vanilla extract.

- Process protocol for osmo-vac dried intermediate moisture ripe banana and jackfruit.

- Process protocol for spray dried fruit juice powders

- Standardization of ethnic health mix based on banana flour, ragi flour and Njavara rice.

- Created a training hall facility for providing hands on training/demonstration to potential entrepreneurs. The centre has conducted more than fifty EDP training programmes to potential stake holders/entrepreneurs.

Agro-processing clusters

Government of India has formulated a new Scheme for creation of infrastructure for Agro Processing Clusters as a sub-scheme of Kisan SAMPADA Yojana. The major objective of the scheme is to develop modern infrastructure to encourage entrepreneurs to set up food processing units based on cluster approach. The scheme will be implemented in identified areas of horticulture/agriculture production through a mapping exercise. These clusters will help in reducing the wastage of surplus produce and add value to the produce which will result in increase of income of the farmers and create employment at the local level. Lack of such an enabling modern infrastructure in the country is hampering the development of food processing sector. Therefore, it has been envisaged to set up at least 5 food processing units with an aggregate investment of minimum Rs. 25 crore in the agro-processing cluster.

Capital intensive common facilities will be created that are required by food processing units irrespective of nature of their processing. These include cold storages, blast freezers, specialized packaging, IQF, warehousing etc. If developed as common facilities, it will encourage entrepreneurs to set up more food processing units with less capital. The basic enabling infrastructure like water supply, effluent treatment and clean surrounding, etc. will also be taken care of. It creates integrated and complete preservation infrastructure facilities from the farm gate to the consumer that enable effective backward and forward linkages by linking groups of producers/farmers to the processors and markets through well-equipped supply chain.

Strategies to boost processing sector

In addition to these specific programmes, the following are some of the possible strategies to be adopted in the area of research, development, and extension, to enhance food processing sector in the country.

- Technologies, industrial plants and machine are to be designed in association with research organisations to suit the processing requirements for specific raw materials
- Strengthen research in frontier areas of post harvest technology
- Industry should make provision for guaranteeing stable prices to the producers and reliable suppliers
- Industry can assure the quality of raw material and consequently the processed product by having a control over farming practices, both in plants and animals. Adoption of good agricultural practices (GAP) can go a long way in reducing the problems of residues and contaminants in food.

- Industry can make use of recent innovations for efficient utilisation of the produce and the wastes. Diversified and economical utilization of commodities and product wastes can multiply the profit of the processing units.

- Industry should be more conscious for human engineering, workers' safety and pollution control, so as to make the food processing industries friendly to the locality.

- Capacity building- potential entrepreneurs may be given hands on training/ demonstration of the food processing equipment at the established business incubators.

- Suitable gender friendly machines in post harvest and processing need to be developed to cater the needs of small, and medium scale food processing industries.

- New technologies in agriculture such as nutrition genomics, bio-fortification, offer unique opportunities for augmenting food and nutrition security through dietary diversification.

- A multi-disciplinary approach has to be adopted to identify critical problems related to post harvest, processing and marketing aspect of commodities and their products. A proper coordination among scientists, growers and industrialists for common strategies is needed.

- Establishment of procurement centres, with facilities for grading, sorting, washing, packing and pre-cooling in centralized locations at production areas can hold effective post harvest care and handling.

- Transportation methods should be standardized, especially for local/rural transport. Refrigerated or insulated trucks and intermediate cooling storage and central godowns with small capacity cold storage at district level needs to be employed for collection and distribution of the commodities.

- Establishment of cooperative societies for proper marketing, distribution and processing and retail outlets at potential consuming areas and multi-raw material distribution and multi-product processing units are essential to process seasonally available commodities.

- Investment in food processing sector should be encouraged by providing subsidies, soft loans and other incentives.

- Food technology coupled with food safety is needed not only for promoting nutrition security but also for economic prosperity. India needs to improve its performance on both the fronts. Food safety evaluation using the HACCP approach is needed from street foods to export-quality processed foods

- Emphasis should be made on utilization of food processing wastes. Presently, some methods are available for the use of food processing wastes but the available technologies are underutilized. Hence, the scope should be widened to meet the need of farmers for on-farm use of wastes.

- As the energy crisis is growing day by day, suitable energy optimisation techniques and use of renewable energy sources should be explored and recommended for unit operations in agricultural processing sector.

Conclusion

Food and other agriculture based enterprises have the potential to take the form of comprehensive occupational schools under the new initiatives and the liberalised economic regime. It can offer rural producers and workers sufficient knowledge, experience, infrastructure, and means to become agribusiness entrepreneurs. This endogenous movement can have far-reaching effects, promoting the overall modernization of primary production, industrialization, and marketing and development of rural areas. More specifically, these programmes create a system that can assist in the identification, adaptation, and commercialization of products from public and private agricultural research institutions and universities.

Inclusive Market-Oriented Development in agricultural sector opens the way for collaboration and joint initiatives aimed at supporting smallholder farmers and entrepreneurs to transform from subsistence to competitive businesses through mentoring and capacity strengthening. It also opens the way for joint resource mobilization, information sharing and event organization in the area of agribusiness. It also helps commercialisation of technologies that could bring in a definite change in the farming system and would create opportunities and prosperity to the smallholding farmers of the country.

References

Awasthi, D. 2011. Approaches to Entrepreneurship Development, The Indian Experience *J. Global Entrepreneurship Res.* 1 (1): 107-124.

Bansal, S., D. Garg & S. K. Saini, 2012. Agri-Business Practices and Rural Development in India - Issues and Challenges *Int. J. Applied Engg. Res.* 7 (11).

Bnerjee, G. D. 2011. *Rural Entrepreneurship Development Programme in India – An Impact Assessment*, National Bank for Agriculture and Rural Development.

Grebmern von, K., J. Bernstein., D. Nabarro., N. Prasai., S. Amin., Y. Yohannes., A. Sonntag., F. Patterson., O. Towey & J. Thompson 2016. *Global Hunger Index, Global, Regional and National Trends*, International Food Policy Research Institute, and Concern Worldwide, Bonn, Washington, DC, and Dublin: Welthungerhilfe, P.12.

Nee, V. 1989. A Theory of Market Transition from Redistribution to Markets in State Socialism. *American Sociological Review* 54(5): pp. 663-681.

Negi, S. 2013. Food Processing Entrepreneurship for Rural Development: Drivers and Challenges. In: *IIM, SUSCON III Third International Conference on Sustainability: Ecology, Economy & Ethics* (pp. 186-197). Tata McGraw Hill Education New Delhi.

Padmaja, G., Sajeev, M. S. & Sheriff, J .T. 2016. Value Added Food Products from Tuber Crops Suitable for Agro-enterprises, In: *Proceedings of VAIGA- 2016- International Workshop on Agro Processing & Value Addition*, held at Thruvananthapuram during 1st to 5th of December 2016. pp. 64-65.

Sherawat, P. S. 2006. Agro-Processing Industries - A Challenging Entrepreneurship for Rural Development. *The Int. Indigenous J. Entrepreneurship, Adv., Strategy Edu.*

Suarez, 1972.Campesino Communitarian Enterprises in Latin America, In: *The Community Enterprise*, USA, IICA.

Sudheer, K. P. 2008. Emerging Technologies in Fruits and Vegetable Processing- An Opportunity for Food Entrepreneurs, In: *Proceedings of National Seminar on Food Security through Innovations in Food Processing and Entrepreneurship Development*, held at Kerala Agricultural University, on 29th and 30th September, 2008, pp.20-27.

Sudheer, K. P. 2016. Prospects of Value Addition and Role of Agri Business Incubation Centre's. In: *Proceedings of VAIGA- 2016- Int. Workshop on Agro Processing & Value Addition*, held at Thruvananthapuram during 1st to 5th of Dec. 2016, pp: 49-50.

Sudheer, K. P. & S. M. Mathew. 2016. *Recent Developments in Post Harvest Technology*, Published by Director of Extension, Kerala Agricultural University, Thrissur.

Sudheer, K. P., S. Saranya & N. Ranasalva. 2017. Entrepreneurship Development in Fruit and Vegetable Processing Sector through Mechanization. In: *Proceedings of National Conference on Horticultural Crops of Humid Tropics - Diversification for Sustainability* at Madikeri, Coorg, Karnataka, May 20-21, 2017, pp: 142-148.

Tan, W. L. 2007. Entrepreneurship as a Wealth Creation and Value Adding Process. *J. Enterprising Culture*.15 (2):101-105.

Tripathi, D. 1971. Indian Entrepreneurship in Historical Perspective: A Re-Interpretation *Economic and Political Weekly* 6 (22): pp. 59-66.

United Nations Food and Agriculture Organization. 2011. The State of the World's Land and Water Resources for Food and Agriculture. *Summary Report*, P. 9.

United Nations in India, 2017. *Nutrition and Food Security*, http://in.one.un.org/.

Verma, S. S. 2016. Indian Cold Chain – An Emerging Industry.*Cooling India.* http://mofpi.nic.in.

Worldbank.org/indicator/SH.DYN.MORT

Worldbank.org/indicator/SN./TK.DEFC.ZS

http://www.indianfood industry.net/ Industry: *Indian Food Industry*, (Accessed on 2017, July 8).

http://www.indexmundi.com/g/r.aspx.

http://www.mofpi.nic.in

2

Developing Entrepreneurship Skills in the Farming Sector: Perspectives and Strategies

Jiju P Alex

Introduction

Entrepreneurship is regarded as one of the most important factors that contribute to economic development of a country. The current epoch of development is marked by structural changes in developing economies to create new opportunities of entrepreneurship. This is largely made possible by liberal policy interventions that foster enhancement of enterprises of all hues. Unlike previously, in every country, an entrepreneur is looked upon as an invaluable contributor to economic growth and development. This is more important in developing economies, as 'entrepreneurship development' has multiplier effects as it invariably fulfils other important pre requisites of development, viz. innovation, technological development, new institutions and human resource development. To put it in other words, without enterprise and entrepreneurs, there would be little innovation, little productivity growth, and few new jobs. This is more relevant in the agricultural sector since the emphasis has now shifted from mere enhancement of productivity to value chain management, through a process of extensive entrepreneurship development. More so, this has been identified as a strategy for sustaining the livelihood options of small farmers who have found it difficult to thrive the perils of price fluctuations, increasing cost of production and several other uncertainties that have plagued the farm sector.

Raising entrepreneurial activity could therefore play a considerable role in promoting economic development. Even while we acknowledge the importance of entrepreneurship in development, most agribusinesses are still encountering challenges that stagnates entrepreneurial activities. This has called for substantive measures to boost the farmers' levels of entrepreneurship capability. There should also be appropriate strategies for producing more entrepreneurs in order to get more agribusinesses to grow. Encouraging and releasing people's entrepreneurial energies are essential keys to the achievement of greater economic prosperity.

However, developing entrepreneurship among farmers is not as easy as it may appear. It involves the challenge of identifying and fostering entrepreneurs who are competent enough to perceive new opportunities and bear the inevitable risks in exploiting them. It also implies evolving new institutional mechanisms and support measures to provide farmers with the environment to test their skills and grow further, either individually or collectively. Empirical studies have shown that sustainable development of agribusiness requires the development of entrepreneurial and organizational competency in farmers. This could be made possible by a two pronged strategy. The first one is bringing in structural changes to amend the social, economic, political and cultural factors and frameworks that hinder entrepreneurial development and the second one is to encourage the farmers by developing their personalities and capabilities to ignite entrepreneurial desire. This chapter examines in detail the objectives of entrepreneurship development and the necessary entrepreneurial skills required by small scale entrepreneurs in the agricultural sector.

Objectives of entrepreneurship development programme

The challenge of entrepreneurs is to break the vicious circle of low income and poverty. In other words, they should be equipped with the skills to try out new means of improving and diversifying their livelihood options. It is important to understand the general objectives of entrepreneurship development programmes in a developing economy, before we set out to design strategies in a given sector. An overview of literature on entrepreneurship development lists out the following objectives.

Generating employment

Entrepreneurs while supporting themselves independently turns out to be a source of direct and indirect employment for many people in a country. This has been regarded as a workable solution to take on unemployment, which is a chronic developmental problem in most of the developing and underdeveloped countries. Entrepreneurial development, particularly through promotion of small business, is looked upon as a vehicle for employment generation. Governments that vouch alleviation of poverty and creation of employment advocate entrepreneurship development as an effective strategy for long term development.

Promoting capital formation

Idle and unproductive funds that are available with people shall be channelized for productive purposes if there is a favourable environment for entrepreneurship development. Policies that facilitate investment in productive enterprises would enhance capital formation even in less commercially advanced regions. The funds which are used by entrepreneurs are mostly a mix of their own resources and borrowed money. In any case, generation of wealth through investment is possible only when the money is ploughed into enterprises.

Enhancing the dynamism of the economy

Small and medium enterprises are always observed to have an enhanced dynamism through quick decision making and smooth communication. This quality of dynamism originates from the inherent nature of the small business, which have to essentially struggle to survive and sustain. Moreover, the structure of small and medium enterprises is less complex than that of large enterprises and therefore facilitates quicker and smoother communication and decision- making. This enables greater flexibility and mobility of small business management. Small enterprises are more often managed by their owners themselves, which make it possible for them to undertake risk and challenges, much more than any employed managers.

Equitable economic development

Small enterprises mostly need relatively low investment which makes it easily implementable in rural and semi-urban areas. This has been found to create additional employment in these areas, preventing migration of people from rural to urban areas. This also gives policy direction on the emphasis on entrepreneurial development. Since, majority of the people are living in the rural areas, entrepreneurial efforts should be directed towards those regions where rural unemployment is rampant. As small enterprises use local resources, there is ample scope for establishing functional linkages with different enterprises. For instance, establishment of an agro processing industry would lead to several linkages with the producers, both forward and backward. This would eventually bring more people into the ambit of economic development. Moreover, flourishing of enterprises in a locality also result in a large number of public benefits like infrastructure facilities for better transportation, education, entertainment etc. Setting up of more industries leads to development of backward regions and thereby promotes balanced regional development.

Fostering innovations

Innovation is the *mantra* of entrepreneurship development. Without this vital element, business enterprises would not perform better. Since small scale enterprises encounter stiff competition and risks, they have a relatively higher necessity and capability to innovate. They also have the advantage of flexibility with technological options. Thus, they are both free and compelled to innovate. However, certain congenial situations are required to foster innovation, particularly in the case of small enterprises. They have to explore opportunities of new markets, new raw materials, new production processes, new technologies, new organisation structures, new linkages etc to improve their performance and take on emerging challenges. It would not be possible for small scale entrepreneurs to get hold of the meaning and gravity of the pre requisites for innovation on their own. This warrants facilitation of entrepreneurs by means of technical advisory and other information support from time to time. It is expected that in India- as we see in China which is surging ahead with stunning innovations and an entrepreneurial boom- an increased number of small firms are expected to come up in the immediate future. Farming community would not be able to reap the benefits of this emerging opportunity unless multitudes of fostering facilities are established across the country in order to capacitate farmers to innovate.

Better standards of living

In the process of achieving a higher rate of economic growth through entrepreneurship development, entrepreneurs would be able to produce quality goods at lower cost and supply them at lower price to the community according to their requirements. The general notion is that when the price of the commodities decreases, the consumers would get the power to buy more goods for their satisfaction, contributing at least partially to improve their standard of living. In fact, this is applicable not only to the consumers, but also for the entrepreneurs, as the spectrum of clientele may increase considerably in this process.

Accomplishing self-reliance

Accomplishing self- reliance in various sectors, particularly agriculture, is very important as they help reduce imports and thereby dependence on foreign countries. It would enhance export of goods and services and foreign exchange of the country. This would bring forth a situation where in import substitution and export promotion would ensure economic independence and self-reliance. This is an important concern in agriculture as sustaining the food security of the country is a major factor that contributes to independence and sovereignty. Historically, it was the over reliance on other countries for food grains, which prompted the country to invest in the agricultural sector for fostering innovations.

Catalysing overall development

Entrepreneurs have multiplier effect on the economy as they catalyse the chain reaction of economic activity through series of interactions among various subsectors. Establishment of enterprises would set in motion a process of industrialisation by generating demand for various types of inputs required by it. There will be so many other units which require the output of this unit. This would lead to overall development of an area due to increase in demand and setting up of more and more units. In this way, the entrepreneurs multiply their entrepreneurial activities, thus creating an environment of enthusiasm and conveying a motivational spirit for overall development of the area.

Even while these objectives are relevant, entrepreneurial success cannot take place in a vacuum. Instead, entrepreneurship exists and evolves in the context of the particular geography; local, national, or even supranational economy and society. Several authors have observed that this naturally involved a mix of attitudes, resources and infrastructure, which shall be termed as an 'entrepreneurship ecosystem'. The health of this system, therefore, is a function of the entrepreneurial attitudes, abilities and aspirations of the local population and the prevailing social and economic infrastructure of the region. It is in this backdrop, an analysis of the traits of entrepreneurs become important.

Traits of entrepreneurs

Traits of entrepreneurs have been studied by several authors and described in different ways. While most of such definitions and explanations have relevance to industrial and commercial situations, traits of entrepreneurs in agriculture would be different from all those conventional definitions to a considerable extent. The traits at the individual level might not differ too drastically. But entrepreneurs in the agricultural sector, particularly those with limited resources and knowledge about the process of entrepreneurship, will have to acquire several core competencies and skills, apart from the naive traits required by an entrepreneur. This has become more relevant at present since new technologies, practices and protocols are being increasingly used in agriculture. Exploring the possibilities of value chain management has opened up new vistas of knowledge based agriculture even in rural areas. The profile of new age agriculture, which is compelled to take on the challenges put forward by market forces, warrants a new genre of entrepreneurs engaged in knowledge intensive and skill bound activities.

Focusing on definite and workable objectives

An entrepreneur is likely to have several notions about the enterprise she has chosen to pursue. This would probably lead to too many objectives and blurred focus. There are also so many distracting forces when trying to build a business. An entrepreneur should set a long term vision and should know how to laser focus on the very next step to get closer to the ultimate goal. An entrepreneur in agriculture should necessarily master this skill as the distractions would be too many and the probability of success is limited. It is always important to focus on the objectives and find out ways by which these objectives are accomplished one by one, through specific action points.

Maintaining resilience

Entrepreneurship in agriculture demands this trait very much due to some of the characteristics of agricultural enterprises cited above. Agricultural enterprises are likely to be affected by several external and internal factors, which are beyond the direct control of the entrepreneur. Uncertain production levels, fluctuation in market prices, perishability of goods etc. add to the woes of the entrepreneurs. There would be seasonal and many times, unseen changes in demand for goods. The ability to endure the ups and downs of any business is an essential trait required by the entrepreneur, since it never goes exactly the way the business plan described it. This skill enables the entrepreneur to keep going when the outlook is bleak. In fact, this trait is dependent not only on the personal quality of the entrepreneur, but also on the congenial situations provided by the policy framework which would suitably ensure the checks and balances required for better prices and other pre requisites.

Anticipating and responding to changes patiently

Most entrepreneurs do not show the patience to wait for things to evolve on their own and focus only on what comes next, rather than where the enterprise needs to go in the long term. In any case, success of an enterprise may take several years. However, in agriculture, the entrepreneur will have to address immediate concerns quite frequently since agricultural production and value addition are susceptible to short term changes much more than any other business process. Entrepreneurs should have short term as well as long term strategies for growth, and maintain an outlook on the trajectory of evolution of the enterprises. More precisely, an agripreneur should anticipate changes in the course of actions every time and respond to them patiently. This would again be possible only if the entrepreneur is provided with congenial situations to thrive and move forward.

Managing human resources and labour

Learning to leverage employees, vendors, consumers and other human resources involved in agriculture is very important to build a scalable enterprise. There is a need to learn to identify appropriate human resources and train them progressively to handle emerging challenges and situations skillfully. This would essentially require human resource management skills at different levels and dimensions. For example, maintaining a group of skilled labourers at the farm is very important for speedy farm operations. Skill training of labourers to handle and maintain machineries has been found to be an essential component of profitable agriculture. The skillsets of different types of human resource have been found to be decisive in enhancing the efficacy of enterprises. Though, agriculture would not have been considered to be a sector that required sophisticated skills till recently, situations have changed drastically. Emerging trends in agriculture warrant skills in handling fairly complex machines. Cultural operations and post-harvest handling have become more knowledge intensive as better standards are now applicable for agricultural products.

Managing information and knowledge

A successful entrepreneur is one who realizes that there are several things that are yet to be known. As stated earlier, agriculture has become more knowledge intensive and information on a plethora of components is vital in decision making. Information on new systems, technology, and industry trends are important inputs in decisions related to farming. Managing the state of art information on agriculture is difficult as it ranges from management of vital natural resources to sophisticated trends in value chain management. In fact, this knowledge intense situation cannot be handled by individual farmers unless there is well organised institutional support for providing them with knowledge support. Still, managing knowledge on agriculture and related sectors by an entrepreneur is not much difficult with the modern information communication technologies (ICTs). Skill to manage ICT tools, however, has turned out to be an important pre requisite if someone wants to try his hands at agriculture entrepreneurship.

Managing the market

Market forces have an unprecedented influence on agriculture in India, consequent to globalisation of the Indian economy. Liberalisation has brought about new consciousness on improving the quality of products. However, this has unnerved the Indian farmer as he has not been duly bolstered by policy instruments. Support to farmers in terms of institutional credit, irrigation, and market access, technical advice etc. have not improved simultaneously. This calls for improving the capabilities of farmers and agri- entrepreneurs to intervene in the market skillfully by understanding the market trends. However, this cannot be undertaken by the farmers themselves as prediction of market trends is a complex process which requires reliable data support and tools for analysis. In spite of such problems, entrepreneurs in agriculture and allied sectors have to be made aware of market, its composition and trends, to name a few.

Contemplating the future and self-reflection

An entrepreneur has to allow downtime to reflect on the past and plan for the future. She should be able to contemplate her future actions based on her experiences. This is nothing but her skill to learn from the surroundings and past experiences. There would be immense scope to improve skills and practices if the entrepreneur is ready to learn from experiences. It is very important that the entrepreneur tries to reflect on herself, with reference to her experiences, knowledge and skills to manage various entrepreneurial functions. This would invariably lead to improvement, and sometimes even innovations. Drawing out a plan for the future would be a rewarding action for the entrepreneur in the farming sector, because it is a daunting task to figure out the future in agriculture, unless one is equipped with deep understanding of trends in the market. Self-reflection for that matter would be a tool for the entrepreneur to examine oneself, in terms of skills, capabilities, opportunities and threats.

Accomplishing self-reliance

One of the most important traits of an entrepreneur is self-reliance. In the cumbersome process of entrepreneurship development, there could be a lot of help to be solicited from different sources for accomplishing an entrepreneurial objective. But in the end, an entrepreneur herself should need to be resourceful enough to be independent. This calls for training oneself on vital skills and core competencies and drawing from experiences as much as possible. Farming offers immense opportunity to explore the ways to be self-reliant. In the case of small scale entrepreneurs, self-reliance assumes more important as they cannot always rely on external support to establish and sustain.

Managing vital resources including time

It has been observed that the quality of activities completed within a given amount of time is what differentiates a highly productive entrepreneur from a mere wishful thinker. Since entrepreneurship is not a solitary activity, there are well defined responsibility to partners, customers, and employees to make the most out of their time. Effective time management skills would help entrepreneurs complete essential tasks expeditiously while still leaving room for additional activities. Some entrepreneurs are found to work with mentors and partners to ensure that they stay focused on their most important objectives. Entrepreneurs who make good use of their time can actualize great achievements while maintaining peace and composure amidst stress. Entrepreneurship in agriculture demands more effective ways of time management for various agricultural operations from seed to seed.

Exploring possibilities and recognising opportunities

It is widely observed that entrepreneurs must be able to recognize opportunities that are mostly ignored by other participants to earn a competitive advantage. It has been observed that intuitive ideas usually do not add much significant amount of value because other entrepreneurs must be making use of them earlier. Innovative entrepreneurs must therefore be willing to verify opportunities by conducting market research, talking to business experts and utilizing intelligence gained from organizational experience. Entrepreneurs can add significant value if they recognize an opportunity that nobody else has perceived. Entrepreneurs should record innovative ideas that they discover whenever they occur. In this regard, ideas that are systematically categorised should be routinely reviewed to eliminate unimportant concepts and highlight opportunities that are worth pursuing.

Formulating and managing institutions

Even with the best information, entrepreneurs could be a baffled lot, lest they do not have the organizational skills to make use of the acquired knowledge. As we might all agree, organizational skills are so essential for managing other people. But how would you exercise these skills? Pursuing development concerns usually end up in debates on institutional arrangements that are required to address the development concerns of the community. This bestows upon the entrepreneur, the responsibility of understanding the dynamics of the community and how best various institutions can be formulated to address the needs. For instance, evolving an institutional mechanism to procure the products of small scale farmers in a region would be a significant step towards entrepreneurship development. Managing an enterprise can be an impossible undertaking without advanced organizational skills to get the most out of available resources.

Rational decision making and critical thinking

Entrepreneurs cannot expect to succeed without their ability to make rational decisions. Unfortunately, most entrepreneurs underestimate the importance of taking important and vital decisions. As a result, many entrepreneurs do make impulsive decisions on the basis of intuition or conjecture. Poor decisions can have enormous consequences that can even drive a prosperous enterprise out of business. Entrepreneurs must consult with experts and evaluate market trends before making any decision that could have significant future consequences. Some decisions shall also be delegated to specialists to facilitate more effective decision-making. Obviously, this is more relevant for the agricultural sector, as entrepreneurship in agriculture demands high degree of decision making in every agricultural operation as well as during post- harvest phases. Since the risk involved is too much, rational decision making becomes inevitable.

The most important entrepreneurial decisions require critical thinking skills. This has been found to be very much essential in agri-enterprises, owing to the peculiar features of this sector. Success of agricultural production depends on several factors, which have to be managed skillfully and cautiously. One cannot casually jump into conclusions based on premonitions and assumptions. Instead, decisions should be taken based on critical analysis of existing situations and review of relevant facts and figures available from authentic sources. In agriculture, choice of technology and other critical inputs can be finalised only on the basis of critical thinking. It is observed that many times individuals do not regularly exercise their ability to think critically about important issues. Most individuals prefer to avoid critical thinking, since it requires significant mental focus and energy. However, entrepreneurs are unlikely to succeed unless they take decision-making seriously and critically think through important issues.

Meaningfully communicating with the stakeholders

It is widely accepted that all business activities are found on the productive basis of adding value to other individuals. Communicating with the stakeholders, who include individuals as well as institutions, is a very important entrepreneurial function. Entrepreneurs with strong communication skills can find it easier to engage with partners, acquire funding, and develop relationships with prospective customers. Oral communication skills can make it easier to manage associates who work for the enterprise. Enterprises in agriculture involve information and technology that are to be deftly handled and communicated. Communicating with the farming community itself is a daunting task, as the target group is highly heterogeneous. Several communication channels and methods will have to be employed to transfer technological information, harness people and prompt them to perform. In agriculture, adoption of technologies would not take place unless technical information is not communicated to the end users effectively. Access to updated information is much required for entrepreneurs as they strive to exist in a highly competitive entrepreneurial environment. It is also important that the entrepreneurs clarify and communicate their thoughts about prospective opportunities. Many types of business communications are most effective when they are conveyed through written materials. For example, entrepreneurs can write a quarterly letter to all customers and suppliers. A well-written letter is a great way to demonstrate appreciation to other business professionals. Writing skills can also enable entrepreneurs to communicate more effectively about advanced topics that might be difficult to comprehend. In today's world of widespread literacy, business professionals have begun to expect that entrepreneurs have exceptional writing skills.

Leading institutions and organisations

The ability to lead can enable entrepreneurs to expand their reach to get more work done through the assistance of others. An entrepreneur's credibility as a leader within an industry or among a group of customers can be a decisive factor contributing to the acceptance of an enterprise. Stakeholders involved in any entrepreneurial venture always look to the entrepreneur for leadership. Therefore, entrepreneurs must possess adequate leadership skills to effectively coordinate the efforts of everyone involved in an enterprise. Leadership skills can be acquired by keenly observing various situations of decision making and learning the behaviour of people.

Motivating and persuading people to perform

Motivation and persuasion are often needed to encourage other individuals to adopt an innovative idea that entails risk. Most individuals are naturally averse to risk or new ideas. Entrepreneurs must be able to ethically convince her aides and associates to accept that a questionable idea might be beneficial in the long-run. Strong persuasion skills are also needed to harness funds, gather resources, develop partnerships, and acquire customers. Motivating the people associated with an enterprise to accomplish better standards and results is the key success factor in any enterprise. Managing business amounts to motivating people to cherish the dreams of the entrepreneur collectively, with harmony of action. Over and above all, negotiation skills are an essential component of persuasion because calculated concessions are often the most effective way of inducing other individuals to accept a position on a particular issue. Entrepreneurs can improve their persuasion skills through critical review of existing situations and exploring possibilities.

Cultivating entrepreneurial skills can help farmers and the rural youth improve their prospects for success in almost any business activity. Most entrepreneurial skills can be improved through a combination of education, motivation, and by imparting practical experience. All the traits described above are important and have to be continuously improved through conscious efforts. Entrepreneurs should make improving their skills a lifelong priority to improve their chances of business success.

Conclusion

A specialised and intense programme to impart entrepreneurial skills to the farming community is very much needed to put the agricultural sector in India on to a growth trajectory. This is because of the dismal performance of the country in this sector as indicated by the negative growth and an increasing eviction of a great majority from the farming sector. The Global Entrepreneurship

Index, which is an annual index that measures the health of the entrepreneurship ecosystems in 137 countries ranks India as 69th among the nations of the world. This calls for a huge, comprehensive and all pervasive drive to inculcate entrepreneurial skills among the rural youth and women with emphasis on agriculture. This, by all means would boost overall economic growth by increasing the growth potential of agriculture and allied sectors. The enormous latent potential of human resources in this highly populated country will not be put to constructive use unless the entrepreneurial skills are imparted on a wide scale. Since India is believed to surpass other countries in terms of demographic dividend by the turn of this decade, this assumes greater importance. The country would not be able to take advantage of this unique condition if our efforts are not directed towards this goal. Creating the right entrepreneurship eco system however does not end with this alone. In fact, simultaneous interventions are required on several fronts to take on the challenge of increasing employment opportunities in the agricultural sector. Development of entrepreneurship skills should be accompanied by congenial policy measures, which would include liberal financial support to entrepreneurs and development of cost effective innovations by research and development institutions. Providing the less endowed entrepreneurs with levelled play grounds to survive and grow is another critical intervention required to enhance entrepreneurship. India as an economy has a long way to go in this direction.

References

Kahan, D. 2012. Entrepreneurship in Farming. *Farm Management Extension Guide*, FAO, Rome, Italy available at: www.fao.org/docrep/018/i3231e/i3231e.pdf.

Kumar, D. 2015. *Entrepreneurship in Agriculture.* SSPH, New Delhi.

3

Business Incubation – Understanding the Process

Preethi Manniledam & Radha Ramanan T

Introduction

The establishment and nurturing of Small and Medium Enterprises (SME's) is a vital input in creating dynamic market economies in the economic and social development of transition countries. Globalization has created a new world order in which entrepreneurship and systematic innovation play important roles by offering competitive advantage for much needed sustainability of SME's. To keep up the momentum, the development process of innovative ideas into products require alignment with the changing business paradigms, market dynamics and prevailing economic scenario.

The concept of business incubation dates back to 1950s and was more of providing a real estate facility to the small scale units to tide over the early years' resource deficiencies. These are called the first generation incubators, which were first established in United States. Later on, the idea of business incubation diffused over different countries, and by the mid of 80s the incubators started providing services like business supports, accelerating the learning curve, including knowledge based services. These are termed as second generation incubators. The third generation of incubators provides the additional facility of networking also to the incubated companies. The importance of networking led to the establishment of virtual incubators or networked incubators, where physical infrastructure has less sanctity. Business accelerators appeared in the mid 2000s, and by now the concept of business incubation is spread all over the world. There exist almost 9000 business incubators in the world.

A Technology Business Incubator (TBI) nurtures the development of technology based and knowledge driven companies helping them to survive and grow during the start-up period (2-3 years) by providing an integrated package of workspace, shared office services, access to specialized equipment and value added services like management assistance, business planning, access to finance, technical assistance and networking support. The main objective of the TBI is to produce successful business ventures that create jobs and wealth in their region (NSTEDB, 2009).

Importance

Deployment of technology in the life of the common man is an indication of the level of development of a nation. Technology based products could be easily developed through start-ups. Technology Business Incubators nurture these start-ups to develop and come out in the market with their products. TBIs are powerful economic development tools. They promote the concept of growth through innovation and application of technology, support economic development strategies for small business development, and encourage growth within local economies while providing a mechanism for technology transfer. By setting up in the premises of higher technology and research institutions and encouraging the technology transfer, the TBIs help to align the education in these institutions more towards the industry and thus promotes industry- academia interaction.

World over, business incubation has attained enough importance as the role of start-ups in the technology development and thus in the progress of the countries are increasing day by day. US, China, Korea, Europe, Japan, etc. are the countries in the forefront as far as the business incubation activities are concerned. In all these countries the incubation industry has reached matured stage. Out of these, the US has the highest number of incubators and has famous ones like Y Combinator. National Business Incubator Association (NBIA) provides the data and guidelines regarding the incubators specifically in US along with other countries.

Indian scenario

In India, technology business incubation efforts are almost three decades old. In these years, incubation has helped to weave many inspiring and motivating success stories of young entrepreneurs, and have helped in creating a positive environment for entrepreneurship in general and for incubators in particular.

There are around 300 Business Incubators in the country in the government sector. Out of this, almost 90 are promoted by National Science & Technology Entrepreneurship Development Board (NSTEDB) of Department of Science and Technology (DST), Government of India (GoI) and the rest are promoted by other Ministry Departments like Ministry of Communication and Information Technology, Ministry of Micro, Small and Medium Enterprises etc and other public and private organizations. Ministry of Communication and Information Technology has set up 27 incubators across the country in the higher educational institutions. These government departments support the incubators with the funds for establishing the common facilities including equipment and the operational expenses for a specific period of time. In 90% of the cases, these incubation centres are set up in the universities with a vision to support the technology based start-ups with the expertise the universities are housing. These institutions primarily have to provide the built up space for housing the incubator in all the government funded set ups. The incubators are expected to build their own business model and become sustainable beyond this period (NSTEDB, 2009).

There are different forms of business incubation in the private sector also. They range from a mere co-working space, where the incubated company will get required number of seats to occupy in a bigger space to a well structured space and associated facilities with common laboratory space etc. In the co-working spaces, normally, the incubators charge on per seat basis. Many corporates have established their incubation activities. They mainly focus on promoting the development of innovative product related to the existing business. This is expected to help the corporates in getting innovative ideas without the hassles of identifying and developing innovative products in a corporate set up. Some of the government departments have taken initiative to set up incubation centres in Public Private Partnership mode also. Start-up Village established in Kerala was such an endeavor of DST and a private sector company MobMe Wireless. New age financing organizations like venture and angel investment firms are also setting up incubation centres.

Depending on the sponsors of the incubator, the goals of the incubators may vary. Typical examples are given in Table 1. (Lalkaka, 2001).

Elements of incubation process

Selection process and its importance

Incubatee companies are selected by the incubator based on policies formed by each incubator. Most of the incubators are sector specific. ie. they provide incubation facility to companies from a particular sector like, agriculture, information technology, biotechnology etc. (Kris et.al., 2007). Mostly, incubators look for start-ups with some innovative products or services. Even though, the criteria for selection to the incubator is based on parameters set up by individual incubators, most of the incubators look for technical viability of the product or service proposed and market potential for the product. Table 2 gives a list of major parameters that an incubator considers for selecting its incubatees.

Table 1: Goals of the incubators with respect to their sponsors

University	Innovation, faculty and student involvement
Research organization	Research commercialization/technology transfer
Public/private partnership	Investment, employment, other social goods
State sponsorship	Regional economic development, employment generation
Private sector initiative	Profit, new products, spin-offs, equity in incubated start-ups, image
Venture capital-based	Winning enterprises, high returns.

Table 2: Parameters for selection of incubatees

1. Uniqueness of product/service
2. Business plan
3. Managerial team
4. Existing/possible cash flows
5. Marketability
6. Growth potential

Most importantly a proper match between what assistance the start-up is looking for and what the incubator can provide matters in the selection process. A proper selection process helps the incubator to choose the right kind of company as its incubatee, which reduces the possibility of failure. Some authors argue that high success rate of start-ups in the incubator is basically because of the selection process (Schwartz, 2013).

Typical services offered by TBIs

Common types of assistances that incubators offer are help in business plan writing, assistance in securing capital, and shared administrative services. But, services vary depending on an incubator's mission and focus as well as on individual clients' needs. These services help incubated companies to grow to a sustainable level and graduate from the incubator after reaching a certain level of growth. This growth is achieved with the help of various services provided by the incubator like physical infrastructure, equipment, laboratory facilities and specific guidance rendered in the technology development and management of the unit as well as the seed fund provided to the incubatee. Let us look into these services in detail to understand how important they are in making of a successful company.

Physical space: Physical infrastructure or built up space for the office and the product development/ production process is a necessity for most types of start-ups. Depending on the nature of business, the requirement of space varies. For example, for the start-ups from IT sector, requirement is for a space with a modern outlook as they will have only computers, software developers and possible client visits to be accommodated in the office. But a start-up from a manufacturing sector requires more space with not so polished nature as they will be dealing more with machinery, other equipment, raw materials and production process.

Equipment: As technology business incubators normally incubate technology based start-ups, there will be typically some product development going on in most of the incubated companies. In many cases the start-ups choose a particular incubator depending on common technical facilities they can offer. The start-ups cannot invest in all necessary equipment and facilities in the initial stage. For solving this problem, typically, the incubator either makes the host institute facilities available to the start-ups or set up separate common facilities in the incubator itself. These common facilities include equipment pertaining to the thrust areas of TBI. Incubators fix up the mode of usage of these facilities and the related usage charges.

Shared office services: Incubator provide common office assistance services like secretarial assistance, telephone, fax, phone and internet connections, housekeeping services, meeting rooms, front office service etc. to the start-ups incubated. In the initial stages of a start-up, it is difficult to set up these services which they don't use frequently. Apart from these services, an important assistance that the start-ups require in early stage is legal and statutory consulting. These services provided by the incubator help the start-up to lay down a proper structure for the company in the initial stage itself and thus help to work in a structured manner. Normally, the incubator will have a legal consultant and chartered accountant or a company secretary, who can guide the start-ups in obtaining statutory compliances including the company registration and tax registrations (Mian, 1996).

Technology consultants: In the technology business incubators, technology consulting plays an important part. When the incubator is attached with an institute of higher learning or research, there will be expertise in different sectors, which may be used by the incubatees. Here, the incubator plays a guiding role in identifying the correct resource person and helping to get the required assistance for the incubatee. In the case of institutes, it could be teachers, students and other technical expertise available with the institute. Whenever required technical expertise is not available with the host institute, normally the incubator tries to get assistance from any of the associated institutes. Thus, the incubator is expected to do proper networking to help the incubatee in getting the assistance. When the incubator is not attached to academic/research institutions, a network of technology sources has to be built by the incubator for providing technical support to the incubatees whenever necessary.

Commercializable technologies/technology transfer: World over, many of the incubators, especially the government funded ones are attached to a university; where a good number of researches are expected to happen. It is envisaged by the government while establishing TBIs that they will act as a mechanism for technology transfer. This will contribute to improve competitiveness of the industry and better utilisation of the facilities available with the technical institutes and research laboratories. Successful transfer of technology is one of the parameters for evaluation of efficiency of the incubator. Successful transfer of new technologies from research laboratory to commercial sector has many benefits: creation of wealth, new jobs and new solutions to society's problems. There are different ways in which the incubators can facilitate the technology transfer process. They can also act as the technology transfer offices which will lead to the dissemination of knowledge created in the academia to the industry (Mian, 1996).

In some incubators attached to the institutes like Indian Institute of Technology, Mumbai, Indian Institute of Technology, Chennai, etc. the faculty or students are encouraged to form the start-ups. In such cases, mostly the research outputs of the faculty/students are getting converted to new products.

Mentors: Mentors play the most important role in the growth of an incubated start-up. Identifying a right mentor for a start-up in its nascent stage itself help the start-up from committing mistakes and help to grow them fast. The mentors could be from different fields, mostly with enough experience. Normally, the mentors are ready to offer their services without any fee. The willingness to share knowledge and spare time for start-up is the most important factor that the incubator should look for in a mentor. Incubators normally have a panel of mentors from different sectors (Ozdemir & Sehitoglu, 2013). Whenever a new start-up is getting incubated, it is ideal to identify a mentor from the panel of mentors of the incubator for that start-up. It is seen from studies that mentors are one of the important aspects which lead to success of the incubated start-ups.

Linking with other organizations: Having linkages with other organizations is an important aspect for the success of an incubator as the incubator or the host institution may not be able to provide all help needed by the incubatees in-house. These could be other incubators, industry associations, institutes which can provide technical assistance or research guidance etc. These associations can bring in a lot of mentors, especially from the industry. Most of the senior businessmen bring in a lot of practical knowledge, which can help the entrepreneurs to avoid mistakes they do in early days of entrepreneurship (Lee and Osteryoung, 2004).

In India, especially in government supported incubators, the idea of co-incubation is well accepted now. In this, a company incubated in one incubator is simultaneously assisted by another incubator, from where it is expected to get help which the first incubator is not able to provide. In such cases, mostly MoU will be entered into, by the incubators.

Business facilitation and marketing assistance: The start-ups may not get access to some of the clients as they have not achieved enough goodwill. In such cases, introduction provided by the incubator gives the start-up enough credibility to enter into the market. Marketing assistances could be provided to incubated start-ups in many ways. It starts from helping in formulating marketing strategy, to directly recommending the product to a possible client. Many of the incubators provide enough opportunities to the incubatees to showcase their products in exhibitions and trade shows.

Ecosystem: The ecosystem that exists in the incubator acts as a motivating factor to the start-ups. Encouragement from peers and support from the incubator management can provide the start-up a suitable background to work for the growth of venture. There could be business deals happening among the incubatees itself. In some other situations, they may enter into joint developments or business. The start-ups which have been there in the incubator for some time provide mentoring support to the start-ups that joins after them. Studies show that these ecosystems act positively and impart confidence in the incubatees (Adlesic and Slavec, 2012).

Seed funding/linking with financial institutions: The start-ups require funds for their operations. Normally, entrepreneurs raise a portion of the fund on their own or through the family and friends. When these funds are not sufficient, the incubator can help the start-up with the seed funds. There are many government schemes through which the incubators are provided with the funds to give it as seed funds to the incubatees. For the start-up, seed fund may be given as loan with an interest rate which is much less than the commercial rate or in some cases the incubator takes equity in lieu of the fund provided. Normally, these shares are held by the incubator until the start-up raises the next round of fund from the market. In the case of equity investment in the incubated companies, the extent of shares held by the incubator is normally limited to a maximum of 10%. Incubators do not involve in decision making activities of the company and mostly exit from the company in 7-8 years. In some cases, incubators take equity in the incubated companies in lieu of the service provided to them. These companies are expected to pay only less or no charges to the incubator for the services offered by them. This model is more common in private sector incubators.

Incubators are also expected to assist the start-ups in raising funds by external investments. Incubator must provide opportunities for finding out suitable angel or venture investors for obtaining sufficient funds for scaling up the operations. Incubator may have to help the start-up in obtaining a proper valuation for the start-up before they go for external investments. Incubator may help the start-up in raising debt financing also by introducing them to the bankers or other financial agencies.

Training: The start-ups require to be trained in technical as well as managerial fields. Most of the entrepreneurs from technical background lack business management and financial skills. Incubation manager in most of the cases identifies this deficiency and provides required training to the incubatees. These inputs are given normally in the beginning of the operations of start-ups. As they go ahead in the product development, entrepreneurs or their employees need to acquire new technologies. Incubator can organize training programmes in these technology areas also for a group of people. It is helpful for entrepreneurs to update skills if incubator is providing training in marketing, financing options, leadership etc. frequently.

Positive image of the host institute: As many of the incubators in government sector are set up with higher educational institutions as host institute, the image of the institute already developed becomes a big boon to the start-ups. When host institute is a technical institute, the start-ups get an easy acceptance for their product in the market, as their customers tend to think that the product receives technical assistance from the host institute. Thus, image of the host institute can positively help the start-ups (Mian, 1996).

Process of incubation

A start-up can be in the incubator only for a specific period. This period varies from incubator to incubator which depends on the policy of the incubator and the sector in which the incubator is operating. For sectors like information technology, electronics etc. duration required for a start-up to become sustainable is comparatively less and hence incubators in these areas normally keep the incubation period as 2-3 years. But, in food processing, biotechnology and pharmaceutical sectors, since it takes longer to make a technology into the product form, incubation period for the start-ups also are kept longer, typically 5-6 years. A typical incubation process is represented in Fig.1.

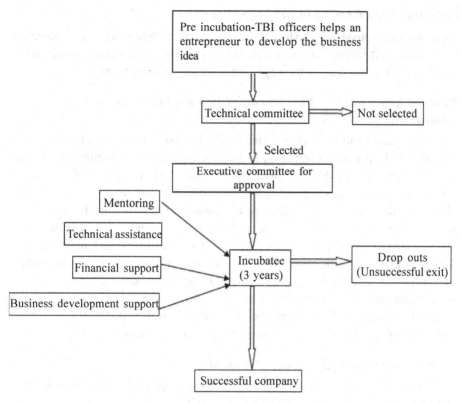

Fig. 1: Typical incubation process

Expected results of incubation/incubators

The major parameters that are used to assess the success of an incubator are discussed here

Number of incubated companies and number of graduated companies in operation after incubation

Number of entrepreneurs benefitted by the incubator is a parameter to measure its effectiveness. Number of incubatees served depends on the sector of the incubator also, as it is easy to accommodate more number of start-ups in sectors like IT, but it is difficult to provide enough facilities for companies from manufacturing sectors in large numbers. When support system from the incubator is no longer available, the start-ups find it difficult to sustain and some of them may close down after the incubation period. Higher rate of sustainability of the start-ups may be taken as an indication of an effective incubation process

Total number of employment generation

Incubators are considered as a tool for the regional development too. Employment opportunities it creates for technical and non technical graduates contribute to employment generation of the region at least to a smaller level.

Revenue generation by the companies and the taxes paid to the government

This is another form of help provided by the incubator to the development of a region which directly contributes to its economic growth. Incubator ensures that the start-ups incubated follow the statutory requirements and pay taxes regularly. Concession on taxes is a benefit that the start-ups get if incubated. Normally, the service based start-ups are exempted from service tax until they reach a revenue level of 50 lakhs per annum. This exemption can be availed for three years during which time the start-up is in the incubator. Under the new Start-up India policy, the start-ups which are approved by government supported incubators are eligible to register in the Start-up India portal after a screening by a committee at the government level. These start-ups which are registered under the start-up India scheme are exempted from income tax for any three years out of the first five years of registration as a company.

Number of technologies commercialized

One of the objectives of setting up the incubators especially in the research and academic institutes are the potential research outputs and technologies for commericalisation. Normally, there will be a gap between the academic research outputs and the industry requirement. Incubator normally plays an important role in finding suitable and easily commercilaisable technologies and often makes the match between the prospective entrepreneur and the technology provider.

Number of other entrepreneurs assisted

Incubators provide assistance to entrepreneurs outside the incubation centre also. This could be in the form of technical consulting, training or help in IP Protection etc.

Awareness creation

Awareness creation on the possibilities of entrepreneurship and the assistance available in the ecosystem is an important factor that will later lead to a pipeline of incubatee companies to the incubator. Hence, the incubator should take up this responsibility of awareness creation in the region. Other networked institutes and the industry associations can act as partners in the awareness creation process.

Monitoring the growth

Main reason for failure of the start-ups generally is the mismanagement. Normally, failure rate in the incubated start-ups is found to be low. This could be because of continuous mentoring provided by the incubators.

Monitoring of the start-ups in the incubator starts from the beginning of the incubation process itself. Normally, while selecting the start-ups itself, the incubator and the start-ups start working based on a growth plan made by the start-up which is approved by the incubator. This plan contains product development and other major milestones to be achieved by the start-up. Incubator conducts periodical reviews of performance of the start-up against the milestones set in the business plan. Some of the milestones set for the incubatees could be the prototype development, pilot production, fund raising etc. Any deviation from the milestones established earlier is analysed and necessary guidance is given to re-orient the goals. A team of experts will be available with the incubation centre to perform this monitoring. Technical and management experts, including those from industry will constitute this team. Apart from this periodical monitoring, the incubation manager is expected to interact with the entrepreneurs on a daily basis and guide them whenever required (Mian, 1997]

There can be five types of outcomes of the incubation process.

- Successful company with large profit
- Successful company with moderate success
- No loss no profit situation
- Failure with moderate loss
- Failure with big loss

For the incubation manager, the close monitoring of the incubatees will help to prevent the last case ie. failure with big loss.

Exit from the incubator

A key feature of incubators is the limited duration of assistance with exit criteria typically specifying that firms should 'graduate' after a fixed period of time. This time period depends on the type of industry sector the incubator is nurturing. For some start-ups, time taken for product development and hitting the market will be less compared to some other sectors. Hence, incubation period varies anywhere from one year to five to seven years. Following factors could be taken as the criteria for exit from the incubator:

- Number of years (this is set by the respective incubators, considering the general industry conditions)

- Size of the company: Sometimes the incubator cannot accommodate a company when number of employee's increases, as the space available has to be given to more number of start-ups. Hence, some incubators fix up the criteria that the start-up has to graduate if it has grown beyond a certain number in terms of size.

- However, in many cases, contact will be retained with 'graduate' companies through the provision of after-care services and/or on-going networking. This networking is proved to be good for the incubator as well as the graduate incubatee. The start-up may still need technical assistance or reference from the incubator to approach other organizations for any assistance. Incubator periodically conducts training programmes which will be beneficial to the graduated companies also.

Incubation management

Group of people or individual managing the incubation centre is generally called as incubation management. Incubation management involves in the activities from selecting the incubatee to the graduation and even beyond in the form of post incubation services. Incubation manager interacts with the start-ups in the form of counseling and networking interactions, and through these, needs of the incubatee are identified. Proper identification of the needs and how effectively these needs are met, leads to satisfaction of the incubatee. Experience and expertise of the incubator team is an important factor in effectively rendering this in-house mentoring process. As the start-ups mostly will be having a day to day interaction with the incubator management, this incubator team plays a crucial role in the sustainability of the start-ups and thus in the success of an incubator.

Future prospects

As the incubation system matures, physical incubation is slowly changing into virtual incubation world over. A number of virtual platforms which provide incubation facilities like mentoring and technology consultancy and many of the services provided by the conventional incubators, have come into existence. This happens mostly in the hi-tech start-up sector. These changes can accelerate the growth of high tech industries in the coming years. In the next few decades when the small, high-tech firms are having a significant role in the growth of nations, the incubators, whether virtual or physical, will certainly have a major role to play.

Conclusion

The Technology Business Incubators are established with the intention of developing the community's economic health. This objective of the incubators is achieved by promoting technology based start-ups, with the assistances brought in from different networked organisations. Moreover, the incubator itself has to be a dynamic model both in terms of financial and operational sustainability. The effectiveness of the services provided by the incubators decides the survival of the start-ups and in turn leads to the sustainability of the incubators. With the increase in number of the incubators and their acceptability as a catalyst to the growth of technology based start-ups, business incubation itself has become an industry and is expected to provide remarkable contribution to the start-up ecosystem.

References

Adlesic, R. V. & A. Slavec. 2012. Social Capital and Business Incubators Performance: Testing The Structural Model, *Economic & Business Review* 14: 201–222.

Kris, A., M. Paul & V. Koen. 2007. Critical Role and Screening Practices of European Business Incubators, *Technovation* 27: 254–267.

Lalkaka, R. 2001. Best Practices' in Business Incubation: Lessons (yet to be) Learned, European Union - Belgian *Presidency International Conference on Business Centers: Actors for Economic & Social Development*, Brussels, 14 – 15 November 2001.

Lee, S. S. & J. S. Osteryoung. 2004. A Comparison of Critical Success Factors for Effective Operations of University Business Incubators in the United States and Korea, *J. Small Business Management* 42(4): 418–426.

Mian, A. S. 1996. Assessing Value-Added Contributions of University Technology Business Incubators to Tenant Firms, *Research Policy* 25: 325-335.

Mian, A. S. 1997. Assessing and Managing the University Technology Business Incubator: An Integrative Framework, *J. Business Venturing* 12: 251-285.

NSTEDB, 2009. Conceptual Document on TBIs: Developing Ecosystem for Knowledge to Wealth Creation, *National Science and Technology Entrepreneurship Development Board.*

Ozdemir, O.C. & Y. Sehitoglu. 2013. Assessing the Impacts of Technology Business Incubators: A Framework for Technology Development Centers in Turkey, *Procedia - Social and Behavioral Sciences* 75: 282 – 291.

Schwartz, M. 2013. A Control Group Study of Incubators' Impact to Promote Firm Survival, *J. Technology Transfer* 38: 302–331.

4

Value Added Products from Fruits and Vegetables Prospects for Entrepreneurs

Sudheer K P & Indira V

Introduction

Fruits and vegetables are very important in a balanced diet. India has become the second largest producer of fruits and vegetables. Though, the production of fruits and vegetables make us believe in our strength for self-sufficiency, a significant qualitative and quantitative loss occurs in the produce from harvest till consumption. For self sufficiency and also for processing, export and to meet additional requirements, a lot of emphasis need to be given to reduce post harvest losses besides increasing production and productivity of fruits and vegetables. Though, processing plays an important role in conservation and effective utilisation of these perishable commodities, only less than three per cent of total production of fruits and vegetables are processed in India. The new industrial policy has placed this sector in the list of high priority areas.

Fruit and vegetable processing industry in the country is in a nascent and primitive stage. Number of establishments in the organized sector is less compared with several developed and developing nations. The technologies adopted for processing, preservation and value addition in medium, small and tiny industries are primitive and outdated. Processing of fruits and vegetables into various products using proper cost effective technology is a viable tool for improving economic status of farmers as well as our country. Fruit and vegetable preservation methods are broadly classified as (i) *Physical methods:* This includes removal of heat, addition of heat, removal of water and irradiation, (ii) *Chemical methods:* like salting or brining, addition of sugar and heating and by addition of chemical preservatives, (iii) *Fermentation* and (iv) Judicious combination of one or more of the above methods for synergistic effect.

Health benefits of fruits and vegetables

Fresh fruits contain 80 to 90% moisture, 3 to 27% carbohydrates and 0.2 to 3.1% fibre. Fruits, especially yellow and orange coloured are rich in β carotene (5 to 500 μg/100 g). The vitamin C content of fruits is high (135 to 600 mg/100 g). Protein (0.2 to 2%), fat (0 to 0.1%) and mineral contents are low in fruits. Avocado contains high amount (22.8%) of fat. Mineral rich fruits include strawberries, cherries, peaches and raspberries. Dry fruits are rich in calcium and iron. High potassium and low sodium give a high dietetic value to fruits and to their processed products. Total sugar content of fruits varies from 3 to 18%. Most of the energy value of fruits is provided by sugar present and are valued as quick sources of available energy. Fruits contain natural acids, such as citric acid in oranges and lemons, malic acid in apples, and tartaric acid in grapes. These acids give fruits tartness and slow down bacterial spoilage. Acidity and sugars are two main elements which determine taste of fruits.

Leafy vegetables contribute mainly vitamins and minerals. They are fair sources of proteins (2 to 7%) and low in fat (0.1 to 1.7%) and carbohydrates (4 to 14%). Important minerals in leafy vegetables are calcium (30 to 500 mg/100 g) and iron (0.8 to 16 mg/100 g). The important vitamins in leaves are pro vitamin A (β carotene) (1200 to 7500 μg/100 g), vitamin C (60 to 140 μg/100 g), folic acid (10 to 30 μg/100 g) and riboflavin. Roots and tubers contain high percentage of carbohydrates (4 to 38%) in the form of starch. They are fair sources of proteins (0.7 to 3%) and poor to fair sources of B complex vitamins and vitamin C. Yellow and orange coloured roots and tubers like carrot and sweet potatoes are rich in β carotene. They are poor sources of calcium and iron. Vegetables which do not come under leafy vegetables and roots and tubers will also provide minerals and vitamins and add variety to diet. They also contribute to fibre content of diet.

Apart from nutritional benefits, scientific community has recognized the health benefits of fruits and vegetables. Beta carotene, an important carotenoid pigment and a precursor of vitamin A present in yellow, orange, red and green vegetables and fruits is an important antioxidant. It protects body against the dangerous free radicals and is beneficial in coronary heart diseases, certain types of cancers, macular degeneration and cataract.

Another carotenoid pigment, lycopene, the predominant pigment in tomatoes, has very powerful disease fighting capabilities particularly in prostate cancer. It is associated with reduced risk of many cancers and gives protection against heart attacks. Lutein present in green leaves prevents macular degeneration, a leading cause of blindness among elderly. As an antioxidant, it helps to prevent cataract and reduce risk of heart disease and gives protection against breast cancer.

Glucosinolates are present in cruciferous vegetables like cabbage, broccoli and brussels sprouts. Glucobrasin, an important glucosinolate is metabolized to produce two important phytochemicals namely isothiocyanates and indoles, which trigger production of enzymes, that block cell damage due to carcinogens. Polyphenols like tannic acid, ellagic acid, vanillin, chlorogenic acid and ferulic acid present in different fruits and vegetables also prevent formation of carcinogens and inhibit mutagens.

Flavonoids present in fruits and vegetables like anthocyanin, qucertin, hesperidin, tangerine, myricetin, kaempferol, resveratrol and apigenin also act as antioxidants. Estrogen like substances namely phytoestrogen present in fruits and vegetables protect against certain cancers.

Antioxidants like vitamin C, flavonoids, carotenoids and polyphenolic substances present in fruits and vegetables prevent fatty deposits in blood vessels and reduce risk of Ischemic heart disease. Folic acid and vitamin B_6 present in vegetables help to lower blood levels of homocysteine, a risk factor for heart disease. Other beneficial effects of fruits and vegetables for a healthy heart are their low calorific value, low sodium and saturated fats and high potassium. Due to the presence of high potassium and other antioxidant phytochemicals, fruits and vegetables also reduce the risk of strokes.

Fibre content and low calorific value of fruits and vegetables reduce risk of diabetes. Fibre slows absorption of sugar into blood and help to reduce blood glucose level. In diabetes mellitus, metabolic state will be altered and produce free radicals which may cause damage to eyes, kidneys, blood vessels and heart. Antioxidants and phytochemicals present in fruits and vegetables help to block free radicals and thus damage caused to different organs. Fibre content of fruits and vegetables are beneficial in lowering blood cholesterol. It is also important as roughage and helps in bowel movement and thus prevents constipation. Fruits and vegetables are also beneficial in controlling body weight by their low calorific value and high fibre content.

Processing of fruits and vegetables

All fruits and vegetables can be processed, but some important factors, which determine whether it is economical are; demand for a particular fruit or vegetable in the processed form; quality of raw material, i.e. whether it can withstand processing; and regular supplies of the raw materials. A particular variety of fruit, which may be excellent to eat fresh, is not necessarily good for processing.

Many ordinary table varieties of tomatoes, for instance, are not suitable for making paste or other processed products. A particular mango or pineapple may be very tasty to be eaten fresh, but when it goes to processing centre, it may fail to stand the processing requirements due to variations in its quality, size, maturity, variety and so on. A processing centre or a factory cannot be planned just to rely on seasonal gluts; although it can take care of the gluts. It will not run economically unless regular supplies are guaranteed. To operate a fruit and vegetable processing centre efficiently, it is important to pre-organize growth, collection and transport of suitable raw materials.

Fruit beverages

Fruit beverages are highly refreshing, thirst quenching, appetizing, easily digestible and nutritionally far superior to synthetic drinks. The fruit beverages include natural and sweetened juice, squash, syrup, cordial, nectar, fruit juice concentrate and ready to serve (RTS) beverages. Fruits most commonly used to prepare beverages are orange, lime, pineapple, grapes, cashew fruit, mango, jackfruit, apple etc. To prepare fruit beverages select fresh, fully ripe, tart, juicy and seasonal fruits.

Fruit juices are products for direct consumption and are obtained by the extraction of cellular juice from fruits. As per FSSAI, in sweetened fruit juice, a minimum amount of 10% Total Soluble Solids (TSS) and 85% juice should be present. However, the unsweetened juice should contain 100% natural fruit juice and the TSS should be natural. The FSSAI specifications for different fruit beverages are given in Table 1.

Table 1: Specifications for fruit beverages (FSSAI)

Beverages	Juice (%) (minimum)	TSS (%) (minimum)	Acidity (%) (maximum)	Preservative (ppm) (maximum) SO_2 or Benzoic acid
Squash	25	40	3.5	350 600
Syrup	25	65	3.5	350 600
Cordial	25	30	3.5	350 600
RTS beverages	10	10	0.3	70 120

(Permitted synthetic colours added to fruit beverages should not exceed 100 ppm)

Squash essentially consists of juice containing moderate quantity of fruit pulp to which cane sugar is added for sweetening. It is diluted before serving. Quantity of different ingredients required to prepare squash is given in Table 2.

Table 2: Ingredients required to prepare fruit squash

Squash	Juice (Kg)	Sugar (Kg)	Water (Kg)	Citric acid (g)	SO_2 (ppm)	Benzoic acid (ppm)
Orange	1	1.7	1	20	350	-
Lime	1	1.7	1	-	350	-
Pineapple	1	1.7	1	20	350	-
Mango	1	1.7	1	20	350	-
Grapes	1	1.7	1	20	-	600

After extracting fruit juice, mix sugar, citric acid, and water and heat till sugar and citric acid are dissolved. Cool sugar syrup and filter through a muslin cloth. Blend the clean syrup with fruit juice. Add preservative after dissolving in a small quantity of boiled and cooled water or cooled squash. Add permitted colour and essence in adequate amounts. Colour and essence should be selected on the basis of the fruit selected to prepare squash. Pour squash in dry sterilized bottles leaving 2.5 cm head space. Close the bottle with pilfer proof closures which should be dipped in 1% potassium metabisulphite (KMS) solution. After sealing, wash and dry the bottles and label. The product keeps well for more than one year without much change in colour, taste and flavour.

To prepare fruit syrup, method described for squash can be adopted. The TSS content of syrup should be at least 65%. To prepare pineapple or grape fruit syrup, for 1 Kg of fruit juice, 2.6 Kg of sugar, 20 g citric acid and 400 ml of water can be used. Mix juice, sugar, citric acid and water and heat till sugar is dissolved. Cool sugar syrup and strain through a muslin cloth. Add KMS @ 610 mg/Kg for pineapple syrup and sodium benzoate @ 700 mg/ Kg for grape syrup. Add colour and essence on the basis of the fruits selected. Bottle and store the syrup as described in squash.

Cordial is a sparkling, clear, sweetened fruit juice from which pulp and other suspended materials are removed completely. It contains at least 25% juice and 30% total soluble solids. It also contains at least 1.5% acid and 350 ppm sulphur dioxide. This is suitable for blending with wine. To prepare lime cordial, Juice is stored in barrels which are lined with microcrystalline wax. KMS is added as preservative@ 1.5 g/Kg of the juice during storage. During storage, the sediment settles and forms a compact layer at the bottom and clear juice remains at the top. Clarification process takes 2-3 months. The clear juice is siphoned off. This method is slow. To make it fast, gelatin and tannin can be added. To prepare cordial, for the clarified 1 kg juice, 1.25 Kg sugar and 1 kg water can be used. Mix sugar and water thoroughly with slight warming. Cool sugar syrup and mix with clarified lime juice. Filter the mixture by means of a filter press or strain the product through a muslin cloth. Add KMS @ of 150 mg per Kg of the finished product. Bottle and store the finished product as described in squash.

To prepare RTS beverages, mix sugar, citric acid and water and heat till sugar is dissolved. After cooling, filter through a muslin cloth and mix with fruit juice. Add colour and essence on the basis of the fruit selected to prepare RTS beverage. The beverage can be consumed immediately. The quantity of different ingredients required to prepare 20 Kg RTS beverage is given in Table.3.

Table 3: Ingredients required to prepare 20 Kg RTS beverage

Ingredients	Pineapple	Orange	Lime	Grapes
Juice (Kg)	2.5	2.5	1.5	2.75
Sugar (Kg)	3	3	3.5	2.5
Water (l)	14.5	14.5	15	14.75
Citric acid (g)	12.5	12.5	-	12.5

If RTS beverage is to be preserved for few days, sulfur dioxide @ 70 ppm or benzoic acid @120 ppm depending on the fruit selected to prepare RTS can be added. If RTS is to be preserved for 1-2 months, heat the beverage to 90°C; fill it hot immediately into clean, warm and sterilized narrow mouthed bottles of 200 to 300 ml capacity, leaving about 2.5 cm head space and seal with crown corks. Hold the bottles in horizontal position for 10 to 15 minutes and allow them to air cool in vertical position.

Use of fruit juices in the preparation of carbonated drinks is practically unknown in India. Mostly, artificially flavoured drinks which have no nutritive value are prepared by this method. Use of fruit juices will increase the nutritive value of carbonated beverages.

One of the most important factors that relates to the taste of bottled fruit juice beverage is carbon dioxide gas content or degree of carbonation. Carbonation is the process of dissolving or incorporating carbon dioxide in a beverage so that when served, it gives off the gas in fine bubbles and has the characteristic pungent taste suitable to the carbonated beverage.

In beverage manufacture, CO_2 not only provides the distinctive taste of carbonated drinks but also inhibits the growth of certain microorganisms. Fruit juices can be carbonated directly or preserved in the form of concentrates for subsequent carbonation. Clarification of such juice is essential prior to carbonation. Carbonated beverage keeps well for about a week without addition of any preservative. For longer storage of carbonated drink, use of preservative (0.05% sodium benzoate) is important.

Juice obtained by removal of a major part of their water content by vacuum evaporation or fractional freezing are defined as "concentrated juice". By concentrating the juice, processor reduces bulk of juice, thereby reducing storage volume requirement and transportation cost. Process starts with pressing fruits and obtaining pure fruit juice. This is then stabilized by heat treatment which inactivates enzymes and microorganisms. Next processing step is concentrating the fruit juice up to 40-65°Brix under vacuum. Many methods like evaporation, membrane concentration and freeze concentration are adopted to concentrate fruit juice.

Production of concentrated juice by evaporation is performed under vacuum (less than 100 mm Hg residual pressure) up to a concentration of 65-70% total sugar which assures preservation without further pasteurization. Modern evaporation installations recover flavours from juice which are then reincorporated in concentrated juice.

In membrane concentration, membranes effectively separate water molecules from other food constituents. Concentration of juice is also possible by using combination of reverse osmosis and evaporation. Specific membranes are used for this purpose. Principle involved is the interposition of a membrane between feed stream and a transfer stream, and establishment of conditions providing a driving force for transport of water across the membrane from feed to transfer stream.

In membrane concentration, since, temperature of processing is less, product quality is maintained. Lower energy requirement, lower labour cost, lower floor space, and wide flexibility are other advantages of membrane concentration.

Reverse osmosis technology is effective in concentrating a low solid juice (7-8°Brix) two or three fold. From there, evaporation technology will be appropriate. Recently, a new reverse osmosis design has claimed effective in achieving a 50-60°Brix concentrate. These fruit juice concentrates are often further stabilised by addition of sodium benzoate and potassium sorbate and are usually stored away from light and are refrigerated or frozen.

Freeze concentration is based on freezing point depression. Pure water freezes at a temperature of 0°C. However, if dry solid is dissolved, freezing takes place at further low temperature. In this process, a freezer is used to produce ice crystals out of fresh juice. Ice crystals are separated from juice slurry by using a centrifuge or filter press. Freeze concentration avoids problems associated with evaporation methods that depend upon heat. It is capable of concentrating most juices to 50°Brix without appreciable loss of taste, aroma, colour or nutritive value.

To prepare fruit juice powders, fruit juice is sprayed as mist into an evaporating chamber and flow of air is so regulated that dried juice falls to the floor of the chamber in the form of dry powder. Powder is then separated and packed air tight. Powder when dissolved in water makes a fruit drink almost similar to its original fresh juice.

Many fruit juices can be dehydrated to powders or crystals for reconstituting into beverages and are available at prices comparable with quality frozen concentrates. These powdered products are available in several sizes of packages. These powdered products are also considered as "sports drink", which require mixing by individuals and avoid inconvenience of transporting large volume of liquid.

These powders are highly hygroscopic in nature and hence require proper packaging. Fruit juice powders from orange, mango, jackfruit, guava etc. have been developed by CFTRI. Powders can be made by vacuum drying, spray drying, freeze drying, drum drying or foam-mat drying. Moisture content of powdered juice varies from 3-5%.

Sugar based products

Preparation of sugar based products like jam, jelly, marmalade, preserve and candy is one of the most important aspects of home scale preservation as well as industrial level processing of fruits. This method of preservation with high sugar concentration is principally based on the reduction in moisture content so as to arrest the microbial spoilage.

Fruit jam

Jam is a product with reasonably thick consistency, prepared by boiling fruit pulp with sufficient quantity of sugar firm enough to hold fruit tissues in position. Jam contains 0.5-0.6% acid and invert sugar should not be more than 40%. The FSSAI specification for jam is furnished in Table 4. In jam, synthetic sweetening agents are not permitted and only permitted colours are used.

Apple, papaya, strawberry, mango, grapes, jackfruit, tomato, pineapple, etc. are used to prepare jam. Various combinations of different varieties of fruits can be often made to advantage, pineapple being one of the best for blending purposes because of its pronounced flavour and acidity.

To prepare jam, select sound fruits and wash in running water. Mode of preparation of fruits varies with the nature of the fruit. For example, mangoes are peeled, steamed and pulped; apples are peeled, cored, sliced, heated with water and pulped; plums are scalded and pulped; peaches are peeled and pulped; berries are heated with water and pulped or cooked as such. For pineapple, remove the crown, peel off the skin and the eyes and cut sound portion into small pieces and pulp the cut pieces to obtain a uniform mass.

Generally, cane sugar of good quality is used in the preparation of jam. Proportion of sugar to fruit varies with type and variety of fruit, its stage of ripeness and acidity. A fruit pulp to sugar ratio of 1:1 is generally followed (Sudheer and Indira, 2007).

Table 4: Specifications for jam, jelly and marmalade (FSSAI)

Ingredients	Jam	Jelly	Marmalade
Fruit content (minimum % in final product)	45	45	45
TSS (minimum % in final product	68	65	65
Preservative (maximum)Sulphur dioxide (ppm) or	40	40	40
Benzoic acid (ppm)	200	200	200

Citric, malic or tartaric acids are present naturally in different fruits. These acids are also added to supplement acidity of the fruits during jam making. Addition of acid becomes necessary as adequate proportion of sugar- pectin-acid is required to give good set to the jam. It is often advisable to add acid at the end of cooking which leads to more inversion of sugar. When acid is added in the beginning, it will result in poor set.

To prepare jam, fruit pulp is cooked with required quantities of sugar and pectin. Concentration of jam is finished at an optimum point to avoid over cooking which leads to economic loss due to low yield. Under cooking will lead to spoilage of jam during storage due to fermentation. End point of jam can be determined by drop test or sheet test.

In drop test, a little quantity of jam is taken from boiling pan in a tea spoon and allow to air cool before putting a drop of it in a glass filled with water. Settling down of the drop without disintegration denotes end point.

In sheet test, a small portion of jam is taken in a spoon, cool slightly and then allow to drop off keeping the spoon or ladle in horizontally inclined position. If jam drops like syrup, further concentration is needed. If it falls in the form of flake or forms a sheet, it indicates end point.

Refractometer method is the most common method used by small and large scale fruit processing industries for jam making. The cooking is stopped when refractometer shows 68 °Brix .

Boiling point method is the simplest and the best to determine finishing point of jam. Jam containing 68% TSS boils at 106°C at sea level.

To prepare pineapple jam, mix 1 kg of pineapple pulp with 1 kg of sugar and keep for ½ to 1 hour. Cook the mixture slowly with vigorous stirring. Add 10 to 12 g of medium rapid set pectin for every 1 kg sugar. Add 5 g of citric acid at the end of cooking. Cook again till the mass approaches jam consistency. Fill hot jam in sterilized dry jars, cool and close the container and store in a cool dry place.

Fruit jelly

A jelly is a semisolid product prepared by boiling a clear, strained solution of pectin containing fruit extract, after addition of sugar and acid. A perfect jelly should be transparent, well set, but not too stiff, and should have the original flavour of the fruit. It should be of attractive colour and should keep its shape when removed from the mold. It should be firm enough to retain a sharp edge but tender enough to quiver when pressed. It should not be gummy, sticky, or syrupy or have crystallized sugar. Product should be free from dullness with little or no syneresis, and neither tough nor rubbery. Amount of ingredients required for jelly as specified by FSSAI is given in Table 4. In jelly also synthetic sweetening agents are not permitted and only permitted colours can be used.

Jelly is usually manufactured from juice obtained from a single fruit species, obtained by boiling in order to extract as much soluble pectin as possible.

Guava, sour apple, plum, papaya, certain varieties of banana, outer rind of jackfruit and gooseberry are used for preparation of jelly. Other fruits can also be used but only after addition of pectin powder due to low pectin content. Pectin is the important component responsible for the texture of jelly.

Select mature, ripe and fresh fruits with good flavour. Avoid overripe fruits as far as possible. Slightly under ripe fruits yield more pectin than over ripe fruits because as the fruit ripens, pectin decomposes into pectic acid, which does not form jelly with acid and sugar. A mixture of under ripe and ripe fruits can also be used to prepare jelly ie., under ripe fruits for their pectin content and ripe fruits for flavour.

After selection, wash the fruits thoroughly. Since jelly is made from aqueous extract, it is not necessary to peel the fruits. For example, guava and apple can be cut into small pieces with the outer skin. However, for orange, the outer flavedo should be removed to avoid excessive bitterness. Most of the fruits are boiled for extraction of juice to obtain maximum yield of juice and pectin. Boiling converts protopectin into pectin and softens fruit tissues. Very juicy fruits do not require addition of water and are crushed and heated to boiling only for 5 min. Firm fruits are cut or crushed and boiled with water for 15- 20 min. The length of boiling varies according to type and texture of fruit. Amount of water added to fruit must be sufficient to give a high yield of pectin. Apples require one half to an equal volume of water, where as citrus fruits require 2-3 volumes of water for each volume of sliced fruits. Only a minimum quantity of water should be added to the fruit to extract pectin. If necessary, a second and even a third extract can be taken and these can be combined with the first extract.

Strain the pectin through bags made of linen, flannel, or cheese cloth folded several times. For large scale production, the fruit extract is made to pass through filter press for clarity. For small scale production, the pectin extract is allowed to settle overnight and the supernatant liquid can be drained off.

Analyse clarified extract for pH, acidity, soluble solids and pectin content by common laboratory methods. To determine pectin content, easiest way adopted is to precipitate the pectin with alcohol. A rapid test for evaluation of pectin content is by mixing one teaspoonful of juice with three teaspoonful of methyl alcohol. Take the pectin extract in a small beaker and add methyl alcohol gently along the sides of the beaker. Carefully rotate beaker and allow the mixture to stand for a few minutes. If the extract is rich in pectin, it will form a single transparent lump of jelly like consistency. If pectin is present in moderate quantity, clot will be less firm and fragmented. Granular clots will indicate insufficient pectin. Insufficient pectin remains in numerous small granular lumps.

Based on pectin test, quantity of sugar to be added is worked out. For the extract, rich in pectin, sugar equal to the quantity of the extract is added. To the extract with moderate pectin, 650 – 750 g of sugar should be added to each kilo gram of extract. For juices rich in pectin, jellification occurs without pectin addition. If pectin content is less, 1-2% powdered pectin can be added to the fruit extract.

The extract is boiled to remove about half of the water. Then, calculated quantity of sugar is added gradually. Remainder of water is evaporated until a TSS of 65% is reached. During boiling, it is necessary to remove foam/scum formed. Product acidity must be brought to about 1% corresponding to pH > 3. Any acid addition is performed always at the end of boiling. Boiling of jelly is performed in small batches (25-75 kg) to avoid excessively long boiling time which brings about pectin degradation.

Since, excessive boiling of jelly results in greater inversion of sugar and destruction of pectin, prolonged boiling of jelly should be avoided. The end point can be judged by sheet test, drop test, refractometer or thermometer test. Methods like sheet test and drop test can be done in similar way as in jam preparation. In refractometer method, the cooking is stopped when the refractometer shows 65°Brix. In temperature test, a solution containing 65% TSS boils at 105°C. Heating of jelly to this temperature will automatically bring the concentration of solids to 65%. After jelly is ready, remove foam, cool slightly and pour into dry glass jars. Cooling is optional and is carried out up to 85°C. After filling in glass jars cool it for jellification.

Marmalade

Marmalade is a fruit jelly in which slices of fruit or its peels are suspended. The term is generally used for products made from citrus fruits like orange and lemon in which shredded peel is used as the suspended material. In the preparation of marmalade, pectin and acid contents are kept on higher side than jelly. Bitterness is regarded as desirable characteristic of marmalade. Marmalades are classified into two: jelly marmalade and jam marmalade.

Good quality jelly marmalade can be prepared from a combination of sweet orange/ mandarin orange and sour orange in a 2:1 proportion. Shreds of sweet orange (Malta) peel are used in the preparation.

Extraction of pectin, filtration/straining of extract and analysis of extract are carried out in the same way as that of jelly preparation. To prepare shreds, peel the outer layer of yellow portion of citrus fruits carefully. Peel is cut into slices of about 2-2.5 cm long and 1-1.2 mm thick. Boiling in water with 0.25% sodium bicarbonate or 0.1% ammonia solution softens the shreds. Before addition to jelly, keep shreds in heavy syrup for some time to increase their bulk density to avoid floating on surface when it is mixed with jelly.

Boil fruit extract before adding sugar. Add required quantity of sugar and during boiling, remove impurities in the form of scum occasionally. When temperature of mixture reaches 103°C, mix prepared shreds of peel at the rate of 5-7% of original extract. Continue boiling till the end point, which is judged in the same way as in the case of jelly. Like jelly, marmalade also contains 65% TSS at 105°C. Boiling should not prolong for more than 20 min, after addition of sugar to get bright and sparkling marmalade.

Cool marmalade to permit absorption of sugar by the shreds from surrounding syrup. If marmalade is filled in hot, shreds may come to the surface instead of remaining in suspension. During cooling, occasionally stir the product gently for uniform distribution of shreds. When marmalade temperature reaches around 85°C, a thin film begins to form on surface, which prevents shreds from coming to surface.

Flavouring is done by adding some flavour or orange oil to the product at the end of boiling to supplement the flavour lost during cooking. Generally, a few drops of orange oil are mixed in marmalade before filling into containers. Like jelly, marmalade is also filled into jars at a temperature of around 85°C.

Jam marmalade is practically made by the method used for preparation of jelly marmalade except that pectin extract is not clarified. Slice orange peel after removing albedo portion and treat in the same way as for jelly marmalade. Mix sliced fruit of orange or lemon after removing peel with little quantity of water and boil to soften. Press boiled mixture through coarse pulper to remove seeds and to get thick pulp. Mix pulp with equal quantity of sugar and cook to a consistency of 65°Brix or consistency of jam. Mix the treated shreds with jam marmalade when it is slightly cool. Add few drops of orange oil also in marmalade before filling into containers. Filling and packaging are done in similar way as adopted for jelly and jelly marmalade.

Preserve

A mature fruit/vegetable or its pieces impregnated with heavy sugar syrup till it becomes tender and transparent is known as preserve. When, fruits are placed in concentrated sugar syrup, water moves out of the fruit and sugar moves into it until equilibrium is reached by osmosis. Apple, cherry, anola, pineapple, pear, mango, papaya, strawberry, ash gourd, cashew apple etc., can be used for making preserve. FSSAI specifications (minimum) for preserve are 68% TSS and 55% prepared fruit.

Preparation involves primary operations like, selection of fruit, peeling, puncturing or pricking (to promote sugar penetration) and blanching. Blanching can be done with or without additives to inactivate natural enzymes and to reduce oxidative discolouration. Blanched fruits are then treated for firming the texture of product. Now, sugar is added, concentrated and packed after addition of preservatives. Different processes employed for the preparation of preserves from fruits and vegetables at commercial level are explained below.

i. Rapid process

Fruits are cooked in low sugar syrup. Boiling is continued with gentle heating until the syrup becomes sufficiently thick. Rapid boiling should be avoided as it makes fruit tough. The final concentration of sugar should not be less than 68% which corresponds to a boiling point of 106°C. This is a simple and cheap process but flavour and colour of the product are lost considerably during boiling.

ii. Slow process

In this method, fruits are blanched until it becomes tender. Sugar, equal to the weight of fruit, is then added to fruit in alternate layers and the mixture allowed to stand for 24 hours. Then, by boiling on second, third, and fourth day consecutively strength of syrup is raised to 70% TSS. A small quantity of citric or tartaric acid is also added to invert a portion of cane sugar and thus to prevent crystallization. Prepared preserve is then packed in containers.

iii. Vacuum process

Fruit is first softened by boiling and then placed in syrup which should have 30-35% TSS. Fruit syrup blend is then transferred to a vacuum pan and concentrated under reduced pressure to 70% TSS. Preserves made by this process retain the flavour and colour of fruit and will be better than the one prepared by other two methods.

Preserve is cooled quickly, drained free of syrup and then filled in dry containers. Freshly prepared boiling syrup containing 68% TSS is then poured into jars/containers which are then sealed air tight. In commercial scale production, it is better to sterilize cans to eliminate any possibility of spoilage of product during storage.

To prepare papaya preserve, select fully mature and unripe fruits. Peel and remove seeds. Cut flesh into pieces of 7.5cm x7.5cm and prick them with stainless steel fork. Immerse fruit pieces in dilute lime water (15g of lime (calcium hydroxide) in one litre of water) for about 3 to 4 hours. Take out the pieces and wash three to four times with fresh water and boil in sugar syrup of 40°Brix. Keep overnight. Next day, drain out the syrup and add enough sugar to raise its Brix to 50°. Repeat the process everyday till the Brix of residual syrup reaches 70 to 75° (with 50% invert sugar). Drain off syrup and cut fruit pieces further to desired shape. Pack the product in sterilized wide mouthed bottles or tin cans and add to them the residual syrup to fill the inter space. Seal the container air tight and label the product.

Candied fruit/vegetable

A fruit or vegetable impregnated with cane sugar or glucose syrup, and subsequently drained free of syrup and dried is known as candied fruit/vegetable. Most suitable fruits for candying are pineapple, cherry, anola, karonda, papaya, apple, peach and orange peel.

Process for making candied fruit is practically similar to that for preserve. Only difference is that the fruits are impregnated with syrup having a higher percentage of sugar or glucose. A certain amount (25-30%) of invert sugar or glucose is substituted for cane sugar. Total sugar content of the impregnated fruits is kept at about 75% to prevent fermentation. It is desirable that the proportion of cane sugar and invert sugar in the final syrup should be approximately equal.

After removing fruit pieces from sugar syrup, drain for about half an hour and remove unwanted pieces. Wipe pieces with a wet sponge or dip for a moment in boiling water to remove adhering syrup. Then, dry in shade or in a drier at about 66°C for 8 to 10 hours until fruits are no longer sticky to handle. Transfer the product into clean, dry, wide mouthed glass jars, seal the container and store in a cool dry place.

Glazed fruit/vegetable

Covering of candied fruit/vegetable with a thin transparent coating of sugar, which imparts them a glossy appearance is known as glazing. The specifications for candied and crystallized or glazed fruit is not less than 70% total sugar and not less than 25% reducing sugar as per cent of total sugar.

Glazed fruits are prepared by passing the dried candied fruits through sugar syrup. Sugar syrup is prepared by boiling a mixture of cane sugar and water (2:1) in a steam pan at 113 to 114°C and skimming the impurities as they come up. Heating is then stopped and syrup is cooled to 93°C. Granulation of sugar is achieved by rubbing syrup with a wooden ladle on side of the pan. Granulated candies are then placed on trays in warm dry room. To hasten the process, fruits may be dried in a drier at 49°C for 2 to 3 hours till they become crisp. These are then packed in air tight containers for storage.

Crystallized fruit/vegetable

Candied fruits/vegetables when covered or coated with crystals of sugar, either by rolling in finely powdered sugar or by allowing sugar crystals to deposit on them from dense syrup are called crystallized fruit/vegetable.

Place candied fruits on a wire mesh tray which is then placed in a deep vessel. Gently pour cooled syrup (70% TSS) over the fruit so as to cover it entirely. Leave the whole mass undisturbed for 12 to 18 hours during which time a thin coating of crystallized sugar is formed. Tray is then taken out carefully from the vessel and drain the surplus syrup. Place the fruits in a single layer on a wire mesh trays and dry at room temperature or at about 49°C in a dryer.

Fruit bar/leather

Fruit bar or leather is prepared from fruit pulps like mango, peach, plum, apricot and papaya. Fruit pulp is taken and its TSS increased to 30°Brix by adding sugar. This pulp is then spread on stainless steel trays smeared with fat which are dried in a mechanical dehydrator at 60 ± 5°C for 2 hours. Usually, five layers are dried one above the other and final product is packed in polythene bags.

To prepare mango leather, take fully ripe, juicy fruits and wash them. Discard over ripe and rotten fruits to avoid unpleasant flavour in the product. Extract pulp and strain it through fine mosquito net cloth to remove coarse particles. Add 2 g KMS for every kilogram of pulp and mix. Spread the pulp in thin layers on stainless steel or aluminium trays. Before spreading the pulp, smear butter on trays to prevent sticking of product. Cover trays with fine mosquito net cloth and place them in sun. When first layer dries, put another layer upon it. Repeat this process till 0.6 to 1.25 cm thick layer of dried pulp is obtained. Cut into small pieces. Wrap the pieces in butter paper and store in glass jars (CFTRI, 1990).

Fruit toffees

This is made by mixing fruit pulp with other ingredients like glucose, milk powder and edible fat. Concentrate the fruit pulp to about 1/3rd of its volume by heating in a steam jacketed kettle or in a sauce pan over a steam bath. Fruit pulp is first concentrated to half its volume. Generally, for one kilo gram of concentrated pulp add 160 g glucose, 320 g milk powder and 200 g edible fat. Heat this mixture to a thick consistency (75- 80°Brix) followed by spreading it as a sheet of one cm thickness on a fat smeared flat tray and cool. Then, toffees are cut into pieces of desired size, wrap and store in cool dry place.

Other products

Chutney

A chutney is basically a mixture containing fruit or vegetable, spices, salt and/or sugar, vinegar, etc. The method of preparation is similar to that adopted in jam except that spices and vinegar are added. Chutney of good quality should be reasonably smooth, palatable, appetizing and have the true single flavour of fruit or vegetable used for preparation.

Select ripe fruit or vegetable, cut into slices or pieces of suitable size and boil in water to make it soft. Then, cook slowly at a temperature below boiling point. Add onion and garlic at the start to mellow their strong flavour. Coarsely powder spices and add to the product. Whole spices, if used, are bruised and tide loosely in muslin cloth before adding to the mixture and removed before bottling. Add vinegar, sugar, and spices before final stage of boiling. This prevents loss of some essential oils of spices due to volatilization. Bottle chutney while hot in clean and warm jars and seal properly.

Sauce

Sauce and chutney are usually made from the same raw materials, spices and flavours. However, difference is that, sauce is sieved and as a result, are thinner and of smoother consistency than chutney. Sieving is done to remove the skin, seeds and stalks of fruits, vegetables, and spices and to give a smooth consistency. Here, cooking process is longer compared to chutney due to use of fine pulp or juice.

Sauces are generally of two kinds, thin and thick sauces. A good sauce whether thin or thick, has a continuous flow with no skin, seeds and stalks of fruit and/or vegetable and spices used for its preparation, and possess pleasant taste and aroma.

Thin sauces mainly consist of vinegar extract of various flavouring materials like spices and herbs. Their quality depends mostly on the piquancy of material used. Some sauces are matured by storing them in wooden barrel or casks. During storage, they develop flavour and aroma. Freshly prepared products have often a raw and strong taste and they should, therefore, be matured by storage.

For preparation of thin sauce of high quality, spices, herbs, fruit or vegetable are macerated in cold vinegar. Sometimes, they are also prepared by boiling them in vinegar. Sauce is filtered through a fine or coarse mesh sieve of non corrodible metal, according to the quality desired. Skin, seeds and stalks of fruit, vegetable and spices used, should not be allowed to pass through the sieve as they spoil appearance of sauce. Usual commercial practice is to prepare vinegar extracts of each kind of spice and fruit/vegetable separately, either by maceration or by boiling in vinegar and then blending these extracts suitably before filling sauce into barrels for subsequent maturation. Soya sauce made from soybeans and Worcestershire sauce made from tamarind are examples of thin sauces.

A sauce which does not flow freely and which is highly viscous is called a thick sauce. On the other hand, thin sauce is less viscous in consistency. Thick sauce also contains more of sugar and less of acid. Generally, spices and colouring added are practically similar to those in thick ketchups and sauces. It should contain at least 3% acetic acid to ensure its keeping quality. Acidity should not exceed 3-4% as otherwise the sauce would taste sharp. Sugar content varies from 15-30% according to the kind of sauce made. Usually malt vinegar is used. In addition to contributing to acidity of sauce, it also improves its flavour. Sweetness is derived partly from dates, raisin, apple and tomato and partly from sugar added. Colour of sauce varies with the raw material used.

Manufacturing process is same as for chutney. Thickening agents are also added to prevent or retard sedimentation of solid particle in suspension in the sauce. The starch obtained from maize, potato, arrow root, sago and rye are used as thickening agents. Indian gum, gelatin, Irish moss, pectin and other similar substances can also be used subject to food laws of the country. Tomato sauce and apple sauce are examples for thick sauce.

Recent developments in fruit and vegetable processing

Major motivations determining trends of development of new, emerging or future food technologies are those which signify responses of food science and industry to demands of consumers due to their changing lifestyles and expectations for fresher and more natural foods. Alternative food-preservation and processing technologies are being developed to a large extent in relation to consumers' requirements. Most of the currently employed preservation techniques act by inhibiting growth of microorganisms, slowing down or completely preventing their multiplication. These techniques are being employed to meet consumer trends mainly by their use in combinations (hurdle technology), by controlled methods of heating that deliver less damage to product quality, and by well temperature controlled cook chill operations. In contrast to many microorganism-inhibitory techniques, few of the currently employed techniques act primarily by inactivating microorganisms in foods (Gould, 2001). By far, the major inactivation technique being heating preservation, the need of the hour is to develop new and improved techniques for elimination of spoilage and pathogenic microorganisms from the most often contaminated foods (Sudheer and Mathew, 2016). It is encouraging, therefore that most of the newer and 'emerging' technologies that are coming into use for food preservation act by inactivating microorganisms. Important methods employed in fruit and vegetable processing are the following.

Irradiation

Irradiation refers to the physical means of exposing food products or bulk food to gamma rays. Food irradiation is also referred as a 'cold process'. Exposure to radiation causes ionizations and structural changes in exposed molecules. In case of living organisms, exposure to radiation causes structural and functional changes in important macromolecules leading to cell death. A dose that causes little chemical changes in food can cause sufficient changes in DNA leading to cell death because of the unique properties of DNA. Radiation damage to components of cells other than DNA, such as cytoplasmic membrane may also contribute to cell injury. Studies on the wholesomeness, microbial safety, nutritional adequacy and lack of mutagenacity, and toxicity of irradiation have reaffirmed its safety.

Different doses (levels) of radiation are used for different purposes. Current FDA limit for irradiation on fresh produce is 1.0 kGy, but to destroy yeasts and moulds that may exist as spores, irradiation levels of 1.5 to 20 kGy are necessary and these levels are damaging to fruit tissues (Sudheer, 2009).

High hydrostatic pressure technology

One of the most promising novel methods of preservation is the use of high hydrostatic pressure. This technology originally used in the production of ceramics, steel and super alloys has now expanded to the food industry. It can perform number of functions such as increase in product shelf life, denaturation of proteins, texturization of proteins, increase in fusion temperature of lipids and decrease of ice fusion temperature of thawing and freezing with little or no heat treatment, so much so that it is called no heat process. By this process, natural fresh taste of fruits and juices is retained and under optimal conditions, colours (i.e., chlorophyll, anthocyanins, carotenoids) are not or very slightly affected (Griet *et.al.*, 2013).

At a pressure of 4000-9000 atmospheres, enzyme and bacteria are inactivated but taste and flavour remain unaffected (Hendrickx *et al.*, 1998). Inactivation of enzymes is due to denaturation of the enzyme protein. Different enzymes are known to respond differently to pressure. To avoid reversible effects and achieve adequate inactivation of all enzymes, pressure above 6000 to 8000 bar is required.

Microencapsulation technology

In microencapsulation process, tiny particles are surrounded by coating to give small capsules. Microencapsulation may be defined as "the technique of packaging minute particles of solid, liquid or gas within continuous individual shells designed to release their contents in a predictable manner under predetermined conditions" (Korus *et al.*, 2003). In other words, micro encapsulation preserves the active material in the divided state and releases it as and when occasion demands, by any desired mechanisms such as disruption, dissociation, dissolution or diffusion and with any desired rate such as instantaneous, delayed, controlled or sustained release. It improves the case of handling of liquids by converting them into dry, free flowing, pseudo solids, which nevertheless have a high liquid content. Micro encapsulation can modify and improve shape, size, distribution and can also alter the solubility, dispersibility, durability and availability of a substance. It can also change weight or volume of a substance.

Oscillating magnetic field

Utilization of oscillating magnetic field to inactivate microorganisms has the potential to pasteurize food with an improvement in quality and shelf life compared to conventional pasteurization process. Magnetic fields are usually generated by supplying current to electric coils. A magnetic field of 5-50 tesla (T) with short pulses of 10-25 ohms-cm and frequency 5-50 HZ are required to inactivate microorganisms. Oscillating magnetic fields of this density can be generated by using (i) superconducting coils, (ii) coils that produce DC fields, or (iii) coils energized by discharge of energy stored in a capacitor.

Technical advantage of inactivation of microorganisms with oscillating magnetic field include minimal thermal denaturation of nutritional and organoleptic properties, reduced energy, potential treatment of foods inside a flexible package to avoid post-process contamination. Oscillating magnetic field treatment ensures no post-process contamination unlike in thermal process.

Ozone technology

Ozone is an allotrope of oxygen and a strong naturally occurring oxidizing agent. In nature, ozone is formed by a high-energy input splitting of the oxygen molecule. The single oxygen thus formed combines with available O_2 to form the very reactive O_3. Ozone is highly unstable and therefore must be generated on site. Hence, artificial methods of ozone generation have to be used. Ozonization is a non-thermal method of killing microorganisms using ozone, through oxidation of their cell membranes. Most of the pathogenic food borne microbes is susceptible to this oxidizing effect.

High intensity pulsed electric field

High- voltage electric field relies on lethal effect of electric field on microorganism. They destroy microorganisms essentially without heating the food. Electric field is applied to fluids in the form of short pulse with pulse duration ranging between few microseconds and milliseconds rather than using electricity directly (Pereira *et al.*, 2009). Minimal heat is generated and the process generally remains non-thermal. Foods are processed within a short period of time and energy lost due to heating is minimal. Foods susceptible to dielectric breakdown are not suitable for electric field processing. The risk of dielectric breakdown limits this technology feasible for liquid foods only.

Microwave processing

Microwave energy for food processing has now become a reality and is currently in commercial use in Europe and USA. Speed of operation, energy saving, precise process control and greater penetration depth are its major advantages. Moreover, rapid temperature rise causes less destruction of nutrients as compared to conventional processing. Microwave energy (a form of electromagnetic energy) is absorbed directly by the material that responds to the rapidly changing high frequency electric field. Direct inductive heating drastically speeds processing time and process efficiency. Relatively high power and low core microwave sources at 0.915 – 2.45 GHZ are now available, making microwave processing a cost-effective industrial and scientific tool.

Ohmic heating

Ohmic heating is also referred to as Joule heating, electrical resistance heating or electro-conductive heating. Food industry has shown a renewed interest in ohmic technology in recent years. Ohmic heating is an advanced thermal process where food acts as an electrical resistor. Design usually consists of electrodes that contact the food, whereby electricity is passed through the substance using a variety of voltage and current combinations. Substance is heated by dissipation of electrical energy. When compared to conventional heating, where heat is conducted from outside in using a hot surface, ohmic heating conducts heat throughout the entire mass of food uniformly. Success of ohmic heating depends on the rate of heat generation in the system, the electrical conductivity of the food and the method by which food flows through the system. Ohmic heating for food industry consists of using electrical energy to heat foods as a method of preservation, which in turn can be used for microbial inactivation for several other processes such as pasteurization, extraction, dehydration, blanching or thawing.

Manothermosonication

A potentially useful synergy of ultrasound with heat was reported for inactivation of bacterial spores (*B. cereus* and *Bacillus licheniformis*), thermoduric *Streptococci*, *Staphylococcus aureus* and other vegetative microorganisms, and for the inactivation of enzymes. However, with increase in temperature, potentiating effect of ultrasound will become less and less, and (for spores) will disappear at approximately the boiling point of water. Important discovery that led to the development of manothermosonication was that this disappearance of the synergism could be prevented if the pressure was raised slightly (e.g. by only the order of tens of megapascals). Combination procedure generally has the effect of reducing apparent heat resistance of microorganisms by about 5-20°C, depending on the temperature, the organism and its z-value.

Manothermosonication represents an interesting combination of low pressure (0.3 MPa) heat treatment and ultra sonic wave treatment that seems effective for inactivation of microorganisms (Sudheer and Indira, 2007). The combined heat treatment can be used in processing fruit juices and other drinks to solve the problem of thermostable enzymes.

Use of pulsed light

The technique finds potential application mainly in sterilizing or reducing microbial population on surface of packaging materials, on processing equipment, fruits and vegetables and medical devices as well as on many other surfaces. The technology uses short duration flashes of broad spectrum white light to inactivate a wide range of microorganisms including bacterial and fungal spores. Light with an energy density of about 0.01 to 50 J/cm^2 and wave length in the range of 170 to 2600 nm is used (Qin *et al.*, 1999; Sitzmann, 2006). The duration range from less than 0.1 second and flashes applied provide high level of microbial inactivation.

Hurdle technology

This concept is a combination of several preservation factors (Hurdles) that should not be overcome by microorganisms present, accounting for the final microbial stability and safety of food. It illustrates the complex iteration of temperature, water activity, pH, heat, chemicals, salt, sugar, additives and so on to the microbial stability of food. Until now, about fifty different hurdles have been identified in food preservation. In hurdle technology microbial stability of product is achieved by selecting few hurdles which in turn selectively disturb the homeostasis of the microorganisms (Selman, 2002). Combination of hurdles may also exert an additive or synergistic effect leading to auto sterilization of microorganisms.

At the level employed in many foods, individual hurdles may not provide adequate protection from spoilage or pathogenic microorganisms. When multiple hurdles are combined, each hurdle plays a role in reducing microbial activity until, eventually, the microbial population is so weakened that it cannot cross any further hurdles and food is protected from spoilage. If hurdles are insufficient to reduce microbial growth, food products may not be adequately protected.

Packaging of fruits and vegetables and their products

Packaging is an integral part of fruit and vegetable processing. It has major influence on storage life and on marketability of fresh as well as processed products. Throughout the entire handling system, packaging can be both an aid and a hindrance to obtain maximum storage life and quality of product. Packaging materials such as trays, cups, wraps, liners and pads may be added to help immobilize the produce. Either hand-packing or mechanical packing systems may be used. Simple mechanical packing systems often use the volume-fill method or tight-fill method, in which sorted produce is delivered into boxes. Packing and packaging methods can greatly influence air flow rates around the commodity, thereby affecting temperature and relative humidity management of produce while in storage or in transit.

Packaging materials

There are two main groups of packaging materials: (i) Shipping containers and (ii) Retail containers. Shipping containers are containers which contain and protect contents during transport and distribution (eg. wooden, metal or fibre board cases, crates, barrels, drums and sacks). Retail containers are consumer units which protect and advertise food in convenient quantities for retail sale and home storage (eg. metal cans, glass bottles, jars, rigid and semi-rigid plastic tubs, collapsible tubes, paper board cartons and flexible plastic bags).

Glass containers are impervious to moisture, gases, odours and microorganisms. They are inert and do not react with or migrate into food products. They have filling speeds comparable with those of cans and are transparent to display the contents. They are suitable for heat processing when hermetically sealed and transparent to microwaves. They can be molded to variety of shapes and colours. Glass containers are rigid to allow stacking without container damage. In addition to this, they have many other advantages viz; their reusable and recyclable property and they are perceived by customer to add value to product.

Main classes of glass receptacles are; jars which are resistant to heat treatments, jars for products not submitted to heat treatment (eg. acidified vegetables), glass bottles for pasteurized products (eg. fruit juices) or not pasteurized (eg. syrups) and jars with higher capacity.

Plastic materials are available in flexible and rigid forms. Modern trend in packaging of food products is introduction of a variety of plastic packaging materials depending on their characteristics as an alternative to metal and glass containers. Plastics are available in different forms like flexible films, laminated films, coated films, co-extruded films, rigid and semi rigid containers (Sudheer and Madhana, 2012). Flexible packaging describes any type of material that is not rigid, but the term flexible film is usually reserved for non fibrous materials which are less than 0.25 mm thick. In general, they are heat sealable, suitable for high speed filling and printing and add little weight to the product. They fit closely to shape of food; thereby wastage of space is less during transportation and storage. In most cases, such films are used in construction of inner containers. Since they are non-rigid, their main functions are to contain product and protect it from contact with air or water vapour. Their capacity to protect against mechanical damage is limited, particularly when thin films are considered.

Flexible films include single films (eg. polyethylene, polyester etc.), coated films (eg: films coated with aluminum), laminated films (lamination of two or more films) and co-extruded films. Flexible films have different values for moisture and gas permeability, strength, elasticity, inflammability and resistance to insect penetration and many of these characteristics depend upon film's thickness. Advantages of plastic films such as cost effectiveness, convenience, lightweight leading to reduction in storage space and transportation cost and easy disposability, have been exploited to encourage replacement of metal cans and glass containers.

Quality of fresh and processed products

Quality is an important factor in the production and marketing of biological products. Quality may be equated to meeting standards required by a selective customer. In this context, customer is the person or organization receiving the product of each point in the production chain. This is important, because quality is perceived differently depending on needs of particular customer. Customer, purchasing fruit/vegetable for consumption usually judges product on the basis of its appearance, including its shape, firmness, and colour, as well as freedom from defects such as spots, marks, or rots. Consumer will also judge the eating quality as well as its keeping qualities in home. Packing shed operator may be much more concerned about percentage of good fruits/vegetables in a batch, and how easy it is to handle and grade. There are many different factors that can be included in any discussion on quality.

Factors influencing product quality

To produce high quality products, processor needs to be aware of quality attributes which consumer discerns as most important and which are most relevant in determining acceptability. Most consumers would initially judge acceptability of products on their appearance, flavour, texture and perceived nutritional benefits. Each of these attributes is a function of biochemical and physico-chemical composition of fruits or vegetables, which is influenced by quality and composition of raw materials, effects of processing and effects of environmental factors, such as temperature, oxygen, light and moisture encountered during storage and distribution, customer handling and use and barriers to these factors provided by packaging.

Food safety is an important factor in quality of processed foods. There are several possible problems due to natural contaminants, synthetic toxicants, microbial contamination etc.

Measurement of product quality

Eating quality can be assessed most accurately by using taste panels. These consist of a selection of customers from market who are trained to assess quality attributes being examined. This is a very expensive and time consuming task. Furthermore, it can only be attempted when product reached eating stage, by which time it is too late to do much about quality. Hence, taste panels cannot be used easily as part of production process. Instead, it is necessary to resort to readily usable test methods, which may involve simple or complex equipment. Reliability and value of these methods depend on how well they correlate with views of consumers, rather than level of scientific objectivity of test.

Another way of measuring product quality is to monitor sales and customer complaints; higher the sales and fewer the complaints, more likely one is to be satisfying consumer requirements. No responsible food manufacturer would rely on this alone as their only method of quality control.

Quality control measures

Some of the important points to be considered for maintaining good quality processed products are as follows:

- Only sound fruits or vegetables of sufficient maturity are to be used for processing.

- Adequate hygienic practices should be followed during processing.

- The inspector must be aware of pesticides and other chemicals used in production of the raw materials. Necessary laboratory analyses can then be arranged to ensure residue levels in final product.

- At commencement of and during processing, inspector should pay attention to the condition of raw materials, preparation of raw materials for processing (peeling, slicing, dicing, blanching, etc.), preparation and density of packing medium (sugar syrup, brine, etc.), state of containers to be used (cleanliness and strength), pasteurization or freezing process (time/temperature relationship), bottle filling and capping and bottle/container for storage.

- People who work in processing plant must maintain a high degree of personal cleanliness and conform to hygienic practices while on duty.

- Persons who are monitoring sanitation programmes must have education and/or experience to demonstrate that they are qualified.

- Plant construction and design shall provide enough space for sanitary arrangement of equipment. Equipment must be self-cleanable as far as possible. Cleaning operations must be conducted in a manner that will minimise possibility of contaminating foods or equipment surfaces that contact food.

- Check final product to ensure vacuum and headspace, packing medium strength and container conditions. Statistically based sampling plans should be adopted for examination of final product to ensure that it meets requirements of export regulations.

- Each processing unit should have its own sufficiently equipped laboratory and staff to carry out physical, chemical and microbiological quality examinations of foods.

• Practice proper sanitary handling procedures. Cleaning operations must be conducted in a manner that will minimise possibility of contaminating foods or equipment surfaces that contact food.

Conclusion

Fruits and vegetables are perishable commodities. Hence, efficient processing methods and storage facilities are required for their preservation. Processing of fruits and vegetables is considered to be the most important sector which adds value to the products. However, it requires professional skills and resources in order to bring benefits to smaller scale stakeholders in developing countries like India. With the adoption of suitable processing technologies, potential entrepreneurs can emerge in this sector. Supportive environments are essential to successful entrepreneurship and are steadily evolving throughout the country. The recent development in the processing methodology and automation/mechanised process lines has improved fruit and vegetable processing facilities. This has increased the capacity as well as hygienic level of processing industries. Adoption of suitable processing technology and improved processing lines will surely lead to the progress of the sector, which ensures economic and environmental sustainability of the country.

References

CFTRI, 1990. Home Scale Processing and Preservation of Fruits and Vegetables. CFTRI, Mysore, p. 91.

Gould, G. W. 2001. Nutritional Effects of New Processing Technologies. *Proceedings of the Nutrition Society* 60: 463–474.

Griet, K., K. P. Sudheer., C. Ines., V. B. Sandy., H. Marc & V. L. Ann. 2013. Isomerisation of Carrot ß-carotene in Presence of Oil during Thermal and Combined Thermal/High Pressure Processing, *Food Chemistry* 138 (2-3):1515- 1520.

Hendrickx, M., L. Ludikhuyze., I. Vanden Broeck & C. Weemaes. 1998. Effect of High Pressure on Enzymes Related to Food Quality, *Trends in Food Science and Technology* 9:197–203.

Korus, J., P. Tomasik & C. Y. Lii. 2003. Microcapsules from Starch Granules. *J. Microencapsulation* 20: 47–56.

Pereira, R. N., F. G. Galindo., A. A. Vicente & P. Dejmek. 2009. Effects of Pulsed Electric Field on the Viscoelastic Properties of Potato Tissue. *Food Biophysics* 4: 229–239.

Qin, B., Q. Zhang., G. V. Barbosa-Canovas., B. G. Swanson & P. D. Pedrow. 1999. Pulsed Electric Field Chamber Design Using Field Element Method. *Transactions of the American Society of Agricultural Engineers* 38: 557–565.

Selman, J. 2002. New Technologies for the Food Industry. *Food Science and Technology Today* 6: 205–209.

Sitzmann, W. 2006. High Voltage Pulse Techniques for Food Preservation. In: *New Methods of Food Preservation*, pp. 236–252. [GW Gould, ed.]. Glasgow: Blackie Academic and Professional.

Sudheer, K. P. 2009. New Vistas in Horticultural Processing- An Emerging Opportunity for Entrepreneur, *Proceedings of the International Farmers Meet* 2009, Cochin Airport on 2nd- 3rd June, 2009.

Sudheer, K. P. & R. S. Madhana. 2012. Innovations in Packaging of Fruits and Vegetables, In: *Proceedings of International Workshop on Strategies in Value Addition and Safety Aspects Pertaining to Dairy and Food Industry* held at Madras Veterinary College, Vepery, Chennai.

Sudheer, K. P. & S. M. Mathew. 2016. *Recent Developments in Post Harvest Technology*, Published by Director of Extension, Kerala Agricultural University, Thrissur.

Sudheer, K. P. & V. Indira. 2007. Post Harvest Technology of Horticultural Crops. New India Publishing House, Pitampura, New Delhi, India.

5

Total Value Addition of Banana Potential Opportunities for Entrepreneurs

Ravindra Naik & Sudheer K P

Introduction

Banana plants are monocotyledonous, perennial and important crop in the tropical and sub tropical world regions. Traded plantain *(Musa paradisiaca AAB)* and other cooking bananas *(Musa ABB)* are almost entirely derived from AA-BB hybridization of *M. acuminata* (AA) and *M. balbisiana* (BB). Plantain and cooking bananas are very similar to unripe dessert bananas *(M. cavendish AAA)* in exterior appearance, although often larger. Main difference in the former is the starchy flesh rather than sweet and they are used unripe and require cooking. Dessert bananas are consumed usually as ripe fruits; whereas ripe and unripe plantain fruits are usually consumed boiled or fried. Large numbers of cultivars totalling about 70 are grown in different parts of the country. Some of the popular cultivars are Dwarf Cavendish, Robusta, Poovan, Nyali Poovan (Elakki Bale or Ney Poovan) Rasthali, Virupakshi, Monthan, Karpooravallli, Nendran or Rajeli. (www.pnbkrishi.com)

Commercial cultivation of Nendran has picked up rapidly in Tamil Nadu in the recent past. Nendran bunch has 5-6 hands weighing about 12-15 kg. Fruits have a distinct neck with thick green skin turning buff yellow on ripening. Fruits remain as starchy even on ripening. Keeping banana as a whole fruit for long time is not feasible due to its poor shelf life. The processed products from banana are also very limited at present and there is a good market potential for banana products such as banana figs, banana powder, banana jam, banana jelly and beverage (Federation of Indian Chambers of Commerce and Industry).

Banana is highly perishable and more often the post harvest losses go unaccounted particularly in the Indian subcontinent. The Indian banana growers are losing around 20 to 30% of their production on account of improper handling practices (Kotecha and Desai, 1998; Davara and Patel, 2009; Anon., 2015). Therefore, in order to have a good return and to avoid glut in market during peak in production, it becomes essential to store the fruit for considerable period without spoilage. Apart from this, value addition of banana is also important. ICAR- National Research Centre (NRC), Banana, Trichy, Tamil Nadu, has developed many value added products such as banana chips, figs, wine, vinegar, jam and confectionery, flour etc. from banana so as to make it available throughout the year.

Health benefits of banana

Bananas are considered to have a lot of nutritional and health benefits. They are listed below.

Nutritional content of banana

- Bananas are an excellent source of potassium. A single banana provides 23% of potassium that we need on a daily basis. Potassium benefits the muscles, helps to decrease blood pressure and reduces the risk of stroke
- Bananas are an excellent source of vitamins including β carotene, B_6 & C
- High carbohydrate content makes it a very good source of energy
- The central core contains high fibre along with minerals such as sodium, potassium, calcium, magnesium and chlorides.

Action of banana against various diseases

- Iron content in banana is useful for persons suffering from anaemia
- High potassium and low sodium help to hit blood pressure
- Potassium packed fruits like banana is helpful to boost memory
- People who suffer from depression feel better after eating banana. This is due to the tryptophan content of banana which is essential to form serotonin, a neurotransmitter
- Drinking banana milkshake sweetened with honey is one of the easiest ways to cure a hangover
- Banana has a natural antacid which gives relief from heartburn
- Taking bananas between meal helps to maintain blood sugar levels
- Swelling and irritation caused by mosquito bite will be reduced by rubbing with inner side of the banana skin
- B complex vitamins in banana helps to calm the nervous system

- Banana reduces the risk of death caused by strokes by as much as 40%
- Bananas can be used to cure ulcer in the intestine because of its soft texture and smoothness.

Scope of entrepreneurship development in banana processing sector

India is the largest producer of banana in the world, contributing 24 per cent to the global production. During the market glut, the excess production of banana can be converted into value added products which could fetch a premium price in the market.

The value added products which can be used by the entrepreneurs can be broadly classified into following categories

- Products from unripe banana and plantains like chips, flour, baby foods, health drink, soup mix, biscuits, flour based snack foods, sauce, raw banana fruit pickle
- Products from ripe banana like RTS beverages, dehydrated fruit (fig), leather (sweet bar), sweet chutney, jam, sip up, wine etc
- Products from banana peel and banana flower
- Products form banana pseudostem
- Utilization of banana leaves.

Value added products of banana

In banana, the value addition through processing can be brought out by converting them to various products like chips, figs, jam, jelly, puree, powder, flour, syrup, health drink, baby food, etc. (Fig.1). Most of these processes do not require large investment in machinery. It can be set up in a small scale sector. Some of the commercial value added products of banana and plantains and preparation procedures are detailed below.

Banana fig

Banana fig is a dehydrated banana product prepared from ripe fruit. Karpuravalli is the best variety suited for fig making, although, other sweet varieties can also be utilized. Banana fig can be eaten as such or can be incorporated in cakes, biscuits, payasam, kesari and ice-creams as a substitute for raisin. Ripe banana fruits are peeled and dipped in 0.1% KMS (Potassium metabisulphite) solution for 2-5 minutes. Excess water is drained and pieces/fruits are dried in sun or hot air oven at 50°C for 48 hours. The moisture content of fig is less than 25%. It is also possible to produce banana figs through osmotic dehydration and microwave drying.

Banana chips

Banana chips/crisps are made by deep frying raw banana slices in a suitable cooking medium. Nendran banana is widely used for preparation of banana chips. However, other varieties like Monthan and Zanzibar or Mindoli can also be used. Peel banana and cut into thin slices of 2-3 mm thickness and fry in any edible oil till the slices turn crisp in texture and yellow in colour. Add 2-3 teaspoonful of 10% salt solution into the hot oil while frying or sprinkle powdered salt. Then, drain the oil from chips. After cooling, pack the chips in polyethylene bags of required capacity, seal and store in dry place. The product has a shelf life of 2-3 weeks under ambient conditions. Packing in nitrogen flushed laminated pouches can extend the shelf life up to 3 months. Semi ripe banana also is being used for making chips which have a sour-sweet taste.

Ripe-banana based products

Fig   Juice/RTS   Jam   Sweet chutney

Banana flour based products

Baby food   Health drink   Soup mix

Banana based pickles

Central core   Peel   Fruit   Flower

Fig. 1: Value added products from banana
(Narayana & Mustaffa, 2007)

Banana slices can be fried in a vacuum frying machine which depressurizes and fry the chips at lower temperature (100-110°C) than the traditional frying methods. The oil content of the vacuum fried product is less (15-18%) when compared to atmospheric fried products (35-42%). At reduced temperature, the oil does not decompose quickly, hence can be reused several times (nearly 50 times). During low heat/pressure frying, banana slices become crisper and retains the nutrients, colour and aroma. In combination with active MAP (nitrogen packaging), it could be stored for longer time.

Banana flour

Banana flour can be made from both ripe as well as unripe fruits. Banana flour made from unripe green banana is an intermediary product which can be used to prepare biscuits, cakes, bread, custard, chapathi, papad, baby food, health drink, etc. Unripe green banana fruits are blanched in hot water at 80°C for 5 minutes to facilitate easy removal of skin and to inactivate enzymes which cause browning. The blanched fruits are peeled and cut into 1-2 mm thick slices using a knife or slicer. The slices are dried in sun or in a hot air oven at 70°C for 6-8 hours. After drying, the slices having a moisture content of 8-10% are powdered in a pulverizer or in a mill and sieved to obtain free flowing flour.

Ready-To-Serve (RTS) banana beverage

Ready-to-serve banana beverage is a drink prepared from clarified banana juice. When carbonated and chilled, RTS banana beverage makes a nutritious and refreshing drink. Banana being a fruit with thick and sticky pulp will not easily form a clear suspension or juice. The pulp will have a tendency to settle and separate. Therefore, it is essential to treat the fruit pulp with pectinolytic enzyme to extract clear juice. Ripe fruit pulp is mixed with commercial pectinolytic enzyme to enhance separation of juice from pulp. Juice is filtered through a muslin cloth or basket centrifuge. Filtered juice is racked or further centrifuged to remove the sediment. Clarified juice is diluted 2.5 times with sterilized/softened water. Sugar and citric acid are added to adjust final TSS to 15°Brix and 0.3% acidity. Juice is pasteurized and poured into sterilized bottles. Sealed bottles are processed in boiling water for 20 minutes and cooled to room temperature before storage. Ready-to-serve banana juice can be stored for 6 months under room temperature. It tastes best when served chilled.

Banana jam

Banana, as an ingredient in mixed fruit jam is common. There is a good scope for jam if exclusively prepared from banana. To prepare banana jam, pulp the ripe fruit and add equal quantity of sugar. Boil fruit-sugar mixture rapidly to concentrate soluble solids to about 68.5% and for the inversion of sugar.

Commercial pectin may be added followed by addition of citric acid, colour and flavour. At the end point, jam should fall as sheet when dropped with a spoon or ladle. The T.S.S should be above 68°Brix. Fill the hot jam in sterilized bottles and close air tight.

Banana sauce

Sauce is a spicy product generally made from tomato and/or other vegetables like pumpkin. Sauce can be of sweet and hot taste depending upon the requirement. Boil banana for 20-30 minutes in water to cook and to facilitate easy removal of peel. Subsequently, pulp the banana into a smooth consistency in a blender. Add water to the pulp in a ratio of 1:4. Add required spices and cook to a thick consistency.

Sweet chutney

Mix ripe banana pulp, water and sugar and boil in thick bottomed pan or steam jacketed kettle. Put all required spices in spice bag and add to boiling fruit pulp and cook for 15-20 minutes. Add required quantity of salt, vinegar, colour and sodium benzoate also and cook to required consistency.

Banana pickle

Banana pickle is made using cooking variety of banana. There are many recipes for preparation of fruit pickles. The ingredients required are gingelly oil, ginger, garlic, chilly powder, vinegar, salt, mustard, asafoetida, fenugreek powder etc.

Banana flower pickle

Banana flower is used for making pickle. Process involves removal of pistil, blanching, grinding and addition of spices and oil. The product is tasty and stable for one year at room temperature.

Banana biscuit

Flour made from green banana is used to prepare biscuits, cakes, bread, custard, chapathi and breakfast agglomerates. High starch, fibre and mineral content of banana flour make the biscuit wholesome and delicious. Biscuit is prepared using 60% banana flour and 40% maida, sugar, baking powder, milk powder, hydrogenated vegetable oil or shortenings and banana essence. This biscuit can be fortified with minerals and vitamins.

Banana leather (bar)

Banana fruit bar is a confectionary prepared by drying ripe banana pulp with appropriate quantities of sugar, pectin and acid. Banana bar is nutritious and tasty which can be popularized among children and adults. Peel ripe banana and pulp in a pulper or mixer. To this, add 10% sugar, 0.3% citric acid, 0.5% pectin and permitted preservative and food colour. Spread the mixture as a thin layer on a tray and dry at 70°C for 24 hours. After 24 hours, turn the material and dry for another 10-12 hours. Cut the dried banana fruit bar into pieces of desired size and shape and pack in cellophane paper or printed laminated pouches. When dried below 20% moisture, bar can be stored for six to seven months without deterioration in quality.

Banana fruit candy

Banana fruit candy is a snack food similar to chips. It is commercially made and marketed in Kerala. It is prepared from nendran banana, jaggery and ginger. Banana true stem (central core) can also be made into candy through osmotic dehydration process followed by sun drying.

Banana wine

Banana wine is produced by fermenting enzyme treated clear banana juice with wine yeast viz *Saccharomyces cerevisiae* var. ellipsoideus. Fermentation is carried out for about 3 weeks followed by filtration, clarification, and bottling. The pasteurized wine is stored in bottles for aging. Alcohol content of banana wine varies from 9-12%.

Banana flour based products - baby food, health, drink and soup mix

During market glut, excess banana can be processed as flour which will form the basic raw material to prepare other value added products like baby food, health drink and soup mix which could fetch a premium price in the market. Banana flour is fortified with milk, green gram flour and sugar for baby food, while for health drink, chocolate powder, barley powder and sugar are mixed with banana flour. For soup mix, corn flour, dried vegetables and spices are incorporated with banana flour in various proportions. All these can be stored up to six months.

New technologies/ innovations in value addition of banana

Research priorities should be focused to more value added products with modern techniques. Application of modern processing tools like non thermal non chemical processing (NTNC) should be looked into. Among the novel non-thermal technologies, application of high-pressure processing (HPP), pulsed electric fields (PEF), high intensity pulsed light, irradiation, ultrasound and modified atmosphere packaging (MAP) for processing and preservation of banana and banana based products are important. Impact of new technologies on nutritional value of developed products also should be studied.

Scope of by-product utilization

Banana stem, which goes as waste or cut and recycled in the field is being utilized for making many value added products and handicrafts (Naik *et.al.*, 2015) as detailed below:

Banana fibre

Banana fibre can be obtained commercially from pseudostem sheath, which yields coarse fiber and fine fiber. Coarse fiber is extracted from outer sheath of banana pseudostem, which is commercially used for making garlands. It can also be used for making yoga mats, ropes, handicrafts like net-bags, fruit baskets, packing materials, etc. Though, all layers of sheath yields fibre, yield of fibre from middle layer sheaths of seven to eight are better than outer and inner layers and are utilized for commercial extraction of fine fiber. It is used to develop value added products such as handicraft items (table-mat, bag, mobile cover, wall hangings and other fancy articles, etc.). Even, shirts and sarees are being prepared from fine fiber by blending with cotton or silk, after extracting thread manually. Quality papers, carry bags, sanitary napkins, currency notes, etc. can be prepared from the fine fiber. Fiber and its handicrafts are being exported to countries like Europe, Canada, etc.

Banana sap and its uses

About 10-15 thousand litres of sap are collected from pseudostem sheath during fiber extraction process. It is rich in potassium. This sap can be used as growth regulator and nutrient solution to commercial crops and vegetables. It is also used as dye mordant in textile industry.

Banana peel

Banana peel is a rich source of starch, protein, crude fat, total dietary fibre, ploy unsaturated fatty acids, particularly linoleic acid and linolenic acid as well as pectin, essential amino acids and micronutrients. Except lysine, all essential amino acids are higher than FAO standard. Banana peel can be used in production of wine, ethanol, substrate for bio-gas and base material for pectin production. Peel ash is used as a fertilizer and a source of alkali for soap production. Peel can also be used in waste water plant.

Banana leaves and sheaths

Leaves are extensively used in weaving basket, mats, food wrapper, covering over food, table cloth, food plates etc. Leaves are a good source of lignin which is higher than banana pseudostem. Leaves can be given to cattle with some protein extract for better digestibility.

Products made out of central core of banana

Banana central core pickle: The product is prepared by making slices of the core stem and by adding spices and preservatives.

Banana central core juice: Juice is extracted from central core stem, which is having property of dissolving kidney stone.

Banana central core soup: It is one of the commercial products made out of central core stem. It is an appetizer taken before any meal to aid in digestion.

Banana central core candy: The core stem part is used for preparing candy, which is sweet in taste and rich in fiber.

Fine fibre extraction from inner layer of banana pseudostem

The fibers are extracted through hand extraction machine composed of either serrated or non serrated knives. The extracted fibers are sun-dried which whitens the fiber. The natural fibre is used in preparing many value added products such as handicraft items (table-mat, bag, wall hangings, and other fancy articles, etc.) (Fig 2). It is a natural fibre and eco-friendly in nature.

Fig. 2: Various products from banana fine fibre

Coarse fibre extraction from outer layer of banana pseudostem

Out of the 14-18 sheaths in the banana pseudostem, outermost 4-6 sheaths yield coarse fibre, outer 6-8 sheaths give soft lustrous fibre and rest of the middle sheaths yield very soft fibres. Quantity of fibre in each sheath depends upon its width and its location in the stem. Ropes from outer sheath of banana pseudostem are in high demand for different applications but is labour intensive with hand spinning or ratt machines. CIAE Regional Centre, Coimbatore in collaboration with ICAR-NRC Banana, Trichy, has developed a package of equipment to mechanize the rope making process easier. Package consists of equipment for splitting the outer sheath of banana pseudostem and twisting and winding of splitted strands (Naik *et al.,* 2016).

Equipment for splitting the outer sheath of banana pseudostem

In this equipment, rollers are to be changed for different width (2mm, 3mm and 4mm) of strands of sheath. Initial cost of the machine is Rs. 20,000 and the capacity of the unit is 3-3.5m strands/min (Fig.3)

Equipment for twisting and winding of splitted strands

Equipment developed has got advantages over manual method of twisting and winding in terms of more uniform twist, lower space requirement, less dependency on skilled labour, cheaper than manual labour and higher output (Fig.4). Initial cost of the machine is Rs. 90,000. Capacity of the equipment is 4800 m strands/day. The twisted rope obtained is used for production of various eco-friendly handicraft materials like bags, window curtains, table mat etc which has huge demand both in local and international markets. Outer sheath of banana pseudostem when processed for rope making would fetch about 2.4 lakh m of banana sheath rope valued at Rs 1.0 lakh/ha

Fig. 3: Equipment for splitting the outer sheath

Fig. 4: Equipment for twisting and winding

Equipment for value addition of banana central core through minimal processing and juice extraction

Equipment for minimal processing of banana central core was developed by ICAR Central Institute of Agricultural Engineering, Regional Centre, Coimbatore in collaboration with ICAR-National Research Centre for Banana, Trichy, Tamil

Nadu. It consists of slicer, dicer, fibre remover, surface water remover, juicer/ grinder and juice squeezer. The details of equipment are summarized.

Slicing unit: Slicing equipment is motorized (1 hp, single phase) which takes the central core to the cutting blade automatically, due to the cam arrangement (Fig. 5). Rotating stainless steel blade is used for slicing banana central core. Cutter blade is fixed inside a casing which can be opened easily for cleaning purpose. Primary fibre accumulated during the

Fig. 5: Banana central core slicing unit

slicing process can be removed by using port available at one side of the casing, holding the cutting blade. Capacity of the equipment is 40 kg/h.

Dicing unit: Sliced pieces are diced into required sizes which are directly collected into the trough having pre-treated water containing preservative to arrest browning reaction. Dicing is carried out by continuous rotary dicing unit and a nylon circular roller, using dicing blades. Equipment is operated by 1 hp single phase motor. Capacity of the equipment is 30-35kg/h (Fig 6).

Fig. 6: Banana central core dicing unit

Fibre removing unit: Once, the central core is diced, secondary fibre is removed from the fibre removing unit. A spindle is fixed to the shaft of a 0.5 hp single phase motor for removing the fibre, which rotates at 200 rpm for 5-7 min. Central core fibres will wind around the spindle arms. Capacity of the unit is 10 kg/batch.

Surface water remover: Surface water in the diced banana central core is removed to increase the shelf life and also to arrest further browning action. It has two cylindrical stainless steel drum. Inner drum has 3 mm diameter perforations around the surface and rotates at a speed of 1440/720 rpm by 0.5 hp single phase motor. The outer stationary drum has an outlet to remove excess water. Capacity of the unit is 7 kg/batch.

Juicer/grinder: For extraction of juice, central core has to be reduced to fine size to squeeze out the juice. This is performed by high speed juicer/grinder (2800 rpm) with 2 hp single phase/three phase motor. Juicer/grinder is completely made up of stainless steel with a set of six blades fixed to the rotating drive shaft. It has a tight lid on the top. Provision has been made to tilt the juicer for easy unloading after grinding. Capacity of the unit is 4 kg/batch and 40-45 kg/h.

Juice squeezer: Ground paste of juice and fibre is fed into juice squeezer to extract the juice. Juice extraction in this equipment is hygienic, without direct contact of human hands and thus helps to reduce contamination. Capacity of this unit is 2 kg/ batch or 40-45 kg/h.

By using this package of equipment, one can save time and labour by 60 to 70%. Banana central core which is wasted mostly can be made into value added products for human consumption, thus helping in generating additional revenue to banana farmers/entrepreneurs/processors. About 5-7 tons of central cores can be extracted from one hectare.

Success stories in banana processing sector

A success story of a small entrepreneur who is involved in minimal processing of banana central core is given in Table 1.

Future scope of research

Production and productivity in banana during the past 2-3 decades has faced a tremendous growth in India and also at global level. Projection for the year 2050 is to be tuned to 50 million tones. With better post harvest handling and processing technologies the farmers could get better profit through value addition. Waste could be converted into by-products to generate wealth from waste. Therefore, new initiatives have to be directed to develop a technology for value added products. Being a seasonal crop and to get a better price during the glut season, better storage, processing and value addition play an important role in marketing and better price realization. There are about 15-20 value added products in banana and plantain utilizing fruit, pulp, peel, flower, unripe fruit etc. The technologies have to be commercialised in different parts of the country

Table 1: Sucess story of a small entrepreneur in minimal processing of banana

S.No	Particulars	Details
1.	Name of entrepreneur	Shri V. Arumugham, Sri Vel foods, 1/65 North Street, Athichanallur, Srivaigundam Thaluk, Toothukudi - 628 621 Phone : 9976737250 Email: arumugamkings@gmail.com
2.	Title of technology	Post harvest mechanization package for banana central core - minimal processing of banana central core
3.	Brief description about technology/ innovation	The details of the equipment are summarized. • **Slicing Unit:** Slicing equipment is used for slicing central core. Capacity of the equipment is 35 - 40 kg/h. • **Dicing Unit:** Sliced pieces are diced into small cubes. Capacity of the equipment is 40 kg/h. • **Fibre removing unit:** After dicing of banana central core, secondary fibre is removed in the fibre removing unit. Capacity of the unit is 10 kg/batch • **Surface water removing unit:** Surface water in the diced banana central core is removed, to increase the shelf life. Capacity of the unit is 7 kg/batch • **Packing equipment:** Pedal operated sealing machine
4.	Investment details	• **Cost of equipment:** 2.50 lakhs • **Building :** Own (invested 2 lakhs for shed, electrical and minor civil work) • **Other miscellaneous expenditure :** 1.0 lakh
5.	Educational qualification	ITI (Air conditioning)
6.	Year of start-up business	2014
7.	Name of product	Minimally processed banana central core
8.	Scale of production	500-1000 kg of minimally processed (diced) banana central core per day
9.	Manpower engaged	2 male and 4 female
10.	Market linkages	Supply of minimally processed banana central core in retail and wholesale market
11.	Present turnover (Rs)	About 3 lakh per month
12.	Net profit (Rs)	Rs 80-90 thousand per month
13.	Financing agency	Support/loan from local bank

for better marketing and profit. In addition, fibre extracted from the pseudostem also has great potential in the local and export market. These need to be tapped further with a concept of value from waste.

Technology in handling, harvesting, transporting, storage, shipment, artificial ripening and packaging needs attention for export trade. Pre and post harvest technologies like maturity index, better harvesting, handling, storage, packaging, and artificial ripening for export in Cavendish bananas need to be further standardized based on location. There is a need to develop technologies for traditional banana varieties. This is essential to popularise traditional yellow bananas grown in India which has great potential for export as "niche" bananas to cater the needs of Indian ethnic population in Middle East, South East Asian countries and Europe. Marketing strategies to popularise Indian niche varieties to the consumers in foreign markets through government intervention should be taken up.

There is a need to develop post-harvest infrastructural facilities like transportation, storage, handling, ripening at the production sites and marketing of bananas on weight basis using carton boxes instead of selling whole bunches which creates environmental pollution. There is a need to develop retail marketing strategies and better out-lets to reduce post harvest losses and also the availability of quality bananas to the consumers. Since, banana being a highly perishable commodity and the trade will fluctuate depending on season and availability of other fruits, development of this sector is very vital for banana industry. There is a need to develop industries for value addition and by-product utilization to minimise the loss occurring due to increased production/glut in the market. There is great scope to develop value added products from banana fibres and other waste materials. Value added products from recent technologies like vacuum packaging, vacuum frying, and smart packaging can also be taken up.

Safety and quality aspects of banana

Safety and quality parameters need to be followed for marketing banana

CODEX standard for bananas (CODEX STAN 205-1997, AMD. 1-2005)

This Standard applies to commercial varieties of bananas grown from Musa spp. (AAA) of the Musaceae family in the green state to be supplied fresh to the consumer after preparation and packaging. Bananas intended for cooking only (plantains) or for industrial processing are excluded. Details of concerned quality are also given in the standards.

Bananas are classified in three classes as detailed below

i. *Extra Class*: Bananas in this class must be of superior quality. They must be characteristic of the variety and/or commercial type. Fingers must be free of defects, with the exception of very slight superficial defects, provided these do not affect the general appearance of the produce, the quality, the keeping quality and presentation in the package.

ii. *Class I*: Bananas in this class must be of good quality. They must be characteristic of the variety. Following slight defects of the fingers, however, may be allowed, provided these do not affect the general appearance of the produce, the quality, the keeping quality and presentation in the package:

   - slight defects in shape and colour
   - slight skin defects due to rubbing and other superficial defects not exceeding 2cm² of the total surface area

iii. *Class II*: This class includes bananas which do not qualify for inclusion in the higher classes, but satisfy the minimum requirements. The following defects, however, may be allowed, provided the bananas retain their essential characteristics as regards to quality, keeping quality and presentation

   - Defects in shape and colour, provided the product retains the normal characteristics of bananas.
   - Skin defects due to scraping, scabs, rubbing, blemishes or other causes not exceeding 4 cm² of the total surface area
   - The defects must not, in any case, affect the flesh of the fruit

Packaging: Bananas must be packed in such a way as to protect the produce properly. Materials used inside the package must be new, clean, and good quality and should not cause any external or internal damage to the produce. Use of materials, particularly of paper or stamps bearing trade specifications is allowed, provided the printing or labelling has been done with non-toxic ink or glue. Bananas shall be packed in each container in compliance with the Recommended International Code of Practice for Packaging and Transport of Fresh Fruits and Vegetables (CAC/RCP 44-1995, Amd. 1-2004). Requirements of the Codex General Standard for Labelling of Pre packaged Foods (CODEX STAN 1-1985, Rev. 1-1991), must be followed.

Hygiene: It is recommended that the produce covered by the provisions of this Standard should be prepared and handled in accordance with the appropriate sections of the Recommended International Code of Practice – General

Principles of Food Hygiene (CAC/RCP 1-1969, Rev. 4-2003), Code of Hygienic Practice for Fresh Fruits and Vegetables (CAC/RCP 53-2003) and other relevant Codex texts such as Codes of Hygienic Practice and Codes of Practice.

Microbiological criteria: The produce should comply with any microbiological criteria established in accordance with the Principles for the Establishment and Application of Microbiological Criteria for Foods (CAC/GL 21-1997).

Marketing standards for fresh bananas issued by European Commission

To sell fresh bananas in the European Commission (EU), one must comply with the EU marketing standards regarding quality, size, tolerance, presentation and marketing. These standards do not apply to plantains, fig bananas or bananas intended for industrial processing.

Across the EU, the import, sale and marketing of most fruits and vegetables is regulated by marketing standards designed to ensure that product is sound, clean, correctly labelled and of marketable quality. Most products are covered by a General Marketing Standard (General MS) and some are covered by Specific Marketing Standards (Specific MS). To comply with the EU marketing standards, product must be quality graded according to either a Specific MS or the General MS. It must also be free from rot and disease, clean, pest-free and undamaged by pests, and in an adequate state of maturity. One must also clearly label packaging with the information required under the Specific MS or General MS applicable to the fruit. This could include country of origin, details of the packer, variety/type, size, weight, and quality class.

AGMARK standards for grading of bananas

According to Agmark standards bananas are classified into following classes (Table 2)

Dole chart

Dole has created a chart that divides the ripening process into seven stages as indicated below. This is popularly called as Dole colour chart

- All green: As received from ware house

- Light green: First colour change during ware house ripening process

- Pronounced break or shading to yellow: 50% green, 50% yellow - Recommended colour for ware house outrun depending on time, temperature, and distance to retail

- More yellow than green: Ready for retail display. Greater product life and lower loss for retailer. Greater colour selection and product life for consumer

Table 2: AGMARK standards for grading banana

Grade	Designation	Grade requirements/Grade tolerances
Extra class	Bananas shall be of superior quality. They must be characteristics of the variety and/or commercial type. The fingers must be free of defects, with the exception of very slight superficial defects, provided these do not affect the general appearance of the produce, quality, keeping quality and presentation in the package.	5% of bananas (by count or by weight) not satisfying the requirements of the grade, but meeting those of Class grade or, exceptionally coming within the tolerances for that class.
Class I	Bananas shall be of good quality. They must be characteristics of the variety and/or commercial type. The slight defects (as listed below) of the fingers, however, may be allowed, provided these do not affect the general appearance and quality of the produce in the package. • slight defects in shape and colour • slight defects due to rubbing and other superficial defects not exceeding 2 sq.cm. of the total surface area The defects must not affect the flesh of the fruit	10% of bananas (by count or by weight) not satisfying the requirements of the grade but meeting those of Class II or, exceptionally coming within the tolerances of that grade.
Class II	This includes bananas which do not qualify for inclusion in the higher classes, but satisfy the minimum requirements. The following defects may be there provided the bananas retain their essential characteristics with respect to thequality and presentation. • defects in shape and colour provided the product remains the normal characteristics of bananas • skin defects due to scrapping, scabs, rubbing, blemishes or other causes not exceeding 4 sq,cm. of the total surface area; The defects must not affect the flesh of the fruit	10% of bananas (by count or by weight) not satisfying the requirements of the grade, but meeting the minimum requirements.

- Yellow with green tips: Maintain colour longer by holding boxes display in cool area. High temperatures increase colour change
- Full yellow: Fruit firm with good eating flavour. Handle with care, display on padded tables only
- Yellow flecked with brown: Ideal eating flavour and consistency. Perfect for puddings and pies.

Packaging

Packaging of fruits is required for efficient handling and marketing, better eye appeal and better shelf life. Proper packaging protects the fruits from pilferage, dirt, physiological and pathological deterioration during further handling. In some cases banana bunches are packed in old gunny bags wrapped with banana leaves. Due to poor packing quality the bananas deteriorate and fetch low price.

High quality bananas are generally exported. Firstly, fingers are removed from the bunch and washed in water. Then, they are washed in dilute sodium hydrochloride solution to remove the latex, dipped in 0.1% of carbendazime solution and finally air dried. These fingers are graded on the basis of their length and girth and packed in plastic corrugated fibre board (CFB) cartons having capacity of about 13 to 14.5 Kg. A suitable packing material like foam may be used. These boxes are kept at 13-15°C temperature and at 80-90% humidity in cold storage. Bananas can be stored in such controlled atmosphere in a cooling chamber for a period of 20-25 days. Bananas are to be exported via cold chain of shipment at 13°C and refrigerated vans in the country.

Qualities of packaging material

Since, a package as a container offers accommodation to the contents for storage and transportation, the packages must have the following basic qualities.

- It must protect quality and quantity
- It must prevent spoilage during transit and storage
- Labelling of package must indicate quality, variety, date of packing, traceability, weight, price etc.
- It must be convenient in handling operations
- It must be convenient to stack
- It must be cheap, clean, hygienic and attractive
- It must be biodegradable
- It must be free from adverse chemicals

- It should be reusable
- It should immobilize the fruits placed inside
- Quality and hygienic cushioning material must be used to protect fruits from impact, injury and compression
- It should offer good ambient conditions to the fruits which are congenial for storage and transportation
- It should meet optimum requirements of ventilation vis-à-vis temperature and relative humidity management.

Precautions to be taken before packaging

- Banana should be plucked at appropriate maturity, keeping in view the time span of the market
- Banana should be sorted and graded as per accepted quality standards before packing. Only sound fruits should be packed
- Before packing, post-harvest treatment with wax and fungicides should be resorted to as a prophylactic measure against pathogenic invasion in transit
- For prevention of bruising/abrasion injuries, paper liners, pads, trays or tissue wraps may be used. As an alternative, cushioning with easily available paddy straw keeps the packing cost minimum. It will maintain a level of R.H. because of moisture absorbing tendency of paddy straw and keep the temperature down
- Careful placement of banana in the cartons is necessary to avoid bruising. The use of telescopic boxes can overcome this problem
- For securing packages, use of adhesive tape (3 to 4 cm) may be used. The packages can also be secured with thin rope of coconut fiber, or polythene sutli.

Precautions during packing

During packing, fruits should never be packed loosely in order to avoid displacement of fruits which leads to friction among fruits surface and thereby causing damage. In cartons, fastening should be done with little pressure so that during transit period when volume of the fruits gets reduced owing to dehydration and adjustment of space due to jerks in transit, the packing does not get unfastened.

It is also observed that during packaging, sharp edges of packing material damage the fruits. Therefore, care should be taken to avoid bruising, puncturing and damaging fruits by sharp edges. Similarly, there should not be too much ventilation which can affect the quality of fruits due to shrinkage, loss in weight, colour, etc.

Parameters of packaging material

Size, type and capacity of the packaging material depend mostly upon locally available raw material, distance of the markets and type of transport to be used. Generally, packing material of different size made of corrugated fiber boxes, telescopic boxes, wooden boxes and plastic crates are used for packing banana.

Preparation of banana for market/processing

Dehanding

- After harvest, dehanding should be carried out with a sharp, clean banana knife, making a smooth cut as close as possible to the stem
- After dehanding, fruits are placed with the crown facing downwards onto a layer of leaves to allow for drainage of latex
- In order to restrict crown disease development, hands should be dipped in a solution of 0.1% Benlate or Thiabendazole.

Stowing

- After harvest, banana bunches are arranged in rows with the cut ends of pedicel upward, called stowing
- Stowing is required at two stages
- Soon after harvest, bunches are stowed in the field usually over a bed of banana leaves
- Before a carriage arrives, harvested bunches continue to remain stowed in this condition which pave way to spread of inoculums to healthy sites
- During transport and at the wholesalers go down, bunches are again stowed before sending them to ripening room
- During stowing, fruits are invariably subjected to mechanical or insect injury in addition to the spread of pathogens carried from field in latent condition or prevalent under local condition of storage.

Packaging

- Arrangement of fruits in box has to be horizontal in two rows keeping crown end towards box side and fruit tips towards the center of the box

- While packing in single layer, hands should be placed in vertical position by keeping their tips up and crown downside
- Cushioning pads or kraft paper should be placed at the bottom of the box and fruit may be covered in LDPE liner of 100 gauges inside the box to create modified atmosphere.

Precooling

- Fruits destined for distant and export market should be precooled considerably to extend storage life
- Precooling of produce should be done within 10 - 12 h of bunch harvesting
- Fruit packed in boxes should be precooled by forced air cooling at 13°C and 85 - 90% RH
- It may take 6 to 8 h to bring the fruit pulp temperature to 13°C from field temperature of 30 to 35°C
- Boxes should be immediately moved to cold rooms for storage purpose.

Storage

- Bananas can be exported successfully by sea-shipment, if guidelines related to harvest maturity are strictly followed
- Storage conditions of 13°C and 85 to 95% relative humidity are required
- Storage temperature below 13°C would cause chilling injury to fruits resulting in surface discolouration, dull colour, and failure to ripen and browning of flesh
- Storage life at 13°C depends on the cultivar and varies from 3 to 4 weeks
- A combination of low temperature with controlled atmosphere storage can further extend storage life
- Unripe green banana (Robusta cvr.) fruits could be stored for 8 weeks under controlled atmosphere storage condition of 5% O_2 + 5% CO_2 at 12 to 13°C with post storage ripening period of 4 to 5 days under ambient conditions.

Ripening

- Green bananas in boxes and or cushioned plastic crates should be loaded into ripening room (lower temperatures can damage the fruit)
- The room should be closed, insulated and airtight and be maintained at 16 to 18°C and 85-90% RH. Temperature is controlled and maintained by thermostat

- Supply ethylene into the room at a concentration of around 100 ppm (0.01%)

- Ethylene act as a catalyst initiating the hormonal process of ripening

- Room is kept closed for 24 hours. At the end of 24 hrs, room should be ventilated to clear ethylene gas and the CO_2 released during initial ripening phase and maintain at 18°C and reduce to 15°C over three to four days.

Transportation

- Harvested bananas from the gardens located in villages are usually transported as head loads, on ponies, as cart loads and as lorry loads. Transport for interstate trade is mainly affected through lorry services and railway wagons.

- Due to difficulties faced by wholesalers to arrange wagons in correct time, lorry transport is more depended in India.

Conclusion

Banana food processing industry is fast expanding and success of banana industry lies with product diversification and value addition. There is a need for diversification of banana industry through agro-processing and value added products of exportable quality. Changing lifestyles, food habits, organized retail and globalization will certainly give boost to processing sector in the years to come. Various Government agencies like Ministry of Food Processing and National Horticulture Board are providing many schemes and financial assistance to establish new units and to modernise existing units. Product diversification and value addition will also create additional rural employment and improve the nutritional and livelihood security of banana farmers and consumers. Thus, banana processing industry can play a vital role in the economic upliftment of agricultural sector.

References

Anon. 2015. Post Harvest Profile of Banana: 2015. *Report by Government of India*, Ministry of India, Ministry of Agriculture, Department of Agriculture and Co-opertion, Directorate of Marketing and Inspection, Nagpur P. 90.

Davara, P. R. & N. C. Patel. 2009. Assessment of Post Harvest Losses Grown in Gujarat. *J. Horticultural Sciences* 4(2): 187-190.

Kotecha, P. M. & B. B. Desai. 1998. Banana. In. Salunkhe, D. K. & S. S. Kadam. (eds.) *Handbook of Fruit Science and Technology. Production, Composition, Storage and Processing*. Marcel Dekker, Inc., New York pp: 67-90.

Naik, R., C. P. Dawn Ambrose., S. J. K. Annamalai & K. N. Shiva. 2015. Mechanization Package for Minimal Processing of Banana Central Core. *Extension folder CIAE/IEP/2015/06*.

Naik, R., S. J. K. Annamalai & K. N. Shiva. 2016. Mechanization Package for Rope Making from Outer Sheath of Banana Pseudostem. *Extension folder CIAE/RC/2016/04*.

Narayana, C. K. & M. M. Mustaffa. 2007. Value Addition in Banana, *Proceedings of National Conference on Banana*, October 25 – 28, 2007, Tiruchirapalli, Tamil Nadu.

6
Prospects in Value Addition of Mango
Sagarika N & Anjineyulu Kothakota

Introduction

Mango (*Mangifera indica* Linn.) is one of the most nutritious tropical fruits, native to Southern Asia and especially Eastern India. Mango belongs to Anacardiaceae family and it was disseminated all over the world in the beginning of sixteenth century, and there are currently around thousand known varieties of mango. It is the dominant tropical fruit variety produced worldwide, followed by pineapple, papaya and avocado, all of which are considered as major tropical fruits. Global production of mango in 2015-16 was estimated to be about 44 million tons, accounting for nearly 50% of world tropical fruit production.

Mango is the most important fruit of India and is known as "King of fruits". Mango is cultivated in 2,312 thousand ha and the production is around 15.03 million tons, contributing 40.48% of the total world production. Main mango producing states in India are Uttar Pradesh, Andhra Pradesh, Karnataka, Bihar, Gujarat and Tamil Nadu. Total export of mangoes from India is 59.22 thousand tons, valuing Rs. 162.92 crores during 2015-16. India exports mango to over 40 countries worldwide. Major importing countries during 2015-16 were UAE, Bangladesh, UK, Saudi Arabia, Kuwait and Bahrain. Important mango varieties are Dashehari, Langra, Mallika, Amrapali, Bombay Green, Banganapalli, Totapuri, Neelam etc.

Health benefits of mango

Mango is an excellent source of bioactive compounds such as carotenoids, vitamin C and phenolics, as well as dietary fibre essential to human nutrition and health. Moreover, mango is known to contain other vitamins, carbohydrates and minerals such as calcium, iron and potassium. It is low in calories and fat. According to Ayurveda, varied medicinal properties are attributed to different parts of mango tree. Mango possesses anti-diabetic, antioxidant, antiviral, anti-inflammatory properties. Various other properties include anti-bacterial, antifungal, antI-helmintic, anti-parasitic, anti-cancer, anti-HIV, anti-bone resorption, anti-spasmodic, anti-pyretic, anti-diarrheal, immunomodulation, hypolipidemic, anti-microbial, hepato protective and gastro protective (Kim *et al.*, 2010).

Entrepreneurship development in mango processing sector

Mango processing generates lot of income and employment opportunities in Chittoor District of Andhra Pradesh which is the largest mango processing area in the State. However, low capacity utilisation and fluctuation in profitability are the important issues often raised by the processing firms. Effort was initiated by a prominent mango grower, Late Mr. Subramanya Reddy who established M/S India Canning Industries, the first merchandised fruit processing unit in the District. Thereafter, various mango pulp units were established in the area. About 11% of the firms have an investment between 10-16 crores falling under the category of large enterprises. While, 18% of the firms have an investment in the range of Rs. 5 to 10 crores and fall in the category of medium enterprises, rest of the 72% of the firms fall in the category of small and micro enterprises. All the processing units in the district are covered under FSSAI.

Mangoes required for processing are supplied directly by the orchards owned by promoters of processing units. Traders located in the mandis also supply the raw materials. Processing industry is forwardly linked with five market channels. Two of these five channels, i.e., (1) Processor————Exporting pulp; and (2) Processor————Exporting agency————Exporting pulp, account for about 85 per cent of the disposal of mango pulp by the processors. Small firms have lesser access to international market and depend heavily on the merchant exporters to dispose product in the international market. Appendix 1 provides information on the capacity and investment of large, medium and small scale firms. The large scale firms process different commodities namely mango, guava, papaya and vegetables. These firms work almost throughout the year unlike medium and small scale firms which are processing only mango and guava fruits and work for three to four months in a year. The large firms adhere better to the quality standards demanded in the international market as compared to small firms.

Value added products of mango

Processing of mango to obtain value added products is important due to seasonality and to prolong shelf life. Mango as a fruit is eaten fresh, and wide range of products can be prepared with the pulp. It can be canned, frozen as concentrates, mashed, dehydrated, minimally processed or processed to make value added products like jam, beverages, puree, chutney etc. Various products that can be prepared with mango are detailed below.

Ripe mango products

Mango pulp

Mango pulp is stored during the peak season of mango crop for subsequent use for various value added products. Pulp is generally standardized to 14 -18°brix and 5 - 6.5% acidity by adding sugar syrup and citric acid respectively. No preservative is used in the canned mango pulp. Pulp is heated to 85°C, filled hot into cans, sealed and processed at 100°C for 20 min (for A2½ cans) and cooled. Addition of 0.1% ascorbic acid while canning mango pulp helps in the retention of colour, flavour and carotene. With the development of ready-to-serve beverage industry, there is an increasing demand for canned mango pulp.

Mango beverages

i) *Mango juice*: It is prepared by adding almost equal quantity of water and adjusting the total soluble solids in between 12 to 15% and acidity in the range of 0.4 to 0.5%.

ii) *Mango nectar*: It contains 20% pulp with sugar adjusted to give 15°brix and 0.3% acidity as citric acid. These beverages are packed in cans. They are heated to about 85°C in heat exchanger, filled hot into cans, sealed, processed and cooled. Mango nectar can also be prepared with pulp from dehydrated ripe mango slices and sugar in the ratio of 4:3 (20% pulp, 15°brix and 0.23% acidity), keeping flavour, colour and texture into consideration.

iii) *Mango squash*: It contains 25% juice, 45% TSS and 1.2 to 1.5% acidity and 350 ppm sulphur dioxide as preservative.

iv) *Mango juice powder*: This is prepared from strained mango juice by vacuum drying, spray drying, freeze drying, drum drying or foam-mat drying. Mango juice powder can be used as a base material for baby foods. Moisture content of powder varies from 3-5% (Sudheer and Indira, 2007).

v) *Carbonated mango beverages*: Mango based beverages are prepared by taking 7% mango pulp, 11.7% sugar and 0.15% acidity. It contains carbon dioxide (3% gas vol.).

vi) *Mango wine*: After washing and peeling, pulp the fruit and dilute with water (1:1 v/v). Increase the TSS to 22°brix by adding sugar. Add 100 ppm SO_2 also to suppress the activity of natural flora. Pecinase @ 0.5% (v/v) is also added. Keep this for five days for fermentation at 20+2°C in BOD incubator. Filter the fermented fluid and store in glass bottles at 20 to 24°C. Clear supernatant is racked and fined with bentonite at 0.1% (w/v). Further, crystal clear supernatant are racked for one month and add 100 ppm sulphur dioxide and bottle.

Mango leather

Select ripe and juicy mangoes and extract the juice. Strain the juice and add 0.6 g potassium metabisulphite for every kilogram of juice. Now, spread the juice on stainless steel trays smeared with mustard oil. Keep it for drying layer by layer. After drying the first layer, spread the second layer and repeat the process until the thickness of the layer comes to about 1 to 1.25 cm. Cut the dried leather into pieces of suitable size, wrap in butter or wax paper and store in a dry glass jar and seal air tight.

Dry mango products

Mango powder

Production of fruit juice powder has gained prominent importance in recent years because of several advantages. Dried mango powder can be used as a good adjunct in ice cream, bakery and confectionery industries. Mango pulp or juice after concentration is mixed with powdered sugar and dried. Mango pulp can be successfully dehydrated in a double drum drier in 6-8 seconds at 141°C yielding a product of golden yellow colour with original flavour. In a patented process, cabinet drier has been used for dehydrating mango pulp at a temperature of 60-63°C.

For foam mat drying of mango pulp, glycerol monostearate (GMS) is the most suitable foaming agent compared to dried egg albumin and fresh egg white. Spread the foam on stainless steel wire mesh tray with 0.5 kg per sqft load and dry in a hot air drier at 70°C with cross flow drying technique. It should be dried to a moisture level below 2%. The product will be yellow in colour, crisp with mild mango flavour.

Through osmotic dehydration, good quality mango slices can be prepared. The dehydrated product on reconstitution will be like canned slices in terms of colour, flavour and texture.

Mango jam

To prepare mango jam, mix mango pulp prepared from dehydrated mango slices with equal quantity of sugar. Boil the mixture till the TSS is 65-68°brix. Add citric acid at the end to get 0.6-0.7% acidity in the final product (Sagar and Khurdiya, 1998).

Mango toffee

Mango toffee is prepared with 53 kg mango pulp, 30 kg sugar, 4 kg glucose, 8 kg skim milk powder and 5 kg hydrogenated fat with essence and colour of suitable choice. Fruit pulp is first concentrated in a steam jacketed kettle to about one third of its original volume. Other ingredients are then mixed and cooked to a final weight equal to about 1.2 times that of fresh pulp taken. Flavouring material is added at this stage. Cooked mass is then transferred to a tray smeared with fat. Spread the product into thin sheet of 1.6 cm thick and allow to cool. Solid sheet is cut into toffees at 50-55°C having moisture content of 5-6% and wrap in a cellophane or tissue paper and pack in air-tight jars.

Mango yoghurt

Mango yoghurt is an excellent substitute for high calorie pudding which contain sufficient amount of curd with cream. Yoghurt can be prepared by taking milk containing 10% fat. Add 8% sugar and 4% mango pulp and homogenize. This results in good flavoured and smooth consistency yoghurt (Balasubhramanyam and Kulkarni, 1992).

Mango custard powder

Custard powder is prepared by mixing mango powder, sugar, skim milk powder, corn starch, lime juice, cream, salt and water.

Mango ice cream

It is a non-conventional product prepared by mixing mango powder and milk in the ratio of 3:10. Ice cream can also be prepared from structured mango pieces modified with natural ingredients.

Mango shake

This can be prepared with mango powder. It is a delicious and popular drink during summer months, when fresh mangoes are not available in the market. Mango shake can be prepared by adding mango powder in sweetened, boiled and cooled milk

Mango lassi

It is also a delicious and popular drink during summer months, which can be prepared with mango powder with curd and other ingredients. Acceptable mango lassi can be made with mango powder and curd in the ratio of 3:10.

Mango cereal flakes, vermicelli and powder

A process for the preparation of mango cereal flakes (fruited cereal) has been developed at CFTRI, Mysore. It was made by adjusting the pH of the pulp to around 5.4 by neutralizing part of the acidity with sodium bicarbonate; mixing with precooked wheat flour and sugar; and drying in a double drum drier at 2-3 rpm and steam pressure of 2872.8-3112.2 N/m^2. Dried product is highly hygroscopic and need moisture proof packing. Finished product is golden yellow in colour and has the characteristic taste and aroma of the fruit. Mango cereal vermicelli and powder have also been prepared by adding mango pulp into cooked wheat flour to obtain dough of suitable consistency for extrusion through vermicelli press or drying as such by spreading on trays. Drying is done in a hot air cabinet drier.

Strained baby food

Highly nutritious baby food can be prepared using strained mango pulp and custard powder. Mango pulp is passed through a 60 mesh sieve to remove fiber. Sugar is added to this pulp to get the desired blend and the mixture is then homogenized and canned. In the preparation of fruit custard, acidity of pulp is partially neutralized to adjust the pH to 5.3-5.6. Then, add sugar, skimmed milk powder and pre-cooked starch. Mixture is then homogenized and drum dried as in the case of mango cereal flakes. Dried product is powdered and packed under nitrogen and stored at 5°C.

Unripe mango products

Mango is one of the few fruits which are utilized in all stages of its maturity i.e. from very young immature unripe stage to the fully mature and ripe stage. During fruiting, mango crop is very much exposed to the adverse weather conditions. About 75% of the fruits are knocked off, right from the flowering stage till ripening. Losses can be minimized to a great extent by utilizing the dropped green fruits in the processing industry for making pickle, chutney, candy, preserve, juice, dried powder (amchur), beverages, jam and other products. Raw mango fruits are valued for their tangy taste, nutritional content and as a therapeutic agent. Various products can be made from unripe mango fruits and the details are given in this section.

Mango amchur

Select mature green mangoes, wash, peel and cut into slices. Dip the slices in 2% salt solution for an hour, drain the pieces and dip in 2000 ppm sulphurdioxide for two hours. Finally, spread these slices on wooden trays and sun dry. Fill the dried slices into air tight containers and store in dry place. Care has to be taken to powder the dried slices before the onset of rain, to prevent spoilage occurring due to high humidity.

Mango pickle

Green mangoes in India are mostly used as pickles and chutneys. Pickles are prepared in almost every Indian home and also commercially. Mango pickles are classified as salt pickle or oil pickle or sweet pickle based on the type of preservation used. They are made from peeled or unpeeled fruit with or without stones and with different kinds of spices. To prepare green mango pickle, select, fully mature, fresh mangoes, wash and cut into uniform size. Cure the pieces in brine at optimal conditions and dry suitably either in sun or in mechanical driers. Clean and dry red chillies, turmeric and mustard and powder separately and mix with the cured and dried mango pieces, pack in polyethylene bags in required quantities and seal. The ready mix is reconstituted in water overnight to get ready-to-eat pickle.

Mango chutney

Mango chutney is prepared from peeled, sliced or grated unripe or semi ripe mangoes by cooking with spices, salt, onion, garlic, sugar/jaggery, vinegar or acetic acid to a thick consistency. Both brined and fresh slices are used for the preparation of chutney. Chutney prepared from mature, but unripe mangoes has good colour and full flavour. Some Indian varieties like Totapuri and Fazli are ideal for chutney.

Green mango powder

From time immemorial, green mango slices have been in use after sun drying. Slices are dried in the sun for about 16 hours at 35-40°C. Green mango powder prepared in an atmospheric double drum drier can form the base material for the preparation of green mango drink, thick mango chutney and other products of acceptable quality. Boil green mango fruits in water for 15 min, peel and extract the pulp using a pulper. Dry the pulp in 152.4 × 163.2 mm wide chrome coated steel drum drier. Pack the dried samples in poly-aluminium foil laminates.

Green mango beverage

Wash the fruits, slice and boil in water for 20 to 30 minutes and prepare pulp. Strain the pulp through a coarse muslin cloth. Measured quantity of sugar and salt are dissolved in water in another container and strain. Mix the strained sugar-salt solution and mint extract with the strained green mango pulp. Add citric acid, powdered cumin and black pepper and mix thoroughly and boil. To prepare beverage, 25% pulp, 40% sugar, 2% salt and 1.5% citric acid can be used. Raw mangoes can also be cooked in pressure cooker for 10-15 min using 10% water. After cooling, peel, destone and recover the pulp. Process the beverage at 100°C for 10 min in glass bottles and preserve.

New technologies/ innovations in value addition of mango

Mango harvester

Locally, mango harvester is made using cloth, bamboo stick and net made of jute thread. Mango harvester developed by IIHR is somewhat heavy and difficult to handle to harvest the fruits by standing on the tree. Two types of mango harvesters which are useful to harvest fruits from ground (for small trees/

young gardens) and on the tree (for old and well branched trees) as alternatives to local harvester have been designed, fabricated and tested. Light weight harvester when used on large, well branched tress gave a harvesting capacity of 625-650 mangoes per hour with Banganapalli variety. The heavy weight model is useful to harvest fruits from ground and has a capacity to harvest 600-615 mangoes per hour. Capacity of the local harvester made of cloth varied from 300-350 mangoes/hour

Mango telescope pole harvester

Harvester basically consists of a cutting device that is attached to an adjustable aluminium pole. Total weight of the pole is 1.55 kg and it can reach a maximum height of 6 m. This cutting device works on a DC motor which is controlled by a remote control device operated from the ground by the operator. The remote control can send a signal up to 10 m. Pole is guided to the fruit and when it touches the pedicel of the fruit, the remote control starts its operation.

Fig. 1: Telescope pole harvester

Operating torque of the motor weighs about 62.6 g. Cut mango falls on the collecting and conveying unit which is made up of nylon net. Harvested mango reaches the ground by gravitational force without any damage by passing through the collecting and conveying unit (Fig.1).

Mango peeling machine

New automatic mango peeling machine (Fig.2) is suitable for peeling mangoes within seconds. Peeling loss amounts approximately 21% depending upon the size of the fruit. Feeding of mangoes into machine is automatic i.e. only mango has to be placed in product holder. Hence, a single person can fill a number of machines and increase the capacity of filling up to 12 mangoes per minute. It mainly works on shearing principle and the capacity of machine is 3 to 4 pieces per minute.

Fig. 2: Mango peeling machine

Mango destoning unit

It has been designed to destone and peel fresh mangoes automatically (Fig.3). By precisely separating stones and peels from pulp, the destoner improves both yield and quality of final product. It has an efficiency of 70% for peel separation. Generally, two models namely super-mango-destoner and jumbo-mango-destoner are available.

Mango grader

Grading is an important post harvest operation. Graded produce fetch more price in the market and it is also convenient to pack the graded fruits in cases/ boxes. The grader is a power operated, differential speed, expanding pitch V-belt apple grader. It consists of six V-belts with 24 wooden pulleys mounted over four shafts (Fig.4). Distance between adjacent belts increases gradually from 2 cm at the feed end to 5.5 cm at the delivery end. Upper portion of the belts between the upper pulleys act as grading section and the whole grading length is divided into three parts to obtain three different grades.

Pasteurizer

A flash pasteurizer (Fig.5) is used to pasteurize mango juice at about 98°C with a contact time precisely maintained. Pasteurizer has its own acid and caustic holding tanks for Clean in place. There are two major sections inside the pasteurizer, one where the product flows and the other where the heating medium flows. Process takes place in a tubular heat exchanger with counter flow, while pressure inside the product line is maintained to prevent contamination in case of holes, pores or other leakages in the tubes. Pasteurizer has a balance tank, 4 levels of tubular heat exchangers, chiller and a heater. Initially, the hot water pump is switched on and the target temperatures are set in between 115°C-122°C. After this, run the pasteurizer for 30min to reach the target temperature before the product valves are opened. First two levels of tubular heat exchangers have hot water medium and the other 2 levels have a chilling medium. After the two levels of heating coils, product passes in a tube to cool down. In this tube, heat sensors are placed and a target temperature is set in between 98°C to 100°C required for complete pasteurization. Then, the temperature of the product is gradually brought down to 90°C, then to 50°C and finally to less than 20°C before it reaches the filler.

Deaerator

Excess air makes the product more susceptible to oxidation, which further leads to browning and increased acidity. Hence, deaerator (Fig.6) is used to remove the air present on the surface of the fruit pulp. Product is maintained at vacuum conditions. It is similar to a product receiving tank used in the sterilization process. Deaerator is used only for juices and pulp but not for concentrates.

Fig. 3: Mango destoning unit

Fig. 4: Mango grader

Fig. 5: Pulp pasteurizer

Fig. 6: Mango deaerotor

Fig. 7: Mango decanter

Fig. 8: Mango sterilzer

Mango decanter

In the decanter (Fig.7) separation takes place in a horizontal cone-cylindrical bowl equipped with a screw conveyor. Centrifugal force causes sedimentation of solids on the wall of the bowl. Mango decanter can also be used for delicate depulping, especially for the removal of black specks.

Pulp sterilizer

In mango processing industry, pulp sterilizer is attached along with deaerator and homogenizer as shown in Fig.8. It is used for sterilization and concentration of pulp. Capacity of sterilizer is about 6000 L/h. Main source of heating medium for sterilizing mango pulp is steam. Temperature for sterilization is about 131°C. Product sterilization temperature varies around 104 to 106°C for a period of 60-90 seconds to concentrate the pulp. Sterilizer consists of four sections i.e. heating section which is maintained at a temperature of 130°C, holding section at a temperature of 80°C, cooling section around 60°C followed by chilling section around 40°C.

Pulp homogenizer

It is used to prepare mango juice. Homogenization is done in two stages. Stage I is applied primarily at 400 psi followed by Stage II at 2600 psi. After homogenization, particle size in the mix is reduced to 3 microns from an initial particle size close to 20 microns. Flow of the mix into the Ready Beverage Tank is stalled by closing the valves till the homogenizer reaches the target pressure. Equipment capacity is about 6000 L/h. This process is essential to obtain a uniform product which remains as a consistent juice without water separation due to settling or clog formation of the ingredients during storage.

High pressure processing system

Packed mango pulp is processed in a batch mode in high pressure processing system. It consists of a cylindrical vessel having a capacity of 2-4 L, with 100 mm inner diameter and 250 mm depth. The pressure inside the system ranges from 400-600 MPa, and isobaric temperatures varies from 40-60°C. The pressure-hold duration is in the range of 5-15 min at a fixed rate of 300 MPa min^{-1} using aqueous monopropylene glycol (30 ml/100 ml) as pressure-transmitting medium (process fluid). High pressure process samples are preconditioned in a water bath, so that a target temperature of around 40°C-60°C can be achieved upon compression. After processing, samples are immediately brought down to a temperature of 2°C to prevent residual effects of processing. This treatment inactivates pectin methyl esterase, polyphenol oxidase, peroxidase and all microbes such as aerobic mesophiles, yeast & mould, total coliforms, lactic acid bacteria, psychrotrophs and also improves the colour and flavour of mango pulp (Kaushik et al., 2016).

Irradiation and fruit quality

Irradiation is recommended to kill or to sterilize the microbes or insects by damaging their DNA. According to the Food and Drug Administration, the approved dosage for irradiation treatment on a fresh produce is 1 kGy (100 krad). However, 1 kGy may not be effective to kill insects, and such high doses negatively affect the quality of almost all fresh fruits. The effectiveness of irradiation on mango fruit quality depends on irradiation dose, cultivar, and fruit maturity. Irradiation doses ranging from 100 Gy to 750 Gy are best suited for mango processing.

Spray dryer

Spray dryer is used for the preparation of mango powder from mango juice. Powder is obtained by operating the spray dryer concurrently which consists of a spray nozzle with an orifice of 1 mm in diameter. Inlet air temperature of 160°C and outlet air temperature of 70–75°C is maintained in the dryer. Feed rate of liquid is maintained at about 10 ml min^{-1}. Flow rate of the drying air is around 0.7 m^3 min^{-1}.

Pulsed light treatment

Pulsed light treatment is carried out using an automatic flash Xenon lamp system, which is composed of eight lamps situated all around the sample. Wave length of the emitted light ranges from UV-C to infrared. Cubes of mangoes are subjected to 4 successive pulses, for a total fluence of 8 J cm^{-2}. Then, cubes are distributed in glass jars of 1 L (H"50 cubes/jar) capacity which allows to maintain high relative humidity at ambient atmospheric conditions. Samples are stored at 6°C for 7 days. Pulsed light treatment maintains the firmness, colour, carotenoid, phenol, total ascorbic acid contents and increase PPO activity after 3 days. Application of pulsed light can be used with fresh-cut mangoes to improve physical quality and also to maintain nutritional properties (Charles *et al.*, 2013).

Success story in mango processing sector

Well-structured juice processing sector can act as a catalyzer for the development of mango value chain. Famous commercial mango processing and products manufacturing companies in Andhra Pradesh are Hindustan Coca-Cola Beverages Private Ltd, Pepsi Co Private Ltd, Jain Farm Fresh Ltd, Priya Pickle Private Ltd and Galla foods Private Ltd.

In India, Andhra Pradesh is the second largest producer of mango holding a cultivation area of about 4.31 lakh hectares and annual production of 43.5 lakh metric tons. Mango crop occupies 68% of the total cultivated area in Andhra

Pradesh and accounts 24% of the total production in India. Mangoes dominate the horticulture resources of Chittoor District of the State. High yield of mango leads to the development of new orchards and adoption of improved crop management practices. Nearly 53 processing units are presently operating in Chittoor District.

Galla food private limited

Sri Ramachandra N.Galla, a non-resident Indian, now settled in India, started Galla food private limited in 2005. Initially, industry started producing mango concentrates from Totapuri and Neelam varieties. Later, the industry started other fruits like guava and papaya with an annual production of 1500 metric tonnes (5.0 tons per day). This was the main raw material for Coca-Cola Beverages Private Limited and Pepsi Co Private Limited for beverage production. From 2014 onwards, they entered into beverage industry and started producing fruit drinks from mango, apple and litchi and fruit nectars and cocktail from mango. Now, they are producing pack house processed products and puree, pack house processed fresh mangoes like Alphonso, Banganapalli, Neelam, Ruman etc. along with papaya, pineapple, guava and pomegranates. Mango puree is prepared using Totapuri and Neelam varieties while for guava both white and pink varieties are used. Papaya products are prepared using both yellow and red varieties. Industry is using high mechanization technology such as mango pulper, mango grader, pulp homogenizer and pulp steriliser for processing. To maintain safety and hygienic conditions HACCP guidelines are followed.

Scope of by-product utilization/effluent treatment/ utilization

Mango is highly valued as a table fruit and substantial quantities of fruits are also processed. During the manufacture of mango products, large quantities of waste are generated which account for 35% –55% of the fruit, depending on the variety. More than one million tons of mango seeds are annually produced as waste, and these are not currently utilized for any commercial purposes. During processing of mango, peel and mango kernel which are good sources of nutrients are discarded. Value addition of by-products can generate revenue and address the disposal issues as well.

Mango peel

It is a major by-product obtained during processing of mango. Mango peel contributes about 20-30% of the fruit. The peel is a good source of sugars and pectin and also contains substantial quantities of protein, tannins, crude fiber, polyphenols, carotenoids, vitamin E, dietary fiber and vitamin C. It can be incorporated into products such as jam and macaroni (Kim *et al.*, 2010).

Mango kernel

Mango kernel is a good source of starch and fat. Seed content of different varieties of mangoes vary from 9% to 23% of the fruit weight. Kernel which is present inside the seed, weighs around 45% to 75% of the whole seed. Mango seed kernel possesses potential antioxidant activity with relatively high phenolic contents. It is a good source of fat and the fat content varies from 8.85% to 16.13%. Lipids of mango kernel are composed of neutral lipids, phospholipids, and unsaponifiable matter. Protein content in mango seed kernel is very low, but it contains most of the essential amino acids such as leucine, valine and lysine. Mango kernel fat alone or when blended with other nut fats like almond, palm stearin can act as cocoa butter replacer with good triglyceride content and thermal properties.

Pectin production from solid (mango peel and kernel) waste

Pectin is extracted from mango solid waste. It is of two types i.e. water soluble and alkali soluble. Water soluble pectin is more in peel of ripe fruits, while alkali soluble fraction is more in peel of immature fruits. This is obtained by using dry peel and kernel of immature and ripe Dashehari mangoes. During the process, 0.05N NaOH, 0.05N HCl and 0.3% sodium hexameta-phosphate are added.

Vinegar production from mango waste

Vinegar is obtained mainly from peel and stones of Totapuri variety of mango. Alcoholic fermentation of these waste products is carried out mainly by two methods. Extract of peel or stone is prepared by mixing with cold or hot water (50%w/w) and saccharin. Then, this extract is treated with 50 ppm SO_2 inoculated with *Saccharomyces cerevisiae* var. ellipsoideus and fermented at 24 ± 4°C for 72h. The base wine obtained from peel and stone washings contain 2.5% and 3% alcohol respectively. Alcohol content is further increased to the desired level for acidification (5%) by alcohol fortification or secondary fermentation by addition of 4% cane sugar and 0.15% yeast extract.

Future scope of research and limitations

Mango cultivators of India are facing grave challenges including small land holdings, non-availability of quality seedlings/saplings for cultivation and ideal varieties suitable for processing. Apart from this, huge post-harvest loss due to dearth of infrastructure, menace of middle men, lack of support by the concerned nodal bodies, lack of cooperative effort among farmers, lack of integration of activities of government departments, nodal bodies and other institutions are other constraints faced by the cultivators. Hence, a co-ordinated, integrated and strategic effort of all the stake holders is essential to turn around this industry. Mango cultivation of India has to undergo a radical shift to address all the above constraints and reap the enormous advantages/benefits/profits which this sector is to offer. Problems and constraints of cultivators and processing industries should be tackled through an integrated and strategic manner rather than adopting piecemeal approach. Cold storage and other amenities at the growers and retailers level can be established to reduce post harvest losses of mango and to increase availability of value added mango products.

Safety and quality aspects of mango products

In the food industry, food safety principles and practices have always been integrated into activities identified within quality assurance or quality control programmes, or within quality management systems; therefore, these programmes and systems can address both food quality and food safety simultaneously. Mango processing industries should strictly follow the rules and regulations of FSSAI and HACCP. ISO specifications are needed to maintain the quality of processed products. For canned mango pulp, specifications are 17.18°BX TSS, 0.5 to 0.7% acidity as citric acid, <4.00 pH, 3:5 brix-acid ratio and bright yellow colour.

One of the most critical aspects of food processing to ensure health and safety of the consumer is hygiene which is the key concern in the food processing industry. With recent food scares and new legislations, demand for clean in Place (CIP) systems in food industry is increasing. Clean in Place is a method of cleaning the interior surfaces of pipes, vessels, process equipment, filters and associated fittings, without disassembly and to free of inorganic and organic contaminants. Clean in place (CIP) is categorized into single-use CIP system and multiple-use CIP system. In mango processing industry they strictly follow the multiple-use CIP system such as 3-step, 5-step and 6- step procedures. The 6-step process includes pre-rinse, hot caustic treatment, water rinse, phosphoric acid rinse, hot water rinse and final rinse. In 5- step procedure, pre-rinse, hot caustic treatment, water rinse, hot water rinse and final rinse are carried out. Hence, the 5-step CIP is also called acid bypass. In the 3- step process only pre-rinse, hot caustic treatment and final rinse are done; hence, this is also called as lye and acid bypass treatment mostly preferred for mango juice preparation.

Packaging of mango and mango products

Raw mango

After harvest, desapping of the fruits should be done by keeping the fruits in inverted position in the desapping nets for about 3-4 hours. Fully harvested fruits are then packed in corrugated fiber board baskets with stock end pointing upward to avoid injury to the fruits.

Modified or Controlled Atmospheres (MA/CA)

The O_2 concentration of around 3–5% and an elevated CO_2 concentration of 5–10% are the suggested atmosphere compositions for successful MA/CA system for mango fruit (Kader, 1994; Yahia, 2009). Application of MA/CA technology allows extending storage life while retaining the overall quality. Higher CO_2 and lower O_2 (than normal air atmosphere) can delay ripening by inhibiting the production of ethylene, delaying the biochemical activities associated with ripening such as; slowing down the changes in skin and flesh colour, flavour, aroma and texture (fruit softening). Most favourable temperature for modified atmosphere are 10-15°C and the composition is 5% O_2 and 5% CO_2 for storing mangoes up to six months.

Mango beverage (maaza)

Poly Ethylene Terephthalate (PET) is widely used as one of the packaging materials for the maaza beverage. PET bottles are widely used as a packaging material due to good hygroscopic nature, high dimensional stability (resistance to reliability, change and displacement of packing material) and good barrier properties for CO_2 and O_2. It is also a better thermal insulator.

Mango pulp /concentrate

Following are the three options where bulk packing of mango pulp is possible i.e. bag-in–box, bag-in-drum or tins. First two are aseptic packaging types which are acceptable by industry standards, although certain buyers may prefer one over the other. Most of the large-scale processors prefer drums (metallic steel drums, 200 kg per drum), which require fewer steps for transfer of the pulp into tanks. Capacity of each box is 25 kg which contains two 12.5 kg bags. Third and lowest-cost packaging option is by using tins. Packaging by using tins is a very good option, if the pulp has to be stored in godown and also for juice making at the same place.

Mango pickle

Mango pickle and chutney can be packed in clean bottles, jars, wooden casks, tin containers covered from inside with polyethylene lining of minimum 250 gauge or suitable lacquered lining.

Mango products

Generally, mango preserve, jam, jelly and marmalade are stored in clean jars, bottles, chinaware jars and aluminium containers.

Conclusion

There exists good scope for an entrepreneur to establish a small or large scale mango processing industry. In order to maintain a well-established industry, proper knowledge about production, processing, mechanization, packaging and quality control are important aspects which need more attention. Considerable amount of by-products like mango stones, peels etc., can be converted to value added products to generate more revenue and to reduce environmental pollution. To realize the industrial process, scale-up studies and shelf-life estimation may be explored further.

References

Balasubhramanyam, B. Y. & S. Kulkarni. 1992. Standardization of Manufacture of High Fat Yoghurt with Natural Fruit Pulp. *J. Food Sci. & Tech.* 28(6): 389-390.

Charles, F., V. Vidal., F. Olive., H. Filgueiras & H. Sallanon. 2013. Pulsed Light Treatment as New Method to Maintain Physical and Nutritional Quality of Fresh-cut Mangoes. *Innov. Food Sci. & Emer. Techn.* 18: 190-195.

Kader, A. A. 1994. Modified and Controlled Atmosphere Storage of Tropical Fruits. Post Harvest Handling of Tropical Fruit. *ACIAR Proceedings*, Chang Mai, Thailand. P. 239-249.

Kaushik, N., R. P. Srinivasa & H. N. Mishra. 2016. Process Optimization for Thermal-Assisted High Pressure Processing of Mango (*Mangifera indica* L.) Pulp Using Response Surface Methodology, *LWT - Food Sci. & Tech.* 69: 372-381.

Kim, H.J.Y.H., D. Moon., M. Kim., H. Lee., Y. S. Cho., A. Choi., Kim Mosaddik & S. K. Cho. 2010. Antioxidant and Antiproliferative Activities of Mango (*Mangifera indica* L.) Flesh and Peel. *Food Chem.*, 121: 429-436.

Sagar, V. R. & D. S. Khurdiya. 1998. Improved Product from Ripe Mangoes. *Indian Food Packer* 52(6): 2731.

Sudheer, K. P. & V. Indira. 2007. Post Harvest Technology of Horticultural Crops. New India Publishing House, Pitampura, New Delhi, India

Yahia, E. M. 2009. Modified and Controlled Atmospheres for the Storage, Transportation, and Packaging of Horticultural Commodities. Boca Raton, FL: CRC Taylor & Francis.

Appendix 1

S.no	Particulars	Scale of production		
		Large scale	Medium scale	Small scale
1.	Capacity (tones)	13000	5200	2000
2.	Capacity utilised (per cent)	75	50	50
3.	Fixed cost (Rs.)	15.0	6.0	2.0
4.	Variable cost (Rs..)	11.0	4.0	1.5
5.	Total cost (Rs)	26.0	10.0	3.5
Employment				
6.	Executives	10	10	2
7.	Managers	5	2	2
8.	WorkersFull timePart time	100600	500	6250
Investment cost (Investment ranging from Rs. 2 crores to Rs. 15 crores)				
9.	Land value(Rs. Lakh)	464	236	56
10.	Buildings(Rs. Lakh)	600	200	100
11.	Machinery(Rs. Lakh)	400	150	40
12.	Effluent treatment plant(Rs. Lakh)	15	10	1
13.	Vehicles(Rs. Lakh)	21	5	3
Break even analysis for mango processing				
		Large firms	Medium firms	Small firms
14.	Internal Rate of Return(IRR) (per cent)	17.50	13.87	19.31
15.	Pay Back Period (years)	5.99	6.89	5.25
16.	Price per ton of processed products (Rs.)	22800	22000	19000
17.	Variable cost per ton of processed products (Rs.)	19000	18500	14500
18.	Money terms (BEP)	113,885,016	38,096,853	11,353,519

7

Scope of Entrepreneurship Developments in Pineapple Processing

Kalpana Rayaguru & Sanjaya K Dash

Introduction

Pineapple (*Ananas comosus* L.) is cultivated commercially in the tropics and parts of the subtropics of the world. It is greatly appreciated by consumers due to its pleasant sweet taste and flavour in addition to its nutritional and health promoting properties. Varieties differ greatly in both taste and shape. Each variety also has local types. Kew/Giant Kew variety is a leading commercial variety. Fruits are big sized (1.5-2.0 kg), oblong and tapering slightly towards the crown. Fruits with broad and shallow eyes become yellow when fully ripe. Flesh is light yellow, almost fibreless and very juicy. Queen fruits are rich yellow in colour, weighing 0.9-1.3 kg each. Flesh is deep golden yellow, less juicy than Kew, crisp textured, sweet with pleasant aroma and flavour. Baby-pineapples are mostly less than 500 g (Singh, 2016).

Pineapple fruits are consumed fresh. However, because of the unique structure of the fruit and the presence of very active protease, fruit does not find extensive use as a table fruit and hence is mostly processed into a number of products. Many types of value added products are obtained from pineapples which have got good domestic and international market. Proper processing and value addition will not only help to provide wide varieties of products to the market, but also help to reduce the losses of pineapple and will provide additional income to farmers.

Health benefits of pineapple

Sensorial characteristics and chemical composition of pineapples depend greatly on the variety and the geographical and climatological plant growing conditions. In general, 100 g edible part of pine apple contains 86 g water, 13 g digestible carbohydrates, 0.1 g fat, 0.5 g fibres, 100 (20-200) IU vitamin A, 30 mg vitamin C, 230 kJ energy and about 40% waste before usage (Carnara *et al.*, 1995).

The fruit is high in the enzyme bromelain. Bromelain is a natural anti-inflammatory molecule that has many health benefits and encourages healing. It also helps in digestion. Pineapple is a good source of dietary fibre essential for digestive system. Pineapple fruit is very low in saturated fat, cholesterol and sodium.

Pineapple is an important source of sugars, organic acids, essential minerals, and other vitamins (A, B-group). Fruit contains considerable amounts of calcium, chlorine, potassium, phosphorus, sodium and other minerals. Pineapple is the only source of a complex proteolytic enzyme used in pharmaceutical market and as a meat tenderising agent. In addition, pineapple contains other bioactive compounds such as polyphenols and carotenoids that have potential health benefits. Main role of these bioactive compounds is their antioxidant, antiviral, anti-inflammatory and anti-cancer activities. However, the functional properties of each bioactive compound depend on its bioavailability and bio accessibility in the gastrointestinal tract. The pulp has very low ash content and nitrogenous compounds.

It is reported that pineapples are beneficial for treatments of diseases like dyspepsia (chronic digestive disturbance), bronchitis (inflammation of bronchial tubes), catarrh (secretions from mucous membranes), high blood pressure, intestinal worms, nausea (includes morning sickness and motion sickness). Pineapple's bromelain stimulates digestion and proper performance of small intestine and kidneys; it helps in detoxification, normalizes colonic flora, helps in hemorrhoid alleviation, and prevents and corrects constipation.

Scope of entrepreneurship development in pineapple processing sector

Fresh whole fruits

Fresh pineapple, in whole or sliced form, has a good domestic and international market (Reinhardt and Rodriguez, 2009). The fruits should be harvested at the ideal time, and this time is dependent upon how the pineapples will be marketed. Fresh fruits destined for local market are plucked when almost ripe. Colour of skin is an important criterion in determining the ripeness of the fruit. Fresh pineapples destined for export are harvested green-ripe or half-ripe (beginning to turn yellow-green at the base of the fruit). Because of their low sugar-content, pineapples harvested too early are unpopular amongst consumers (pineapples do not ripen afterwards).

After harvesting, fruits are graded according to size, shape, maturity, and freedom from diseases and blemishes. The cut surface is treated with a suitable fungicide to control fungal decay (UNCTAD 2003). Proper choice of packaging materials and containers minimizes damage and losses during storage and transit. In addition to containing and protecting the fruits, it enables retailers and fruit vendors to enhance sales of both fresh and processed pineapples.

For local markets, fruits are packed in bamboo baskets lined with paddy-straw. First layer of fruits is arranged in such a way that they stand on their stumps. Second layer is arranged on the crowns of the first layer fruits. Each basket weighs 20-25 kg. For distant markets, fruits are wrapped individually with paddy straw and then packed. For export purpose pineapples are packed into fibreboard or wood containers. Product stacking will depend on the type and size of container and must be carefully planned to minimize physical damage. Fruits can be placed vertically or horizontally in container. Pineapples should be fixed inside the box, in order to avoid wounds in the shell and/or the crown. The interspace between the fruits should be filled with straw and firm lining all around the container. Reusable plastic containers (RPCs) and corrugated containers (CCFs) are also commonly used for handling and distribution of fresh pineapples throughout the supply chain. An important characteristic in this stage is that the boxes should have holes with lengthened form in all sides for ventilation, because it allows a quick exit of heat of the fruit. Packing measures for pineapple are not standardized, but are guided with international packing norm for agricultural products according to size (Wang and Shu, 2017).

A careful crop handling and postharvest conditions contributes to the maintenance of quality of the products (IFOAM, 2002). When fruits are transported for long distances or need to be stored for several days, refrigerated transport is required to slow down ripening process. In tropical areas, partially ripe, healthy and unbruised pineapple could be stored for almost 20 days at 10-13°C with 85-90% RH. Ripe fruits can be stored at 6-7°C and 90-95% relative humidity for up to 2 weeks. Exposure of pineapples to temperatures below 7°C results in chilling injury. Signs of chilling damage are: black-brown spots appearing in the pulp, loss of skin gloss, formation of brown to black stripes under the skin and around the woody central cylinder (endogenous brown spot), watery flesh, insipid taste, and susceptibility to rotting and loose crown leaves. At temperatures above 10°C, the crown leaves have an increased tendency to bolt. Bolted crown leaves impart a tired appearance to the fruit and diminishes its value. In addition, tendency to fruit rot (black rot) also increases, often occurring as butt rot above the stem but also arising in the crown. Controlled atmosphere storage (3-5% O_2 and 5-8% CO_2) can also be used to delay senescence and extend shelf life (Medina and Garcia, 2005).

Cool chain is essential during transport of export quality commodity all the way from the farm to the customer, both for fresh and fresh cut products. This helps in maintaining temperature inside the box at the same low level as in cold storage. Various stages of cool chain are:

- Cold store at the farm
- Refrigerated truck from farm to the airport
- Cold store at the airport
- Off loading direct into a cold store in the receiving country
- Refrigerated truck to the customers.

Transit time should be calculated so that fruits are at the optimal ripening conditions right before reaching the consumers. In view of the chilling injury occurring to the fruits, transport system is designed to maintain appropriate fruit temperature.

Radiation processed products

Under radiation processing, food is subjected to radiation by exposing it to a source of ionising radiation. This ionising radiation usually is in the form of gamma rays from a source of cobalt-60 or from a non-radioactive source like electron beam generated from electricity. A dose of 2 kGy has been found to be safe which also does not significantly affect the nutritional value as well as the sensory quality of minimally processed pineapple.

Safety and quality aspects of fresh pineapple

As discussed above, fruits intended for export should be harvested half-ripe, just when the colour begins to change on their base. Juice squeezed out of the middle of the fruit should be at least 13°Brix. After harvesting, fruits are cleaned, stalks cut to 2 cm, sorted, classified and packed.

Chilling damage arises in some varieties as "Queen" when they are stored for 14 days at temperatures of less than 7°C. Spoilage may occur as a result both of inadequate ventilation (danger of rotting) and of excessive ventilation (drying-out, weight loss). During transportation, respiration process must be controlled (release of CO_2, water vapour, ethylene and heat) in such a way that the cargo is at the desired stage of ripeness.

Pineapple is a non-climacteric fruit which does not improve in quality including sugar accumulation and flavour once they are harvested. Pineapples harvested at more advanced maturity stages (ripened on the plant) have much better flavour and sweetness than the less mature fruit. Immaturity and poor quality are indicated by a hollow thud.

Pineapples are selected or sorted manually according to the quality criteria agreed upon by the people involved in the trade. It is a common practice to grade pineapple based on variety, weight, shape, maturity and other factors. Prices are influenced mainly by variety and fruit size. Method is very subjective and can vary according to market or location. It is therefore important to incorporate proper grading and standardization in pineapple handling especially for export. Worldwide Codex Standard for Fresh Pineapple (Codex Stan. 182 – 1993) provides some standard guideline for pineapple (Appendix 1).

Fresh cut pineapple

Fresh-cut pineapple has a good market demand. Good manufacturing practices and hygiene procedures should be strictly followed to meet the consumers' requirements in terms of safety and quality. Fresh-cut pineapples can be put under modified atmosphere packaging or controlled atmosphere packaging.

Another innovative strategy is using edible coatings. Edible coatings can help to prevent juice leakage and act as a carrier of substances such as antimicrobials. Edible coatings and modified atmosphere packaging (MAP) reduce juice leakage, retain colour changes and inhibit microbial growth of fresh cut pineapples.

Active packaging such as oxygen scavengers, carbon dioxide releasers/emitters, ethylene scavengers and antimicrobial packaging may be used to improve quality and safety of fresh cut pineapples and pineapple products. Intelligent packaging through the use of indicators and sensors can be used to inform about the product's quality, whereas time–temperature and radiofrequency identification (RFID) tags are increasingly used to monitor quality and safety of particularly fresh and frozen products through the supply chain.

Value added products

Pineapple can be converted to many value added products. Raw fruits are utilized for preparation of products like chutney, pickle, sauce, beverage, etc. while ripe ones are used in making pulp, juice, nectar, squash, leather, slices, etc. Major export products include dried and preserved canned fruits, jam, jelly, dehydrated, frozen fruits, pulp and freeze dried products. Besides, there are other products as minimally processed chunks and slices, dried chips, cocktail-type drinks and powder. Any of these products can be taken up as an entrepreneurial activity in order to cater to the market demand (Sudheer and Indira, 2007).

Canned product

Canning basically involves exposing the product to a high enough temperature for specific time to kill microorganisms and thus extend the shelf life. Types of packages may differ. General canning process for pineapple is shown in Fig. 1.

In a primary processing facility for pineapple canning, fruits are unloaded from trailer trucks to conveyor bands where crowns are removed and then fruits are placed in a washing vat. Fruit should be washed very carefully as it can easily be damaged. Washing can be done at this stage by chlorinated water. Residual chlorine level in wash water should be 20-40 ppm.

Pineapples are then moved by an elevator to a roller sorter. Objective of roller sorter is to separate fruits in to different sizes. Usually, fruit is sorted into two sizes and two slides take the fruit to different packing lines. Peeler removes the peel and forms a cylinder with the peeled fruit, by a circular blade that spins at high speed. Peeling can be done manually, or with knives, yet sometimes the skin is loosened with steam and then subsequently rubbed away mechanically. For removal of the peels and cores and preparing the rings, small tools made up of stainless steel/ plastics are available.

The cylinder is sent by another slide where they are placed manually in another machine that removes the core. The coreless cylinder is then moved to the dicer and if required the product is diced. These dices are ready for being canned. Finally, fruits are sorted again to remove any blackened pieces, bits of peeling, seeds, etc.

Receiving
↓
Removal of crowns
↓
Washing/ cleaning
↓
Sorting and grading
↓
Peeling/coring/ slicing
↓
Can filling
↓
Syruping
↓
Lidding or clinching
↓
Exhausting
↓
Sealing
↓
Can washing
↓
Heat processing
↓
Cooling
↓
Drying
↓
Labelling
↓
Storage/ Marketing

Fig. 1: Canning of pineapple

In case, canning is not to be done immediately, or the canning plant is located at a distance, then the dices are packed in buckets holding 10 to 15 kg. Product is frozen in a chamber set at -20°C and stored or transported for further processing.

Canned products can have different forms as whole fruit, slices, dices, bars, flakes and cubes. The shape of the cut fruit must be given on the can (slices, diced, pieces etc.). Pulp is also often canned. In that case peeled fruits are pulped, and sugar added. They might also be mixed with water or fruit juice.

Most important types of containers used in canning of foods are metal cans, glass jars/ bottles, flexible pouches and rigid trays.

Cut pieces are now filled into jars or cans and covered with syrup. The exhauster maintains steam at 3 bar and 120°C to eliminate air and create vacuum. Then, cans are sealed and sent to the autoclave. Cans are then pasteurised (temperatures above 80°C), or sterilised (temperatures above 100°C). After the heating process, canned fruits are first cooled to 40°C, and then subsequently down to storage temperature. After they have been cooled, canned fruits are labelled and stored. In order to be exported, slices/pulp/juices can be packed into single or wholesale packages (bulk) consisting of glass jars, tin cans or polyethylene or polypropylene bags, and also filled antiseptically into 'bag-in-boxes'.

Use of fruit varieties especially suited to canning and careful observance of established commercial size and quality grades are the two most important fundamentals in commercial canning process. Fruit must be gathered at proper stage of maturiy, i.e. it should not be so ripe or so soft as that can be eaten fresh, yet it should have the flavour characteristics of the ripe fruit. Overripe fruits are more prone to be infected with microorganisms and underripe fruits shrivel and toughen on canning.

Dried pineapple

Drying is a common method for preservation of fruits and vegetables which reduces the activity of microorganisms and enzymes and other biochemical reactions and thus rate of spoilage of the product is reduced. Scientifically, most of the free water available in the fruit is removed, which reduces the water activity. The final moisture is near 5%, and this allows the dried fruit to have a long shelf life as long as proper packing is provided (to prevent absorption of moisture during storage) and storage is done in an appropriate place. Usually, chunks or slices are prepared for better presentation and make handling easier. Dehydrated pineapple has the longest shelf-life when compared to either canning or freezing.

Dehydration process for pineapple is shown in the Figure 2. Thinner slices can be dried in less time. Different types of commercial dryers are available for drying fruits and vegetables. The common is the cabinet dryer or tray dryer (Figure 3). For large capacities, belt conveyor dryer or tunnel dryers can be advantageous. Temperature of drying should be kept below 55-60°C. Drying time is normally 12-14 hours.

Other innovative drying techniques as microwave, heat pump, ultrasound (Jaeger *et al.*, 1998) spray or freeze drying are also used in a limited scale for producing dehydrated products. Drying under sun gives inferior quality product in terms of colour and flavour and has low consumer acceptance.

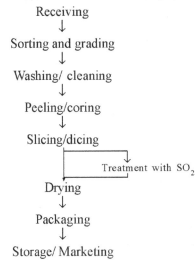

Fig. 2: Process flow chart for dehydrated pineapple

Osmo-dehydrated pineapple products

Osmotic dehydration is a process used for the partial removal of water from plant tissues by immersion in a hypertonic (osmotic) sugar or brine solution. Water is removed by evaporation at atmospheric pressure and temperature near the ambient temperature.

Kew variety pineapples are often used for the preparation of osmotically dehydrated pineapple slices or rings. Different steps involved in the preparation of osmo-dehydrated products are given in Figure 4.

First of all remove the peel and core of pineapple. The core of these fruit pieces are removed by stainless steel corers (Figure 5). Prepare the rings as mentioned under canning. Put these rings in 59°B sugar syrup for 10 hours. During this time, some portion of water is removed from the fruit pieces. Next, remove the pieces from sugar syrup and wipe excess syrup from the surface of these pieces. Dry the pieces in tray dryer at 60°C for 24 hours. Osmo-dehydrated pineapple rings as obtained from this process are shown in Figure 6.

Left over sugar syrup can be used for preparation of nectar and squash. After removing rings from pineapple slices for making the osmo-dehydrated product, some flesh still remains with the skin. These portions can be used for juice extraction, which can be used to prepare pineapple squash, nectar, RTS etc.

In addition, some novel food processing methods as freeze drying, spray drying, osmo-dehydro freezing and other methods to produce powder, pineapple pulp glazings, dried pineapple slices, fruit cocktails, nectar, flavoured yoghurt and fruit sauces and ice cream are also available.

Ultrasound drying is a recent development in which the ultrasound treatment is a pre-treatment for drying of pineapple. Ultrasound pre-treatment consists of immersing fruit pieces in water or in an osmotic solution and to subject the fruit and solution to ultrasonic waves (at frequencies ranging from 18 to 40 kHz) for some time (usually less than 60 min). Ultrasound has shown to have higher influence on fruits with high water content (pineapples, melons) and high content of fibers and phenolic cells.

Fig. 3: Commercial tray dryers

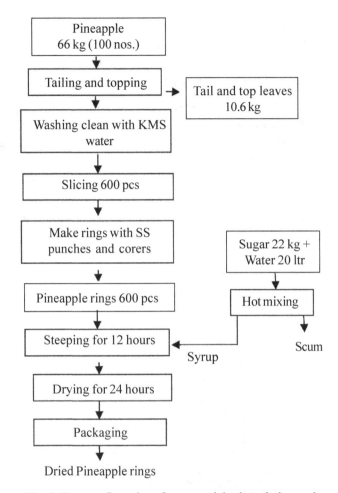

Fig. 4: Process flow chart for osmo-dehydrated pineapple

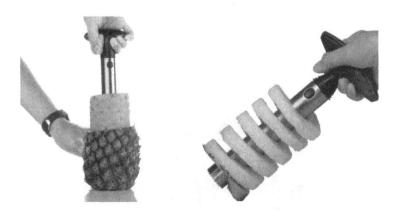

Fig. 5: Pineapple corer

Vacuum fried pineapple chips

Vacuum frying is a technology where food products are deep fried under vacuum or near vacuum condition to reduce the fat content compared to normal deep frying (Figure 7, 8). The process produces healthy fruit snacks (pineapple chips) which also helps in partially preserving the fruit's original colour and nutrients and have a high hydrophilic antioxidant capacity.

Fig. 6: Osmo-dehydrated rings

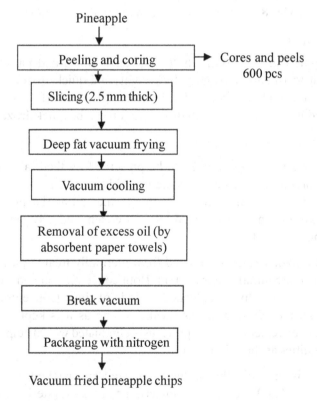

Pineapple

Peeling and coring ▸ Cores and peels 600 pcs

Slicing (2.5 mm thick)

Deep fat vacuum frying

Vacuum cooling

Removal of excess oil (by absorbent paper towels)

Break vacuum

Packaging with nitrogen

Vacuum fried pineapple chips

Fig. 7: Preparation of vacuum fried pineapple chips

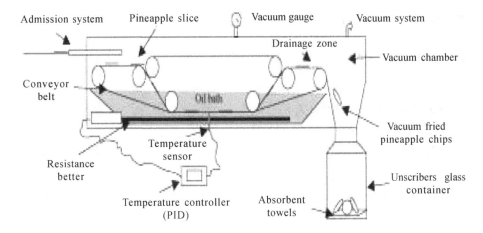

Fig. 8: Equipment for vacuum fried pineapple chips

Frozen product

Ripe pineapple can be frozen whole or peeled, sliced and packed in sugar (1 part sugar to 10 parts pineapple by weight) and quick-frozen in moisture-proof containers. Diced flesh of ripe pineapple, bathed in sweetened or unsweetened lime juice to prevent discoloration can be quick-frozen.

Pulp

Peeled pineapple is crushed and can be preserved by thermal treatment, by addition of preservatives. Pulp is stored in either small packages or in bulk packages and sold to other industries for further industrial processing and formulations as frozen juice, nectar, drinks, ice cream mix, jelly, jam, sodas, fruit cheese, concentrate etc.

Concentrated frozen pulp is prepared from thermally treated pulp to remove at least 50% of the initial water content. Both concentration and freezing help the preservation of pulp for extended periods of time. Concentrated pulp is stable without the addition of chemicals as long as it is kept frozen. Upon reconstitution (by replenishing the previously eliminated water) pulp should have the same qualities as the original pulp.

Aseptic pulp is the pulp that is heat-sterilized and packed aseptically; no chemicals are added. Whole processing is done in aseptic (sterile) environment and the product has a long shelf life. Specific equipment is available to perform this process and it is considered to be at the cutting edge of technology.

Juice

Pineapple juice is obtained from crushing fruit pieces and proper physical separation of solids (Figure 9). Juice is then pasteurized and packed to extend its shelf life and a preservative or refrigerated storage is used as additional barrier to microbial spoilage (McLellan & Padilla-Zakour, 2005).

No juice should reach the market if it has fermented or mixed with water. Packing is done in plastic bottles or bags, coated cans, multi-laminate (plastic, paper, metal foil) or several newer materials. The pH values of the product must be controlled to keep it acceptable for human consumption. It is a common practice to blend batches of juices to attain proper acidity and sensory qualities. Juice from other fruits can be blended with pineapple juice and the mixture

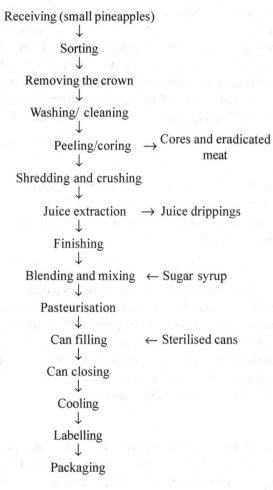

Receiving (small pineapples)
↓
Sorting
↓
Removing the crown
↓
Washing/ cleaning
↓
Peeling/coring   → Cores and eradicated meat
↓
Shredding and crushing
↓
Juice extraction   → Juice drippings
↓
Finishing
↓
Blending and mixing   ← Sugar syrup
↓
Pasteurisation
↓
Can filling   ← Sterilised cans
↓
Can closing
↓
Cooling
↓
Labelling
↓
Packaging

Fig. 9: Preparation of pineapple juice

makes novel products (Arthey, 1995). Proper canning/ packaging after pasteurization are important for these blended products.

Frozen juice is prepared from pressing fresh clean ripe pineapples. There is no concentration, dilution or fermentation and the juice is strained through a 0.5 mm mesh. Juice is then centrifuged, homogenized, air removed, pasteurized and aseptic packaged prior to freezing.

Product does not contain added sugar or preservatives and has a minimum of 12°Brix, 20 to 40% solids, 0.9% acidity and pH values from 3.6 to 3.8. This should be packed in steel containers and to be transported at -18°C or lower.

Concentrated frozen juice is also available at approximately –20°C. It is prepared by direct application of heat to pineapple juice to reduce its water content. Concentration is done at moderate temperature and pressure to yield good nutritional and sensory qualities. Preservation methods are similar as described for concentrated pulp in which no chemical additives are used. Chunks are also filled in syrup into cans or bulk containers and frozen. Osmotic damage is likely to occur during slow freezing whilst structural damage occurs during fast cooling; both conditions lead to significant changes in product quality. A pre-treatment before freezing, such as osmotic dehydration is beneficial to reduce freezing time.

Nectar

It is the product of blending juice with a certain amount of solids from the pulp containing the same amount of °Brix as the original fruit. Normally, nectars are prepared by diluting fruit pulp to 30°Bx. Methods of preservation and packing are similar to those described for juice.

Jelly

As for any other fruit jelly, pineapple jelly is also prepared by mixing 45 parts of fruit and 55 parts of sugar. This mixture is cooked until final solid contents reach 65 to 68%. It is hot-filled for better stability. Final textural firmness is dependent on pectin which is added under controlled acidity and solid content. To assure proper shelf life at ambient temperature, preservatives may be added. These chemicals are mainly used to control mould growth; but once the jar is opened, it shall be stored under refrigeration.

Marmalade

This is also considered as a fruit preserve using the same proportions of fruit and sugar, and cooked until the same solid content as jelly. Consistency is semi-fluid and not a gel as jelly. Preservation criteria and shelf life considerations are similar as for jelly.

Sauce

Pineapple sauce can be prepared by concentrating 1 kg strained pulp by adding 20 g sugar to 1/3 of its original volume in the presence of suspended spice bag containing 50 g chopped onion, 5 g garlic, 50 g ginger, 10 g powdered spices and 5 g red chillies. There is a need to press out spice bag occasionally and then squeezing it out finally to obtain maximum spice extract. Thereafter, add 15 g salt and remaining 40 g sugar and cook the mix to thick consistency. Subsequently, after adding 450 ml vinegar, cook the mix again to end point. Add preservative after dissolving in minimum quantity of water. The mix is then heated to boiling and hot packed.

Fillings

Pineapple pieces mixed with bakery cream may be used as cake fillings for institutional service and large scale production of bakery goods. Stability of the product depends on cleanliness and hygiene of the manufacturing process. Product may be packed in plastic bags, plastic containers or metal bins. If no additive is used, the fillings must be kept refrigerated. Due to its elevated nutrient and water content, shelf life is not very long.

Preserve and candy

Pineapple candy have good consumer acceptance. Prepare rectangular slices (4x1 cm) or suitable sized cubes from fully mature, ripe, washed, peeled fruits after removing inedible portions. Keep the slices in 1.5% lime water for 3-4 hours and then drain and wash 3- 4 times in plain water. Prepare sugar solution by dissolving 400 g sugar in 600 ml hot water and then filtering. Keep the pieces in sugar syrup, boil and keep overnight. Next day, drain the syrup and raise the syrup strength to 50°Brix. Again, add the slices in sugar syrup, boil and keep again. Repeat this process until Brix reaches 70-75°. Keep the product for a week after which drain the syrup, fill the pieces in dry jars and for preserve, cover the slices with freshly prepared sugar syrup of 70°Brix.

For the preparation of candy, the Brix of syrup is raised to 75°, and keep for a week. Drain the product, and dry the pieces under shade. Dip the pieces often in boiling water to remove adhering sugars and then drain, dry and pack.

Toffee

Concentrate one kg sieved pulp to 1/3 volume and cook with 600 g sugar, 100 g glucose and 100 g hydrogenated fat till a speck of the product put into water forms compact solid mass. Prepare a thick paste of 100 g skim milk powder in minimum quantity of water and mix with boiling mass. Spread the cooked mass as 1-2 cm thick layer over stainless steel/aluminum trays smeared with fat.

Flavoring material is added at this stage, if necessary. Then, cool the product, cut and wrap in butter paper.

Plant and machinery

Major equipment used in a pineapple processing facility are listed below. List can be cut short after deciding the specific types of products to be prepared (Reinhardt and Rodriguez, 2009).

- Cutting table/ Crown removing machine
- Fruit washer/ Washing tanks
- Sorting or inspection conveyer/ Sorting tables
- Peeler/ Cores/ Punching machines/ Slicers
- Canning equipment set (Can filling machine, Exhaust box, Lidding machine, etc.)
- Bottle washers and Bottle brush, Crown cappers
- Steam jacketed kettle, Boiling pans/Pasteurisers, Retorts
- Pasteurising kettle, Pressure cooker and Autoclave/ Sterilizer
- Fruit crushers, Fruit presses
- Blanchers/ Sulphuring cabinet
- Cutting boards, Working tables, Packing table
- Deep fat fryers
- Electric heating element
- Dryers
- Hydrometers, Thermometer, pH meters, Refractometers
- Clean water filters, Filter clothes or pads
- Gas burners and Gas cylinder
- MS trolleys, Food grade drums, Plastic barrels, Crates, Knives, Heat sealers, Weighing scale etc.

By-product utilization

Depending on variety and shape of the fruits, total waste generated is 40–60% of the fruit. Important waste components include crown, stem, core, leaves, and mill juice. Peel represents the largest portion (30–42% w/w) followed by core (9–10%), stem (2–5%, w/w) and crown (2–4%, w/w). Different value added products can be prepared from processing wastes that occurs in the field, at packing houses and processing industries. The stems and leaves of pineapple plants are also a source of fibre that is white, creamy and lustrous as silk. Parts of the plant are used for silage and hay for cattle feed (Dev and Ingle, 1982). Processing wastes in the form of shell, core materials and centrifuged solids from juice production are also used as animal feed.

Mill Juice

Most of the pineapple waste components can be milled and pressed to extract milled juice. A number of value-added products can be prepared from milled juice. A typical composition of soluble solids of milled juice is 75–85% sugar, 7–9% citric acid, 2% malic acid, 2.5–4% protein, 3.5% gum and several mineral constituents. Properly prepared high quality sugar syrup from mill juice has a high market demand. High acidity of the juice is removed by treating with either lime or a suitable ion exchange resin. Deacidified and clarified juice syrup can be used successfully in canned pineapples.

Feed

Pineapple wastes are recommended as good sources of organic raw materials and are potentially available for conversion into useful products such as animal feeds. Pineapple waste contains high amount of crude fibre and suitable sugars.

After extraction of mill juice, the compact residue that is obtained also has so many uses. This is further dried to produce pineapple bran. Bran so prepared has about 6–8% crude protein and 22–28% crude fiber. It can be a potential source of feed for ruminants.

Wet or partially dried pineapple bran can be ensiled with rice straw, maize meal, dried sweet potato chips, molasses, and urea in various combinations and proportions. Conservation methods and feeding value of these pineapple bran mixtures for ruminants have been established and when adequately supplemented, bran has been found to be a valuable and cheap feeding stuff well suited to use in a variety of canneries.

Solids from the centrifuge from juice production may be used as feed for pork.

Bromelain

Among all the pineapple by-products, an important by-product is bromelain, a proteolytic enzyme. A process for the production of bromelain from pineapple has been patented in India. Enzyme is extracted into dibasic potassium or sodium phosphate solution or potassium phosphate buffer (pH 5.5–7.5). Extracted bromelain is precipitated with ethyl alcohol, acetone, or *n*-butanol and dried at reduced pressure. In another process developed in China, mill juice is clarified and the enzyme is precipitated with alcohol. Settled bromelain slurry is spray dried to obtain crude enzyme. Another method (Tisseau, 1986) has been developed for bromelain extraction based on milled juice clarification followed by ammonium sulfate precipitation.

Vinegar

Pineapple canary waste juice, a good source of sugar (7–10%), is a good substrate for vinegar production (Fig.10). Vinegar is prepared by an acetic fermentation of alcohol solutions derived from sugar or starchy materials (fermentable sugar content of 8-20%). A method for the production of good-quality vinegar from pineapple waste juice involves fermenting the extracted juice with *Saccharomyces cerevisiae* after adjusting the final Brix to 10° and then passing the alcoholic fermented juice through a generator for acetic fermentation with *Acetobacter* sp. Culture (Satyavati *et al.*, 1972). Other methods are also available.

Vinegar must be pasteurized once it is prepared and bottled. It is stable at ambient temperature.

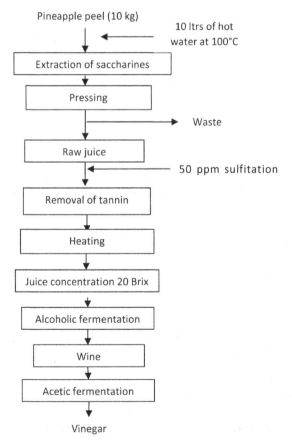

Fig. 10: Preparation of vinegar from pineapple peel

Other products

Pineapple by-products are a potential source of important compounds such as sucrose, glucose, fructose, cellulose, fibre and phenolics. In addition to these, there are also methods to prepare sucrose and molasses containing glucose and levulose from pineapple juice by hot precipitation of extract with lime and filtering to produce clarified juice. After demineralization with ion exchange resins juice is concentrated to produce syrup containing crystals of sucrose, which is then recovered by centrifugation.

The underflow from centrifugation of pineapple juice, which is a viscous fluid, has been used to prepare potentially useful by-products, including oleoresin having a strong pineapple flavour.

Pineapple peel is rich in cellulose, hemicellulose and other carbohydrates. Peel flour has very good prebiotic potential to support probiotic bacteria in the gut. Fermentable sugars and other nutrients make pineapple waste extracts excellent media to produce enzymes, single cell proteins, bacterial cellulose and organic acids. Waste can also be utilized to extract bioactive compounds that can be used in food, pharmaceutical or allied industries. Waste also has the potential to act as alternative source of energy.

Pineapple industry waste is also used for producing fibre and paper. Process of obtaining fibre consists of removal of green tissue by scraping, soaking of fiber and sun drying. Pina, the delicate and expensive fabric of the Philippines is made from this fiber. Waste material left during extraction of fiber is suitable for paper making. This fiber is as thin as the finest quality of jute but about 10 times as coarse as cotton (Chakraverty et al., 2003).

Ensilaging of pineapple peels produces methane which can be used as a biogas. Anaerobic digestion takes place and the digested slurry may find further application as animal, poultry and fish feeds.

Future scope of research and limitations

According to various studies available, evidence indicates that pineapple contains the enzyme (protease) which has several therapeutic properties including thrombus formation and control of malignant cell growth, inflammation, diarrhoea, dermatological and skin debridement. Further research is required for confirmation that it is well absorbed with its therapeutic effects being enhanced in a dose dependent manner. Thus, if incorporated in foods, it could become more acceptable as a nutraceutical and functional product (Rashida et al., 2017).

More research is needed for full-scale commercialization of novel technologies with respect to microfiltration (Tran and Le, 2011) process optimization, packaging, and consumer acceptance.

Marketing of processed products

Marketing of processed products from pineapple is the biggest challenge. There are several popular brands in the market, which act as an entry barrier for new food products. Indian market is also flooded by imported products. Therefore, newcomers need to give maximum importance to marketing of its products. It is always better to engage a professional agency for product branding and marketing. The units should also allocate reasonable budget for advertisement and promotion. Due to changing lifestyles, hassles of making these items, consumers prefer their availability throughout the year from market. There are many variants of these products. New comers may introduce new flavour with certain change in the ingredients to alter taste of the products. It is imperative to cater to regional palate and ethics. Apart from domestic market, there is good demand for processed products in export market. Nowadays Indian products are widely accepted throughout the world for commercial as well as household consumption. Of late, large numbers of Indians have migrated to various countries and higher number of Indians visits other countries. Therefore, demand for Indian food products is on the rise in many countries. New entrants may initially join hands with existing merchant exporters to get entry into such markets. Thereafter, vast export potential for such products can be tapped.

The processed products should follow the Food Safety and Standard Authority of India (FSSAI) Act 2006. FSSAI Act is applicable PAN India for all food products. It prescribes minimum standards operating procedures, food safety norms, packaging and labelling norms. The new units need to take a license called FSSAI number from Food Safety and Standards Authority of India. Licensing procedure is given at FSSAI website linkhttp://foodlicensing. fssai.gov.in/UserLogin/Login.aspx?ReqID=99887766.

Conclusion

Pineapple is a tropical fruit which is consumed fresh or in a processed form. It contains nutrients which are good for human health. As pineapple contains protease, it can be incorporated into foods which are an advantage to human health when taken orally. World pineapple demand has been expanding rapidly. Consumer acceptance of new foods is based largely on their convenience, appearance and reproducibility as well as nutritional, sensory and economic values. Processed pineapples are consumed worldwide and processing industries are trying out or using new technologies to retain nutritional quality of pineapple fruit. This is to meet the demand of consumers who want healthy, nutritious and natural products. Moreover, some of these preserved products such as canned pineapple, fruit juices, dehydrated products and frozen fruits are gaining popularity in the foreign market and are good foreign exchange earners. Thus, there is considerable scope for establishing pineapple processing industries in India, which in turn will help in the development of horticulture and in earning more foreign exchange.

References

Arthey, D. 1995. *Food Industries Manual*, Chapman & Hall, London, UK.

Carnara, M., C. Diez & E. Torija. 1995. Chemical Characterization of Pineapple Juices and Nectars. Principal Components Analysis. *Food Chem.* 54:93–100.

Chakraverty, A., A. S. Mujumdar., G. S. Vijayaraghavan & S. H. Ramaswamy. 2003. *Handbook of Post Harvest Technology Cereals, Fruits, Vegetables, Tea, and Spices*, Marcel Dekker, Inc. New York, USA.

Dev, D. K. & U. M. Ingle. 1982. Utilization of Pineapple By-products and Wastes: A Review, *Indian Food Packer* 15:15-19.

IFOAM, 2002. *Proceedings of the 14th Organic World Congress*, Compiled by R. Thompson. Canadian Organic Growers, Ottawa.

Jaeger, de Carvalho L. M., C. A. Bento da Silva & A. P. T. R. Pierucci. 1998. Clarification of Pineapple Juice (*Ananas comosus* L. Merryl) by Ultrafiltration and Microfiltration: Physicochemical Evaluation of Clarified Juices, Soft Drink Formulation, and Sensorial Evaluation. *J. Agric. Food Chem.* 46:2185–2189.

McLellan, M. R. & O. I. Padilla-Zakour. 2005. Juice Processing, in *Processing Fruits: Science and Technology*, D. M. Barrett., L. P. Somogyi & H. S. Ramaswamy, (eds.), CRC Press, Boca Raton, pp. 73–96.

Medina, J. D. & H. S. Garcia. 2005. Pineapple: Post-Harvest Operations. Danilo Mejia, (ed.), *Agricultural & Food Engineering Technologies Service (AGST)*, FAO of United Nations.

Rashida, R. T. A., P. P. Joy., R. Anjana & M. Rini. 2017. *Pineapple as a Functional Food*. Pineapple Research Station (Kerala Agricultural University), Vazhakulam, Kerala, India.

Reinhardt, A. & L. V. Rodriguez. 2009. Industrial Processing of Pineapple – Trends & Perspectives. *Acta Horticulturae* 822:323-328.

Satyavati, V. K., A. V. Bhat., G. Verkey & K. K. Mukherjee. 1972. Preparation of Vinegar from Pineapple Waste, *Indian Food Packer* 26(3):50.

Singh, D. 2016. *Pineapple: Production and Processing*, Daya Publishing House, New Delhi.

Sudheer, K. P. & V. Indira. 2007. *Post Harvest Technology of Horticultural Crops*, New India Publishing Agency, Pitampura, New Delhi, India.

Tisseau, R. 1986. Utilization of Pineapple-Canning Waste: Potential for Extraction of Bromelin. *Fruits* II(12): 703-710.

Tran, P. P. T. & V. V. M. Le. 2011. Effects of Ultrasound on Catalytic Efficiency of Pectinase Preparation during the Treatment of Pineapple Mash in Juice Processing. *Int. Food Res. J.* 18:347–354.

UNCTAD, 2003. UNCTAD/DITC/COM/2003/2. Organic Fruit and Vegetables from the Tropics, Market, Certification and Production Information for Producers and International Trading Companies. *United Nations Conference on Trade & Development*, New York .

Wang, C. & H. Shu-Fang. 2017. Case Study on Entrepreneurship Value Of Tainan Pineapple Products, Taiwan - Based On TQM. *The International Journal of Organizational Innovation* 9(3):1-11.

Appendix 1

CODEX STAN 182-1993 (REV. 1-1999)

1. Definition of produce

This standard applies to commercial varieties of pineapples grown from *Ananas comosus* (L.) Merr. of the Bromeliaceae family, to be supplied fresh to the consumer, after preparation and packaging. In the worldwide Codex Alimentarius Standard for Pineapple, it is stated that the total soluble solid (TSS) content for harvesting must be higher than 12°Brix.

2. Provisions concerning quality

2.1 Minimum requirements

In all classes, subject to the special provisions for each class and the tolerances allowed, the pineapples must be

- whole, with or without the crown
- fresh in appearance, including crown, when present, which should be free of dead or dried leaves
- sound; produce affected by rotting or deterioration such as to make it unfit for consumption is excluded
- clean, practically free of any visible foreign matter
- free of internal browning
- practically free of pests affecting the general appearance of the produce
- practically free of damage caused by pests
- free of pronounced blemishes
- free of damage caused by low and/or high temperature
- free of abnormal external moisture, excluding condensation following removal from cold storage
- free of any foreign smell and/or taste.

When a peduncle is present, it shall be no longer than 2.0 cm, and the cut must be transversal, straight and clean. The fruit must be physiologically ripe, i.e., without evidence of unripeness (opaque, flavourless, exceedingly porous flesh) or overripeness (exceedingly translucent or fermented flesh).

2.1.1 The pineapples must have been carefully picked and have reached an appropriate degree of development and ripeness in accordance with criteria proper to the variety and/or commercial type and to the area in which they are grown.

The development and condition of the pineapples must be such as to enable them

- to withstand transport and handling, and
- to arrive in satisfactory condition at the place of destination.

2.2 Maturity requirements

The total soluble solids in the fruit flesh should be at least 12°Brix (twelve Brix degrees). For the determination of Brix degrees a representative sample of the juice of all the fruit shall be taken.

2.3 Classification

Pineapples are classified in three classes defined below.

2.3.1 "Extra" Class

Pineapples in this class must be of superior quality. They must be characteristic of the variety and/or commercial type. They must be free of defects, with the exception of very slight superficial defects, provided these do not affect the general appearance of the produce, the quality, the keeping quality and presentation in the package.

The crown, if present, shall be simple and straight with no sprouts, and shall be between 50 and 150 per cent of the length of the fruit for pineapples with untrimmed crowns.

2.3.2 Class I

Pineapples in this class must be of good quality. They must be characteristic of the variety and/or commercial type. The following slight defects, however, may be allowed, provided these do not affect the general appearance of the produce, the quality, the keeping quality and presentation in the package.

- slight defects in shape
- slight defects in colouring, including sun spots
- slight skin defects (i.e., scratches, scars, scrapes and blemishes) not exceeding 4 per cent of the total surface area.

The defects must not, in any case, affect the pulp of the fruit. The crown, if present, shall be simple and straight or slightly curved with no sprouts, and shall be between 50 and 150 per cent of the length of the fruit for pineapples with trimmed or untrimmed (Trimming consist in tearing some leaves off the top of the crown).

2.3.3 *Class II*

This class includes pineapples which do not qualify for inclusion in the higher classes, but satisfy the minimum requirements specified in Section 2.1 above. The following defects may be allowed, provided the pineapples retain their essential characteristics as regards the quality, the keeping quality and presentation.

- defects in shape
- defects in colouring, including sun spots
- skin defects (i.e., scratches, scars, scrapes, bruises and blemishes), not exceeding 8 per cent of the total surface area.

The defects must not, in any case, affect the pulp of the fruit. The crown, if present, shall be simple or double and straight or slightly curved, with no sprouts.

3. Provisions concerning sizing

Size is determined by the average weight of the fruit with a minimum weight of 700 g, except for small size varieties (Victoria and Queen) which can have a minimum weight of 250 g, in accordance with the following table.

Size Code	Average Weight (+/-12%) (in g) -with crown-	Average Weight (+/-12%) (in g) -without crown-
A	2750	2280
B	2300	1910
C	1900	1580
D	1600	1330
E	1400	1160
F	1200	1000
G	1000	830
H	800	660

Significant volumes of pineapples in international trade are packaged and sold by count per box.

Boxes are packed to minimum weight expectations e.g. 10 kg, 20 lbs, 40 lbs, appropriate for the various markets. Fruits are segregated for packaging by weights which approximate the above size codes, but may not consistently fall within a single size code, but would retain the uniformity required by the code.

4. Provisions concerning tolerances

Tolerances in respect of quality and size shall be allowed in each inspection lot for produce not satisfying the requirements of the class indicated.

4.1 Quality tolerances

4.1.1 "Extra" Class

Five per cent by number or weight of pineapples not satisfying the requirements of the class, but meeting those of Class I or, exceptionally, coming within the tolerances of that class.

4.1.2 Class I

Ten per cent by number or weight of pineapples not satisfying the requirements of the class, but meeting those of Class II or, exceptionally, coming within the tolerances of that class.

4.1.3 Class II

Ten per cent by number or weight of pineapples satisfying neither the requirements of the class nor the minimum requirements, with the exception of produce affected by rotting or any other deterioration rendering it unfit for consumption.

4.2 Size tolerances

For all classes, 10 per cent by number or weight of pineapples corresponding to the size immediately above and/or below that indicated on the package.

5. Provisions concerning presentation

5.1 Uniformity

The contents of each package must be uniform and contain only pineapples of the same origin, variety and/or commercial type, quality and size. For "Extra" Class, colour and ripeness should be uniform. The visible part of the contents of the package must be representative of the entire contents.

5.2 Packaging

Pineapples must be packed in such a way as to protect the produce properly. The materials used inside the package must be new, clean (recycled material of food-grade quality), and of a quality such as to avoid causing any external or internal damage to the produce. The use of materials, particularly of paper or stamps bearing trade specifications is allowed, provided the printing or labelling has been done with non-toxic ink or glue. Pineapples shall be packed in each container in compliance with the Recommended International Code of Practice for Packaging and Transport of Tropical Fresh Fruit and Vegetables (CAC/RCP 44-1995).

5.2.1 Description of Containers

The containers shall meet the quality, hygiene, ventilation and resistance characteristics to ensure suitable handling, shipping and preserving of the pineapples. Packages must be free of all foreign matter and smell.

6. Marking or labelling

6.1 Consumer packages

In addition to the requirements of the Codex General Standard for the Labelling of Pre-packaged Foods (CODEX STAN 1-1985, Rev. 2-1999), the following specific provisions apply:

6.1.1 Nature of produce

If the produce is not visible from the outside, each package shall be labelled as to the name of the produce and may be labelled as to the name of the variety and/or commercial type. The absence of the crown should be indicated.

6.2 Non-retail containers

Each package must bear the following particulars, in letters grouped on the same side, legibly and indelibly marked, and visible from outside, or in the documents accompanying the shipment.

6.2.1 Identification

Name and address of exporter, packer and/or dispatcher. Identification code (optional).

6.2.2 Nature of produce

Name of produce if the contents are not visible from the outside. Name of variety and/or commercial type (optional). The absence of the crown should be indicated.

6.2.3 Origin of produce

Country of origin and, optionally, district where grown or national, regional or local place name.

6.2.4 Commercial Identification

- Class
- Size (size code or average weight in grams)
- Number of units (optional)
- Net weight (optional).

6.2.5 Official inspection mark (optional)

7. Contaminants

7.1 Heavy metals

Pineapples shall comply with those maximum levels for heavy metals established by the Codex Alimentarius Commission for this commodity.

7.2 Pesticide residues

Pineapples shall comply with those maximum residue limits established by the Codex Alimentarius Commission for this commodity.

8. Hygiene

8.1 It is recommended that the produce covered by the provisions of this Standard be prepared and handled in accordance with the appropriate sections of the Recommended International Code of Practice – General Principles of Food Hygiene (CAC/RCP 1-1969, Rev. 3-1997), and other relevant Codex texts such as Codes of Hygienic Practice and Codes of Practice.

8.2 The produce should comply with any microbiological criteria established in accordance with the Principles for the Establishment and Application of Microbiological Criteria for Foods (CAC/GL 21-1997).

8

Entrepreneurship Ventures in Cashew Processing

Sobhana A & Mini C

Introduction

Cashew (*Anacardium occidentale*) is valued as one of the most important commercial horticultural crops in India, belonging to the family Anacardiaceae and widely grown in tropical areas. The name *Anacardium* refers to the shape of the fruit, which looks like an inverted heart (*ana* means "upwards" and - *cardium* means "heart"). It is native to Brazil, introduced to India by Portuguese for afforestation and soil conservation and now grown widely in the traditional and non-traditional regions of our country. Cashew is well adapted to warm humid climatic conditions. It is restricted to altitudes below 700 m, where temperature does not fall below 20°C for prolonged periods.

India is the second largest producer, processor as well as consumer of cashew. Yet, nearly 50% of the requirement of cashew processing industries is met through import. As against the processing requirement of approximately 20 lakh MT per annum, the domestic production of raw cashew nut is only about 9 lakh MT per annum. Major cashew producing states in India are Maharashtra, Andhra Pradesh, Orissa, Kerala, Karnataka, Tamil Nadu, Goa, West Bengal and interior tracts of Chasttisgarh, A & N islands, Gujarat, Jharkand and North Eastern regions. Though, there is limited scope for expansion of cashew producing area there is scope to increase income through cashew based commercial enterprises.

The economic importance of cashew cultivation is mainly depended on the production of raw nut and associated production of cashew apple, cashew nut shell liquid and to a small extent the timber and firewood obtained from the crop. It has a unique role in industrial and economic performance, livelihood support to farming communities, export earnings, agricultural and rural development and sustainable agriculture.

Cashew nuts, unlike other oil tree nuts, contain starch to about 10% of their weight. This makes them more effective than other nuts in thickening water-based dishes such as soups, meat stew, and some milk-based desserts. Many Southeast Asian cuisines use cashews due to this characteristic, rather than other nuts. The shell of the cashew nut is toxic which is removed before its use. Cashew nuts are commonly used to garnish sweets or curries, or ground into a paste that forms a base of sauces for curries, or some sweets. It is also used in powdered form in the preparation of several Indian sweets and desserts.

Cashew nuts can also be used in its tender form when the shell has not hardened and is green in colour. The shell is soft and can be cut with a knife and the kernel extracted can be soaked in turmeric water to get rid of corrosive material before use.

Ripe cashew apple is very juicy, spongy, somewhat fibrous, having a unique smell, and has a very thin skin that gets easily bruised. Cashew apple is the pseudo fruit, attached to the nut which develops from the pedicel. Astringency in cashew apple produces a rough, unpleasant and biting sensation on the tongue and throat, which is a major drawback of the fruit. Astringency of cashew apple is determined to a large extent by the tannin content, a phenolic compound, and its content varies from 0.06 to 0.76 g per 100 g.

An overview of the multiple uses of cashew apple is given in Fig. 1. Figure depicts the large number of uses of cashew apple unlike other common fruits. Traditionally, several products are prepared from cashew apple, including those with medicinal properties.

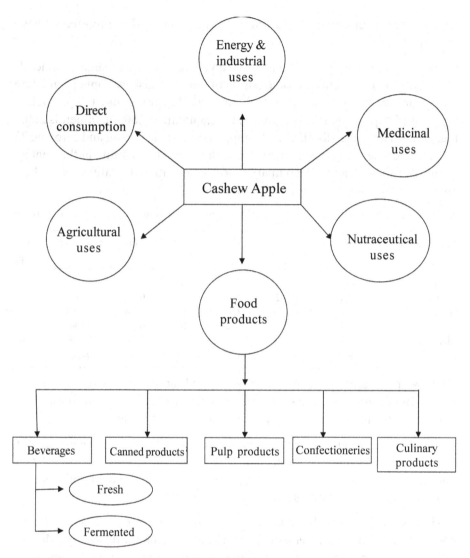

Fig. 1: Multiple uses of cashew apple

Health benefits of cashew

Cashew, as with other tree nuts, is said to be a good source of antioxidants and also a source of dietary trace minerals like copper, manganese, magnesium and phosphorus. Many parts of the plant are used in the traditional medicine to treat snakebites. Nut oil is used to cure cracked heels and used as an antifungal agent. Fruits, bark, and leaves are also used to treat many ailments including sores, rashes and diarrhea. They have anti-fungal and antipyretic activities also. Leaf extract with petroleum ether and ethanol is reported to inhibit growth of several species of bacteria and fungi. Cashew shell oil is reported to have bactericidal action.

Cashew apple is highly nutritious and is a valuable source of sugars, minerals and vitamins. The chemical composition of 100 g of cashew apple is moisture 86.3 g, protein 0.2 g, fat 0.1 g, carbohydrate 12.3 g, crude fibre 0.9 g, calcium 10.0 mg, phosphorous 10.0 mg, iron 0.2 mg, vitamin C 180mg, minerals 200.0 mg, thiamin 0.02 mg, riboflavin 0.05 mg, nicotinic acid 0.4 mg and carotene 23 µg. Thus, cashew apple is comparable with several other fruits in the content of most of the nutrients. A comparison of the nutritional qualities of cashew apple with the commonly used tropical fruits is given in Table 1.

Table 1: Comparison of nutritive value of cashew apple with common tropical fruits (Per 100g)

Name of fruit	Moisture (g)	Protein (g)	Fat (g)	Calcium (mg)	Iron (mg)	Carotene (µg)	Vit. C (mg)
Cashew apple	86.3	0.2	0.1	10	0.2	23	180
Apple	84.6	0.2	0.5	10	0.66	0	1
Banana	70.1	1.2	0.3	17	0.36	78	7
Orange	87.6	0.7	0.2	26	0.32	1104	30
Mango	81.0	0.6	0.4	14	1.30	2743	16
Pappaya	90.8	0.6	0.1	17	0.50	666	57

Cashew apple liquor is used as a cure for cold, body ache, fever, toothache, fresh wounds and cuts. Cashew apple powder lipids are rich in unsaturated fatty acids, the major ones being palmitoleic and oleic acids.

A valuable by- product that can be obtained from cashew apple waste is pectin. Pectin is used in manufacturing jam, jelly, marmalade etc. It is useful as a thickening, texturising and emulsifying agent and finds numerous applications in pharmaceutical preparations and cosmetics.

Cashew apple juice also has medicinal properties. For instance, its high tannin content makes it a suitable remedy for sore throat and chronic dysentery (Morton, 1987). It is reported to have anti-bacterial, anti-fungal and anti-tumor properties (Kubo et al., 1993, ; Kozubek et al., 2001) as well as antioxidant (Melo-Cavalcante et al., 2003) and anti-mutagenic (Cavalcante et al., 2005) properties.

Scope of entrepreneurship development in cashew apple processing sector

Generally, cashew nut is considered as the only economic produce from the crop. Research studies as well as experiences running in India's first cashew apple processing unit at Cashew Research Station, Madakkathara, Thrissur, Kerala, revealed that the cashew apple, weighing about 8-10 times that of the nut, is an equally valuable produce from crop, if it is economically exploited. However, cashew apple is almost completely wasted in India, without any commercial exploitation, except in Goa, where it is used for making *Feni*. Currently, 60-70 lakh tones of cashew apple are wasted annually in the country. In spite of its high nutritive value, it is quite unfortunate that the country is wasting such an excellent fruit causing economic loss both to the farmers and the nation.

Cashew Research Station, Madakkathara, Thrissur, Kerala, has succeeded in commercializing eight value added products from cashew apple (Fig.2). Efforts bestowed in this line enabled the commercial production of different fermented and non fermented products viz; syrup, RTS beverages, Jam, pickle, chutney, candy, wine, alcohol, vinegar etc (Vijayakumar, 1991; Mathew *et al.*, 2008; Mini & Mathew, 2007).

Processing of cashew apple is an economically viable enterprise in cashew growing tracts. Women Self Help Groups can very well take up this enterprise, thereby effectively contributing to the cause of women empowerment. If legal permission is available for production of fermented products like alcohol and wine, it can substantially enhance the income from cashew apple processing. There is also a vast untapped export market for cashew apple products.

Infrastructure development is the primary requirement for starting any processing unit. Cashew apple production being seasonal, ensuring availability of raw material round the year for full utilization of infrastructure capacity is the basic necessity. Processing of locally available fruits and vegetables along with cashew apple can solve this problem. Economic analysis for establishing a fruit processing

Fig. 2: Cashew Apple Products

unit including cashew apple is given in Appendix-1. The social dimension of this investment is evident by the level of employment generation from an otherwise wasted fruit or by- product utilization. Cashew apple based processing units can be an effective driving force for rural development.

Economics of processing cashew apple for syrup production has been worked out. By processing one tons of cashew apple, a net profit of approximately Rs. 12,000/- can be obtained. Considering the average yield of nuts as 800 kg/ha, a production of 6.4 t/ha of cashew apple can be anticipated. A production of about 2 t/ha of good cashew apple can be ensured, taking 30% of the total production for processing. Thus, additional income from a hectare of cashew orchard from the processing of cashew apple is worked out to be approximately Rs. 24,000/-, if a farmer or farmers' groups can venture into this endeavor. The income can be enhanced by processing cashew apple for high value products like alcohol and wine. Compared to other fruits, the advantage of cashew apple is that it is available free of cost and hence the price of cashew apple beverages can be fixed at a lower rate than that of conventional fruit drinks like mango and pineapple.

Total production of cashew apple in the country is estimated to be around 60 lakh tons. At least a minimum of 30% of the total quantity can be economically utilized for production of value added products, working out to 18 lakh tons. With a net profit of Rs. 12,000/- by processing one ton of cashew apple, the total national income that can be obtained through cashew apple processing is estimated to be around Rs. 2160 crores, which will significantly contribute to national economy.

Nut processing

At present, the farmers are directly selling their cashew nut to the traders without even attempting to process it due to which the farmers are not getting the proportional benefit of increased price of cashew kernels in the market. Farm level processing is advantageous to farmers for getting better returns from cashew nut. This can be undertaken by *Kudumbasree,* SHG or such working groups, thus leading to entrepreneurship development. Kerala State Cashew Development Corporation Ltd., based at Kollam, Kerala, is preparing plain, roasted and salted kernels under the name "CDC Cashews". They also developed cashew nut products namely Cashew Vita, Cashew Powder, Cashew Soup, Choco-Kaju, Milky Kaju and Cashew Bits and are being sold. Their new developed products ready for launch are pepper coated cashew and garlic coated cashew. The immature as well as mature kernels could be used for the preparation of highly relished curries and products like spiced and salted kernels, burfi etc. Cashew nut shell liquid is another marketable product from cashew nut which has medicinal value.

Value added products of cashew apple

Beverages

Fresh apple beverages: Clarified and cloudy juice, juice concentrate, syrup, squash and ready- to- serve drink are some of the nutritious and refreshing beverages that can be made from the unfermented juice of cashew apple by adding varying concentrations of sugar, citric acid and preservative. Kerala Agricultural University has standardized the technique for the preparation of juice, syrup, carbonated drink and ready to serve beverage.

Fermented beverages: Cashew apple can be utilized for the manufacture of fermented products like wine, vinegar, liquor and alcohol. Cashew apple vinegar can be prepared by alcoholic and subsequent acetic fermentation of juice, which is perhaps the oldest known fermentation product. Cashew liquor is made by distillation of pure juice of cashew apple without addition of any extraneous matter. One litre of 60-62% ethyl alcohol can be obtained from eight liters of cashew apple juice. Kerala Agricultural University has standardized the method of producing four different grades of liquor from cashew apple.

In Goa, cashew apple is used in the manufacture of fermented beverages. Cashew apple juice is extracted and kept for fermentation for a few days. Fermented juice is then double distilled and the resulting beverage is called *feni* or fenny. *Feni* has about 40-42% alcohol. The single-distilled version is called *urrac*, which has about 15% alcohol.

In southern region of Tanzania, cashew apple is dried and stored. Later it is reconstituted with water and fermented, then distilled to make strong liquor often referred by the generic name, *gongo*. In Mozambique, cashew farmers commonly make a strong liquor namely *agua ardente* (burning water) from cashew apple.

Cashew wine is a product of fermentation of hexose sugar of cashew apple juice by intact yeast cells to form ethyl alcohol and carbon dioxide. Kerala Agricultural University has developed methods for producing four grades of wine such as soft, medium, hard and sweet, based on the alcohol percentage and sweetness.

Cashew apple wine can be mixed with fresh juices of orange, pineapple, tomato, grape and cashew apple as well as tender coconut water to produce wine coolers to serve as good health drink as they contain both wine with its medicinal properties and fruit juices with high amount of nutrients and minerals.

Products from cashew apple pulp

Jam is the most important pulp product of cashew. It can be prepared by boiling the cashew fruit pulp with sufficient quantity of sugar and a pinch of citric acid to a reasonably thick consistency. Mixed fruit jam can also be prepared by mixing cashew apple pulp with equal quantity of banana pulp, mango pulp or pineapple pulp. Madakkathara Centre is commercially producing Cashew apple-Mango mixed jam named *Cashewman*.

Fruit bar having 80°brix can be prepared by heating layers of fruit pulp after mixing with pectin, sugar, glucose and potassium metabisulphite and citric acid to 90°C and drying to 15% moisture. Different layers of cashew apple paste are sun dried and cut into required size after placing one on top of the other to form leather. The layers, after smearing sugar syrup and pressed together, can be eaten like fruit wafers.

Confectionery products

Candied fruit is prepared from cashew apple by impregnating with cane sugar with subsequent draining and drying. One kilogram of cashew apple on processing gives 745 g candies. Madakkathara Centre is commercially producing cashew apple candy. Syrup left over from the candying process can be used for sweetening chutneys, in vinegar making or for candying another batch of fruits. Cashew apple can also be utilized for the preparation of tutty fruity. One kilogram of cashew apple on processing gives 715 g tutty fruity. Whole fruit can also be processed in to nutritious toffee, a feasible dessert item with extended shelf life.

Cashew apple juice can be used for preparing frozen desserts and dairy/ confectionery items by optimization of juice concentration and spray drying. Only constraint here is the large capital investment required for spray drier equipment.

Ready- to- serve beverage mix, fruit–milk/milk shakes, ice creams, ice candy mix, etc can be prepared from clarified juice by homogenization, spray drying and mixing with fruit/milk powder as required.

Osmotic dehydrated cashew apple is a novel value added product developed from cashew apple. Sugar has been completely replaced with honey in preparation of this product, hence having medicinal property with no side effects of sugar. Thus, it is possible to make the seasonal fruit available to the consumers throughout the year. One Kg of good quality fresh cashew apple on processing gives about 200 g of osmotically dehydrated cashew apples.

Culinary products

Sliced raw green fruit can be used to prepare pickle using chilly powder, gingelly oil, fenugreek powder, asafoetida, turmeric powder, garlic, mustard powder, a pinch of sodium benzoate and salt to taste. Chutney can be prepared from sliced cashew apple using sugar, onion, ginger, cumin, pepper, cardamom, cinnamon, coriander powder, salt, vinegar etc.

Several traditional culinary preparations are in vogue in cashew growing areas using both unripe and ripe cashew apples.

Post harvest handling of cashew apple

Desired qualities of cashew apple for processing are

- Medium to large fruit size
- More than 70% juice
- Juice containing more than 11% sugar and 0.39 - 0.42% acidity
- Sugar: More than 13% in varieties Kanaka, Dhana, Raghav and Damodar
- Varieties with over all suitability for processing: Madakkathara-1, Poornima, Damodar, Madakkathara-2, Dhana and Kanaka

Collection and sorting of cashew apples

Harvest of cashew apple at full ripe stage and separation of nut with minimum damage to the apple are essential for its effective utilization. Crisp, firm, tight and full colour developed apples are to be collected and used for processing purpose; riper the fruit, the sweeter it will be. The physical and chemical properties of apple are optimum during 44 to 46 days after fruit set and at this stage, apple becomes suitable for processing and it falls to ground along with the attached nut. This period seems to be the best time for collecting apples without spoilage. Apples are to be collected every day when it falls to the ground and if the apples are left un-gathered for some time, rotting of cashew apples occur. Once damaged, apples may ferment and deteriorate rapidly. Additional losses may also occur when apples are taken away by birds and animals. These losses can be minimised by spreading polythene sheets under the tree. After harvesting, fruits are to be sorted to select the best quality ones. Selected fruits are washed with water in different ways, such as soaking or washing with cold or hot water sprays.

Extraction of cashew apple juice

Depending upon the quantity of juice required, extraction can be done either manually or mechanically. For domestic purposes, extraction by hand pressing will serve the purpose. Mechanical extraction by hydraulic press or screw press, or juice expeller, saves labour and ensures increased recovery of juice and hence is preferred for commercial processing.

Astringency in cashew apples and its removal

Presence of astringent and acrid principles in cashew apple produces a rough, unpleasant and biting sensation on the tongue and throat. Astringency of cashew apple is determined to a large extent by the presence of phenolic compounds, such as tannins, oily substances and anacardic acid and cardol. Removal of components responsible for astringency, from whole or sliced cashew apples (ripe/green) is known as De-tanning. Clarification is the process of removal of astringency from cashew apple juice.

De-tanning of cashew apples: Efficient method of de-tanning of whole ripe cashew apple for making jam, candy, chutney is as follows. Cleaned ripe whole cashew apples are immersed in 5% salt solution for 3 days. To ensure full immersion, keep weight of glass or stainless steel. Salt solution is to be changed daily. The fruits, which are taken out on the fourth day are washed thoroughly in water and can be used for product preparation.

De-tanning of whole green cashew apples: For pickle preparation, de-tanning of green cashew apples is done by cutting fruits into small pieces, washing, and keeping in 8% salt solution for 3 days. The salt solution should be changed daily. Fruits taken out on the fourth day can be used for product preparation after washing with water.

Clarification of cashew apple juice: Clarification of cashew apple juice can be done using either 2.5 to 4.0 g gelatin dissolved in hot water or 125 ml of rice gruel per litre of juice. Calcium hydroxide or pectin can also be used for clarification. Any one of the above materials is added, stirred and allowed for settling for required time ranging from 2-3 hours to overnight. Decant the clarified clear juice without disturbing the sediments.

Utensils for cashew apple processing

Stainless steel vessels are to be used for cooking. Iron, copper, aluminum and brass containers should never be used as it will blacken the products. Food grade plastic barrels and containers are to be used to store cashew apple juice. Plastic buckets and trays are to be used for cleaning, sorting and de tanning of cashew apples. Jars and bottles of clear white glass, which can withstand heating, are preferred for storage of raw materials or products. Though, glass containers are fragile and require extra care in handling, being visible, the contents can be easily displayed.

Storage of pulp and juice

In view of the seasonal nature of production, long term storage of cashew apple is necessary either as juice or pulp for off-season processing. Rapid deterioration of both pulp and juice demands its immediate preservation even for short term processing. To store as pulp, make de-tanned apple into pulp and add preservatives. Juice can be stored either by sterilizing or treating with preservatives. Heat treatment adversely affects flavour and imparts cooked taste and hence not preferred. Potassium metabisulphite (KMS) can be used as preservative.

Storage of cashew apple pulp: The de-tanned cashew apple is steamed for 10-15 minutes in low pressure taking care not to become too soft. The black spots and parts of pedicel are removed from cooked fruit and made into pulp by thorough agitation using mixer or pulper. Add 2.5 g potassium metabisulphite (KMS) and 5.0 g citric acid for every kg of pulp and store in air tight food grade plastic barrel or glass bottle.

Storage of cashew apple juice: Add 2.5 g potassium metabisulphite (KMS) and 5.0 g citric acid during clarification process and stir well. Dissolving potassium metabisulphite and citric acid in little quantity of juice and then adding to bulk juice ensures better incorporation. Treated juice after separating the tannin can be stored for long period, even up to one year.

Value added products of cashew apple developed at Kerala Agricultural University, Thrissur

Cashew apple syrup

Clarified juice is siphoned out as the raw material for preparation of syrup (Fig.3). Sugar and citric acid are added to the clarified juice in required quantity to produce syrup. Taste is better if served chilled. Syrup has a storage life of one year.

Fig. 3: Cashew apple syrup

Cashew apple drink

The drink is an RTS (Ready - To -Serve) beverage. It is prepared by adding water and sugar to required quantity of clarified juice (Fig.4). Drink is marketed both in glass bottles and in attractive food grade pouches. Pasteurized drink in glass bottles has a storage life of three months under ambient storage conditions.

Cashew apple carbonated drink

Cashew apple carbonated drink (soda) is made by adding chilled carbonated water at 100 psi pressure (Fig.5). Drink is marketed both in glass bottles and in attractive food grade pouches. Pasteurized drink in glass bottles has a storage life of three months under ambient storage conditions.

Cashew apple- Mango mixed Jam

Ripe apples are selected, cleaned and soaked in salt solution for three days to remove tannin. Apples are again washed in water, cooked, made into pulp and is mixed with equal quantity of mango pulp. Pulp is mixed with sugar and citric acid to prepare jam. It is marketed under the trade name *Cashewman* Mixed Jam (Fig.6).

Fig. 4: Cashew RTS   **Fig. 5:** Carbonated drink

Fig. 6: Cashew apple jam

Cashew apple candy

It is a sweet product and quality apples with good shape are selected for candy preparation (Fig.7). As in jam preparation, tannin is removed from apples, cooked, pierced using fork and dipped in sugar solution. Concentration of sugar solution is gradually increased so as to reach 70°brix. After two weeks of soaking, sugar solution is drained out and candy is dried in shade. It takes about 2-3 weeks for making the final product.

Fig. 7: Cashew apple candy

Cashew apple pickle

Mature but unripe cashew apples are collected directly from plantations without disturbing the flowers and tender nuts. After cleaning, fruits are cut into small pieces and astringency is removed by immersing in salt water. After removing from salt water, it is again washed and pickle is prepared using oil, chili powder, fenugreek powder, turmeric powder, ginger and garlic (Fig.8).

Fig. 8: Cashew apple Pickle

Cashew apple vinegar

Vinegar preparation involves two stages, alcoholic fermentation followed by acidic fermentation. In the first stage, add a little sugar and yeast to cashew apple juice and keep for one week for alcoholic fermentation. In the second stage, add mother vinegar (three times that of alcoholic ferment) to the alcoholic ferment and again keep for 2 weeks. After this period, check the acidity and if it is 4-5%, then vinegar is ready. Mother vinegar contains acetic acid forming bacteria.

Cashew apple chocolate

Cashew apple powder is used to make chocolate, by adding sugar, milk powder and ghee in appropriate proportions.

Cashew apple biscuit

Cashew apple biscuit is prepared using cashew apple powder, ghee, sugar and *maida* in required quantities and further baking.

Cashew apple wine

Wine is a fermented product from cashew apple. This is made by mixing cashew apple, sugar, clove and luke warm water, after adding starter solution. Starter solution is prepared by adding 5 g yeast to 10 g sugar in warm water and keeping for half an hour. Keep the mixture in glass bottles or clay pots for 21 days with daily shaking. Then, strain the solution and again keep for 21 days by which time the wine is ready.

Cashew apple squash

For making squash, add sugar and citric acid to water and boil. After cooling the sugar solution, add clarified juice of cashew apple, strain, bottle and seal. This can be diluted three times for drinking.

Cashew apple bar

Cashew bar is prepared from pulp by adding 40% sugar.

Cashew halva

Halva is made from cashew apple pulp by adding coconut milk, sugar and ghee.

Cashew apple pudding

Cashew apple pudding is another confectionery which is prepared by mixing cashew apple powder, sugar, milk and gelatin along with vanilla essence.

Osmodehydrated cashew apple

The process involves dehydration of cashew apple in three stages. Osmotic dehydration, in which fruits are subjected to osmosis by dipping them in aqueous honey syrup under specific conditions followed by dehydration wherein the fruits are dried to 15-20% moisture by air and oven drying and then packaging in flexible laminated pouches filled with nitrogen. It is a value added product from cashew apple with complete replacement of sugar with honey and a ready to eat snack.

New Technologies/innovations in value addition of cashew apple

Removal of tannin (clarification/detanning)

Presence of tannin interferes with the taste of cashew apple and the processed products. Cashew Research Station, Madakkathara has developed effective low cost and organic technology for removal of tannin by using sago.

Clarification and storage of juice: Strain the extracted juice through a muslin cloth. Add powdered and cooked sago @ 5 g/litre of juice for clarification. Then, add 2.5 g potassium metabisulphite (KMS) and 5 g citric acid to every litre of juice. Keep for 12 hours to allow the tannin to settle and decant the upper layer of clear juice carefully without mixing the sediments. This clarified juice can be stored in well sterilized air tight food grade plastic barrels for one year and used for off season product preparation.

Detanning and storage of cashew apple: Dip cleaned apples in 5% salt solution for three days, changing water every day. Fourth day, the fruits are taken out and wash thoroughly in fresh water. Steam the fruits in pressure cooker for 10 -1 5 minutes, pulp it and store in air tight glass bottles after adding 2.5 g KMS and 5 g citric acid for every kilogram of pulp.

Success stories in cashew apple processing sector

Cashew apple is not commercially exploited anywhere in the country except in Goa where it is used for the preparation of *feni*. However, many institutions in India, particularly Kerala Agricultural University, have developed technologies for the preparation of various nutritious and refreshing products from cashew apple. Cashew Research Station, Madakkathara of Kerala Agricultural University is profitably running a commercial unit of cashew apple products. This success story is to be replicated elsewhere in major cashew growing areas of the country, which will give additional revenue to farmers and can generate considerable employment for unemployed youth and women. An FPO licensed (Now converted to FSSAI registration) cashew apple processing unit has been established at Cashew Research Station, for the manufacture of unfermented cashew apple products. It is the first ever unit established in India for cashew apple processing. The unit is undertaking commercial production of syrup, cashew drink (RTS), mixed cashew apple- mango jam, pickle, chocolate, carbonated drink, vinegar and candy from cashew apple. Other products developed are halva, fruit bar, tutty fruity, wine, chutney, cookies, pudding, blended squash with cashew apple and pineapple in 50: 50 ratio as well as blended RTS beverages with cashew apple and pineapple in 60: 40 ratio.

The model unit serves to present a comprehensive idea to the trainees and visitors regarding the technical, marketing, economic, infrastructural and licensing requirements for running a commercial unit and a positive response to the cashew apple products. Several entrepreneurs intending to start cashew apple processing units are visiting the model unit and get convincing ideas about its technical and commercial aspects.

First ever cashew apple processing unit in private sector using Madakkathara technology has been established at Iritty, Kannur District, Kerala under the trade name "TOMCO PRODUCTS" and they are marketing cashew apple syrup and candy. Success of the SHG units largely depends upon the support of the State and Central Governments. Being a processed product, cashew apple products are also charged Valued Added Tax. This is a major impediment in selling cashew apple products at attractive price. Extending financial support for establishing cashew apple processing units from Government can encourage entrepreneurs to start new units. Cashew apple products can create great demand in market through its unique traits.

By-product utilization/effluent treatment

Conversion to vermicompost: Pomace or waste after the extraction of juice can be converted to vermicompost with the help of earth worm, *Eudrillus euginae*, with high manurial value of 1.69% N, 0.44% P and 0.58% K. it is also a good ameliorant for acidic soils, since the pH is 8.9.

As animal feed: Ripened apple or residue can be converted to cattle, pig and poultry feed. Apples/residues are dried and subjected to anaerobic ensiling and are preserved as cattle feed for rainy season/long storage.

Biofuel: There is a great scope to utilize cashew apple for production of alcohol to be used as a biofuel. The residue, after extracting juice for *feni* preparation, is used as fuel in liquor industry in Goa.

Biogas: Ripened fruits as well as fruit waste can be used as raw material for biogas plant.

Tannin extraction: Cashew peel can be used for the extraction of tannin.

Future scope of research and limitations

Following aspects need further research
- Varietal difference in product development from cashew apple
- Spray dried products from cashew apple
- More honey based products from cashew apple

- RTS in tetra packaging with more shelf life from cashew apple
- Blended products with cashew apple
- Diversified products from cashew apple
- Storage studies with fresh cashew apple for raw consumption - Large scale availability of cashew apple is a problem due to scattered plantations. Apples have to be transported without affecting the quality.
- Nutraceuticals from cashew apple.

Safety and quality aspects of cashew apple products

Apple and nut fall together when ripe, and in commercial plantations only nuts are collected and apples are left on the ground which are being grazed by cattle or other animals, birds etc. Cashew apples should be processed within two or three hours of picking, since they undergo rapid deterioration when kept for a longer time. Perishable nature of cashew apple is a limitation to the development of processing option and consequent difficulties in transportation from growing areas to distant processing plants. In areas where labour costs are not very high, apples may be gathered and taken to markets or processing factories.

Packaging of cashew products

Cashew apple products are being packed in pet bottles, glass bottles, plastic bowls, flexible laminated pouches, etc. Tetra packing will give more shelf life.

Conclusion

Processing of cashew apple is to be considered as a programme of agricultural waste utilization, adding income to the growers. Excellent qualities of cashew apple offer immense opportunities for its processing to various value added products. Commercial exploitation of cashew apple is the need of the hour considering its vast potential in enhancing the income from cashew plantations. It is one of the prime areas of utilizing the indigenous fruit and opens up wider market possibilities and hence tremendous scope for commercialization. Running of cashew apple processing unit at CRS, Madakkathara under Kerala Agricultural University for the commercial production clearly demonstrates the economic viability of cashew apple processing. However, the support of State and Central Governments are vital in promoting processing of cashew apple, including financial and policy support, such as exemption of VAT for cashew apple products, at least in the initial years. Extending financial support for establishing cashew apple processing units under National and State Horticulture Mission, National Horticulture Board, and Rashtriya Krishi Vikas Yojana, NABARD etc can encourage entrepreneurs to start new units.

References

Cavalcante, A.A.M., G. Rubensam., B. Erdtmann., M. Brendel & J.A.P. Henriques. 2005. Cashew (*Anacardium occidentale*) Apple Juice Lowers Mutagenicity of Aflatoxin B1 in *S. typhimurium* TA102. *Genet. Mol. Biol.* 28: 328-333.

Kozubek, A., R. Zarnowski., M. Stasiuk & J. Gubernator. 2001. Natural Amphiphilic Phenols as Bioactive Compounds. *Cell Mol. Biol. Letters* 6: 351-355.

Kubo, I., H. Muroi., M. Himejima., Y. Yamagiwa., M. Hiroyuki., K. Tokushima., S. Ohta & T. Kamikawa. 1993. Struture-Antibacterial Activity Relations of Anacardic acids. *J. Agric. Food Chem.* 41: 1016-1019.

Mathew, J., G. Zachariah & C. Mini. 2008. Economic Potentials of Tuber Crops for Intercropping in Young Cashew Plantations. *J. Plantation Crops* 36 (3): 366-367.

Melo-Cavalcante, A. A., G. Rubensam., J. N. Picada., E.G. Silva., F. J. C. Moreira & J. A. P. Henriques. 2003. Mutagenic Evaluation, Antioxidant Potential and Antimutagenic Activity against Hydrogen Peroxide of Cashew (*Anacardium occidentale*) Apple Juice and Cajuina. *Environ. Mol. Mutagen.* 41: 360-369.

Mini, C. & J. Mathew. 2007 . Multi uses of Cashew Apple. *Proceedings of 6th National Seminar on Indian Cashew in the Next Decade- Challenges and Opportunities*, 18-19 May 2007, Raipur. pp: 45-52.

Morton, J. 1987. *Cashew Apple: Fruits of Warm Climates*. James Morton and Co. Ltd., Miami Florida, pp: 239-240.

Vijayakumar, P. 1991. Cashew Apple Utilization - A Novel Method to Enhance the Profit. *The Cashew* 5: 17-21.

Appendix-1

Economics of establishment of cashew apple processing unit

Proposed products- Cashew apple squash

Cashew apple jam

Cashew apple pickle

Cashew apple vinegar

Cashew apple candy

I. Establishment cost

1)	Renovation of building with wiring and water connection	: Rs. 5.00 lakhs
	Total	: Rs.5.00 lakhs
2)	Machinery and Equipment	
	1. Fruit pulper with 1.5 HP motor	: Rs. 0.93 lakhs
	2. Electric tray drier with 6 stainless steel trays and control cubicle and 2 KW heater coil.	: Rs. 1.06 lakhs
	3. Juice extractor	: Rs. 1.50 lakhs
	4. Refractometer	: Rs. 0.24 lakhs
	5. Glass wares	: Rs. 0.50 lakhs
	6. Vessels	: Rs. 0.50 lakhs
	7. Containers	: Rs. 0.50 lakhs
	8. Refrigerator	: Rs. 0.25 lakhs
	Subtotal (Rs)	: Rs. 5.48
3)	Other expenses	
	1. Label-designing, licensing, transaction costs : Rs 0.30 lakhs	
	2. Computer	: Rs. 0.50 lakhs
	3. Telephone	: Rs. 0.05 lakhs
	Subtotal (Rs)	: Rs.0.85 lakhs
	Grand total	: Rs. 11.33 lakhs

II Recurring expenses

A	Raw materials:	: Rs. 1,36,000
	Other ingredients(Table2)	:Rs. 2,55,462
	Subtotal	: Rs. 3,91,462
B	Labour	
	A. Fuel charges	: Rs 90,000
	B. Packing (bottles)	: Rs. 1,20,000
	C. Labels/ gum etc. / telephone charges	: Rs. 50,000
E	D. Selling cost	: Rs. 60,000
	Other expenses	: Rs. 12000
	Subtotal	: Rs. 3,32,000
	Total	**: Rs. 7,23,462**

Table 2: Raw material (per year)

Raw material	Quantity (Kg)	Cost/ Kg (Rs)	Total cost (Rs)
Cashew apple	17000	8	1,36,000

Ingredients (per year)

Sl.No	Ingredients	Quantity(Kg)	Total Cost (Rs)
1.	Sugar	5300	2,12,000
2.	Citric acid	50	27500
3.	KMS	1	470
4.	Gingely oil	40	6000
5.	Salt	54	432
6.	Ginger	10	600
7.	Green chilly	10	200
8.	Mustard	5	380
9.	Chilly powder	15	3000
10.	Garlic	5	600
11.	Turmeric	1	50
12.	Fenugreek powder	1	130
13.	Asafoetida	1.5	1600
15.	Miscellaneous		2500
	Total (Rs)		**2,55,462**

III Production and Returns (per year)

Item	No. of bottle	Quantity in the bottle/container	Quantity	Price (Rs)
Squash (Rs 90/ bottle)	14000	700ml	9800litre	1260000
Jam (Rs 85/ bottle)	5370	500g	2685kg	456450
Pickle (Rs 100/bottle)	1450	500g	72500kg	145000
Vinegar(Rs.25/bottle)	5000	500ml	2500litre	125000
Candy(Rs.20/container)	15000	50g	750kg	3,00,000
Grand total (Rs)				22,86,450

IV. Economics of the enterprise

Sl. No	Item	Amount (Rs. in lakhs)
I	a) Interest on fixed investment @ 12%	
	b) Depreciation 10% of fixed investment	1.36 1.13
II	Total recurring expenses	7.24
III	Recurring expenses' interest @ 7%	0.51
III	Total cost of the enterprise	
	Total expense per year(including interest on fixed investment and depreciation and recurring expenses)	10.24
	Total expenses	10.24
IV	Benefit cost analysis	
	1. Total returns	22.87
	2. Expenses	10.24
	3. Net returns	12.63

9

Underutilised Fruits and Vegetables- Scope for Value Addition

Mini C , Geethalekshmi P R & Manjunath Shetty

Introduction

Modern agricultural systems that promote cultivation of a very limited number of crop species have relegated indigenous crops to the status of neglected and underutilised crop species. This has caused a decline in crop diversity in agricultural systems across the world and current research efforts have identified these underutilised crops as having potential to reduce food and nutrition insecurity. This is because of their adaptability to low input agricultural systems and better nutritional composition. Underutilised crops are otherwise known as undervalued or neglected or development opportunity crops (DOCs) (Kahane *et al.*, 2013). They are already cultivated, but underutilised regionally or globally giving relatively low global production and market value. These underutilised crop species have also been described as "minor", "orphan", "promising" and "little-used". Many of these traditional crops grown for food, fiber, fodder, oil and as sources of traditional medicine play a major role in the subsistence of local communities. With good adaptation to often marginal lands, they constitute an important part of the local diet of communities providing valuable nutritional components, which are often lacking in staple crops. Global Facilitation Unit (GFU) for Underutilised species define underutilised crops as, those plant species with under-exploited potential for contributing to food security, health (nutritional/medicinal), income generation and environmental services.

Apart from their commercial, medicinal and cultural values, traditional vegetables are also considered important for sustainable food production as they reduce the impact of production systems on the environment. Many of these crops are hardy, adapted to specific marginal soil and climatic conditions, and can be grown with minimal external inputs. Value addition by applying appropriate production and post harvest techniques ensures that high quality produce reaches the market and satisfies consumer expectations.

Health benefits

Underexploited fruits and vegetables are known for their therapeutic and nutritive values. Because of their curative properties, they have been used in traditional Indian systems of medicine such as Ayurveda and Unani since time immemorial. Many have excellent flavour and very attractive colour. Colour in fruits and vegetables are gaining importance nowadays as they play a vital role in human health and nutrition because of their antioxidant properties. Underutilised fruits and vegetables are rich sources of carbohydrates, fats, proteins, energy, vitamins-A, B_1, B_2, B_3, B_6, C, folic acid, and minerals such as Ca, P, Fe, and dietary fiber. Value addition to such fruits would not only promote economic status of people but also promote their health and nutritional status. Health and nutritional benefits of certain underexploited fruits and vegetables are detailed below.

Health benefits of underutilised fruits

Phyllanthus emblica Linn. (syn. *Emblica officinalis*), commonly known as Indian gooseberry or amla, is highly nutritious and is an important dietary source of vitamin C, amino acids, minerals and phenolic compounds. All parts of the plant are used for medicinal purposes, especially the fruit, which has been used in Ayurveda and in traditional medicine for treatment of diarrhea, jaundice and inflammation. Bael (*Aegle marmelos*) or wood apple fruits are rich sources of vitamins, minerals, protein, fibre, carbohydrates and phenols. Bael is used in 60 patented drugs (Nandal and Bhardwaj, 2015). Sweet drink prepared from the pulp of fruits produce a soothing effect on the patients who have recovered from bacillary dysentery.

Ber (*Zizyphus lotus*), is richer than apple in protein, phosphorus, calcium, carotene and vitamin C. Ber pulp contains 13.6% carbohydrates of which 5.6% is sucrose, 1.5% is glucose, 2.1% is fructose and 1.0% is starch (Jawanda *et al.*, 1981). Plant is also rich in polyphenols, cyclopeptide alkaloids, dammarane saponins, vitamins, minerals, amino acids, and polyunsaturated fatty acids, which are supposed to be responsible for most of the biologically relevant activities including antimicrobial, anti-inflammatory, hypoglycemic, antioxidant, and immunomodulatory effects. Bilimbi (*Averrhoa bilimbi*) fruits has 100% edible portion and contains 94.4 g moisture, 0.5 g protein, 0.3 g fat, 3.5 g carbohydrate, 15 mg calcium, 10 mg phosphorous, 1.2 mg iron, 32 mg ascorbic acid and 19 Kcal of energy per 100g (Radha and Mathew, 2007).

Carambola (*Averrhoa carambola*) fruits contain 9.4% carbohydrates, 0.4% protein, 0.9% fibre, 0.3 to 0.5% minerals, 570 to 700 IU of vitamin A with 0.5 to 1% acidity (Radha and Mathew, 2007). Singh *et al.* (2014) studied prophylactic role of carambola against hepatocellular carcinoma and suggested that it can be a good chemopreventive natural supplement against cancer. Ripe custard apple (*Annona reticulata*) fruit contains about 1.6% protein, 0.02% calcium, 0.02% phosphorus, 0.007% iron and 0.7% minerals. It is also high in carbohydrates, especially fructose and it has large amount of vitamins C, B_1, and B_2 and possess anticancer, antiparasitic and insecticidal activities (Moghadamtousi *et al.*, 2015).

Guava (*Psidium guajava*), often labeled as "super-fruit" is low in calories and fats but carry several vitamins, minerals, and antioxidant polyphenolic and flavonoid compounds. Fruit is a very rich source of soluble dietary fibre (5.4%), which makes it a good bulk laxative. Guava fruit is an excellent source of antioxidant vitamin-C. 100 g fresh fruit provides 228 mg of this vitamin, more than five times the required daily recommended intake. Fruit is a very good source of β-carotene, lycopene, lutein and cryptoxanthin, which are known to have antioxidant properties and therefore essential for optimum health. 100 g of pink guava fruit provides 5204 μg of lycopene. Fresh fruit is a very rich source of potassium, which is an important component of cell and body fluids that helps in controlling heart beat and blood pressure.

Jackfruit (*Artocarpus heterophyllus*) is a fleshy fibrous fruit. Every 100 g of ripe flakes contains 287-323 mg potassium, 30.0-73.2 mg calcium and 11-19 g carbohydrates. Simple sugars like fructose and sucrose in fruit make it an energy fruit. High fibre prevents constipation and smoothen the bowel movements. Jamun (*Syzygium cuminii*), is a summer fruit, which is available abundantly, but for a short period. Fruit is known for its higher content of anthocyanins responsible for its purple colour. Jamun fruit is rich in many bioactive compounds such as anthocyanins, glucoside, ellagic acid, isoquercetin, kaempferol and myrecetin.

Health benefits of figs (*Ficus carica*) come from the presence of minerals, vitamins and fibre in the fruit. Figs contain a wealth of beneficial nutrients, including provitamin A, vitamin B_1, vitamin B_2, calcium, iron, phosphorus, manganese, sodium, potassium and chlorine. Figs are rich in phenol and omega 6 fatty acids, which are natural heart boosters helping to reduce the risk of coronary heart diseases. Figs are known for their soothing properties, and they are also low in fat and sugar. *Garcinia indica*, commonly known as kokum, with a unique reddish colour is used as a natural food colouring agent. Kokum is rich in dietary fibre, low in calories and contains absolutely no cholesterol and saturated fats. It also contains high levels of vitamin C that acts as a powerful antioxidant. Kokum is loaded with magnesium, potassium and manganese that protects against heart diseases. Kokum is widely used to combat digestive problems like constipation, acidity and flatulence. It contains hydroxyl citric acid (HCA) that acts as an appetite suppressant. Its anti-helmintic properties help in removal of worms from the stomach. HCA acts as a hypo-cholesterolaemic agent, which suppresses the activity of the enzymes responsible for conversion of calories into fat. It induces weight loss by hindering with the process of lipogenesis (fatty acid synthesis). Kokum enhances the immune function by acting as a powerful antioxidant and anti-inflammatory agent.

Karonda *(Carissa carandas)* has high concentration of iron, which in combination with high vitamin C makes it a healthy fruit. Fruits of Karonda act as a remedy for improper digestion, stomach pain and constipation. Lovi- lovi *(Flacourtia inermis)* fruits are rich in β carotene (303 µg/ 100g), iron, protein (4.2%), fat (3.6%) and fibre (5.7 %). It is also rich in 2, 3- dihydroxybenzoic acid, an important antimicrobial agent (George *et al.*, 2011).

Fruit juice of malayan apple *(Syzygium malaccensis)* known as malay rose apple contains minerals like calcium, phosphorus, iron and vitamins like nicotinic acid, vitamin C and riboflavin. Mangosteen *(Garcinia mangostana)* contains 79.2% water, 17 mg phosphorus, 19.8 g carbohydrate, 0.3 g fibre, 0.09 mg thiamine, 0.06 mg riboflavin, 66 mg vitamin C per 100 g edible portion (Chadha, 2001). Its juice is a health tonic rich in antioxidants and has antibacterial and antifungal properties (Kevilli, 2005). Mangosteen is considered as an energy tablet because of high content of reducing sugar which imparts instant energy on consumption (Radha and Mathew, 2007).

Papaya *(Carica papaya)*, the 'wonder fruit of tropics' is valued for the nutrititive value as well as for the proteolytic enzyme, papain, which has varied applications in the industrial sector. The awareness of multifold uses of papaya for table, processing and papain extraction is growing steadily and papaya has the potential to emerge from the status of a homestead crop to that of a commercial crop. Papaya cultivation has good economic potential especially due to its multifarious uses as fresh fruits, processed products, production of papain and carpaine alkaloid. Ripe papaya contains high quantities of carotene, the precursor of vitamin A, fair quantities of vitamin C, riboflavin and niacin. The amount of carotene in papaya is the second highest among tropical fruits and comes next to mango.

Yellow passion fruit, *Passiflora edulis*, is known for its exotic, unique flavour, aroma natural sweet taste and medicinal purposes. Passion fruit contains high amount of provitamin A and vitamin C both of which are strong antioxidants. Passion fruit has a significant content of iron, potassium, zinc and manganese. Passion fruit contains high amount of fiber which reduces cholesterol levels in blood and soluble fiber cleanses toxin stored in the colon by facilitating healthy and regular bowel movement. Purple passion fruit is a fair source of ascorbic acid and riboflavin, a good source of niacin and a fair source of minerals. Total sugar in 100 g fresh passion fruit is about 7.21-14.45 g, which includes glucose, fructose and sucrose.

Pomegranate (*Punica granatum*) is considered as highly nutritious. Edible portion of pomegranate is aril which is nearly 68% of the total fruit, containing 78% moisture, 0.1% fat, 1.6% protein, 14.5% carbohydrate, 5.1% fibre and is rich in thiamine, riboflavin, niacin, vitamin C, calcium, magnesium, potassium, phosphorus, iron and ellagic acid. Pomegranate can help to prevent or treat various disease risk factors including high blood pressure, high cholesterol, oxidative stress, hyperglycemia, and inflammatory activities. Antioxidant potential of pomegranate juice is more than that of red wine and green tea, which is induced through ellagitannins and hydrosable tannins. Moreover, pomegranate fruit extract prevents cell growth and induces apoptosis, which can lead to its anticarcinogenic effects.

Pummelo (*Citrus grandis*), the juicy fruits are eaten raw or used for juice extraction. A single tree can produce 70 to 100 fruits per year. Flesh of the fruit is segmented and constitutes 23 to 24 per cent of the fruit. The segments are filled with juicy sacs and the juice content varies between 30 and 75 ml per 100 g fruit. Besides high juice yield, pummelo has many nutritional and medicinal properties. The fruit contains vitamin C (16 to 102 mg per 100g of fruit) and B vitamins. The flowers, fruits and seeds are used for medicinal purposes. Outer soft and thick fruit skin is an excellent source of pectin and provides longer shelf life to the fruits. Rose apple (*Syzygium aqueum*) fruits contain minerals such as phosphorus (30 mg), calcium (10 mg), magnesium (4 mg), iron (0.5 mg), 1.2 g fibre, 0.7 g protein, 0.2 g fat per 100 g of pulp and several vitamins (Radha and Mathew, 2007). Rose apple is rich in vitamin C which prevents the damage of free radicals, pollutants and toxic chemicals.

India is the largest producer of sapota (*Manilkara achras*), a highly delicious dessert fruit which enjoy much popularity throughout India as "chikku". Sapota fruit is a good source of digestible sugar which ranges from 12-18%. Composition of ripe fruit per 100 g of edible portion is 73.7 g moisture, 21.4 g carbohydrate, 0.7 g protein, 1.1 g fat, 28 mg calcium, 27 mg phosphorus, 2 mg iron, 6.0 mg ascorbic acid and 10.9 g total digestible fibre. Sapota fruits are rich in calories and contain sugars, acids, phenolics, carotenoids and possess high antioxidant properties. Sapota is considered to be one of the healthiest fruits to alleviate micronutrient malnutrition.

Health benefits of underutilised vegetables

Underutilised vegetables are considered essential for well balanced diet since they supply vitamins, minerals, dietary fiber, and phytochemicals. Each vegetable contains a unique combination and amount of these phytonutriceuticals, which distinguishes them from others. In the daily diet, vegetables are strongly associated with improvement of gastrointestinal health, good vision and reduced risk of heart disease, stroke, chronic diseases such as diabetes and some forms of cancer. Health benefits of certain underutilised vegetables are discussed below.

The leaf extract of agathi (*S. grandiflora*) contains linoleic acid and aspartic acid, which were found to be the major compounds responsible for the anti-glycation potential of the leaf extract. Compounds with anti-glycation activity offer therapeutic potential in delaying or preventing the onset of diabetic complications. Bathua (*Chenopodium album*) leaves are store houses of protein, dietary fibre and vitamin C. It is a good appetizer and also helps to improve the hemoglobin level.

Moringa oleifera or drumstick leaves are rich sources of many micronutrients – β carotene, folic acid, calcium, iron and vitamin C and it also provides good quality protein. Pods are rich sources of proteins, vitamins, minerals and bioactive compounds. *M. oleifera* leaf powder holds some therapeutic potential for chronic hyperglycemia and hyperlipidemia (Mbikay, 2012). Ash gourd is ideal for diabetic patients due to its low calorie content and it acts as a good detoxifying agent. Ash gourd juice is beneficial for those suffering from peptic ulcer. Due to its strong 'antacid' action, it helps in maintaining the body pH.

Bottle gourd (*Lagenaria siceraria*) is very valuable as a medicine in urinary disorders. Its use during summer prevents excessive loss of sodium, quenches thirst and helps in preventing fatigue. Ivy gourd (*Coccinia indica* L.) is a rich source of carotene and good source of protein and fibre and many important minerals such as calcium, phosphorus, iron, copper and potassium. It also contains a fair amount of ascorbic acid. It helps in control of diabetics. Pumpkin provides valuable source of carotenoids. Consumption of foods containing carotene helps to prevent skin diseases, eye disorders and cancer. Moreover, the antidiabetic properties of pumpkin have generated interest in consuming this fruit (Ravani and Joshi, 2014).

Scope of entrepreneurship development in processing sector

Availability of diverse major fruits and vegetables in India has resulted in limited scope for expansion of other minor fruits and vegetables, though they are nutritious and are the main source of livelihood for the poor. Most of the underutilised fruits and vegetables of the tropics are often available only in the local markets and are practically unknown in other parts of the world. A large number of these crops can be grown under adverse conditions and are also known for their therapeutic and nutritive value and can satisfy the demands of the health-conscious consumers. However, some of these fruits are not acceptable in the market in fresh form due to their acidic nature and astringent taste.

They are categorized as underutilised as they lack recognized orcharding and little is known about their utilization and value addition. Many have not undergone any conscious phase of domestication and human selection. These crops are cultivated, traded, and consumed locally. These crops have many advantages like easier to grow and hardy in nature, producing a crop even under adverse soil and climatic conditions. There are many underutilised food crops in India and majority are not well known or well documented. Only 40% of the cropped area is under irrigation in India, while the remaining 60% is rain fed and has a vast tract of arid and semiarid land. Most of these minor crops could be grown in such areas where there is not enough water for field crops. Therefore, there is a great potential for increasing the area under minor crops and it is worthwhile to look into the organized cultivation in dry land conditions where the delicate fruit and vegetable plants cannot be grown satisfactorily, so that their utilization can be maximized.

A number of acceptable value added products can be prepared from underexploited fruits and vegetables of our country retaining their nutritional and medicinal properties. Many such preparations have been standardized, nutritional properties studied and storage requirements have been formulated to enable commercial exploitation. There is always demand from consumers for new, delicious, nutritious and attractive food products. To satisfy this demand, there is a constant effort to develop products from diverse sources. Modification in traditional value addition methodologies can also be tried for quality improvement. The potentiality of processed products from some minor fruits in the country is still untapped and remains to be unknown in world market which needs to be popularized. This is the great challenge faced in the processing and value addition scenario of underutilised fruits and vegetables. Drying of underexploited fruits and vegetables is an efficient processing method, which can be adopted as a rural based simple technology by small entrepreneurs, home scale industry and also by self-help groups in close association with NGOs. Small entrepreneurs can adopt this process on large scale.

As underutilised fruit and vegetable wealth can contribute towards food safety, health and nutritional security and uplift socio-economic status of the vulnerable communities, tapping their potential through processing and value addition remains the need of the hour. Identifying the market and promoting new brands for the products, taking advantage of brand consciousness and diversifying the product palette, developing and maintaining strong relationships within the market chain and improving labour efficiency and using quality packaging materials can improve the underutilised fruit and vegetable processing enterprises in the state.

Value added products

Analysis of post harvest handling, processing and value addition of fruit and vegetable crops reveals that emphasis in the past has been given only to the major fruit and vegetable crops in the country and the wealth of indigenous fruits and vegetable crops has not been brought to the forefront. In the present scenario of changing food habits, job profile and health awareness, new and improved processed products are in demand world over. There is always a demand from consumers all over the world for new food products which are nutritious and delicately flavoured. To satisfy this demand, there is a constant search and effort to develop novel products from hitherto little sources. India offers exciting possibilities of adding new dimensions to the food processing industry. Consumers today are becoming increasingly conscious of the health and nutritional aspects of their food. The tendency is to avoid chemicals and synthetic foods and choose therapy and nutrition through natural resources. Although, some fruits such as kokum are explored for commercial processing, the nutritive and therapeutic advantages of minor and underutilised fruits and vegetables are to be exploited at their full potential by the processing industries. Many investigators have suggested value addition strategies for several underexploited fruits and vegetables, some of which are discussed below.

Value added fruit products

Aonla is considered as wonder fruit for health. It has an astringent taste and therefore, not popular as a table fruit. However, it shows great potential for processing into various quality products which can have great demand in domestic as well as international markets. Juice, pickle, jam, syrup, preserve, candy, pulp, murrabba etc., are prepared from Aonla. Aonla powder can be prepared by mincing the fresh fruits and powdering slices after dehydration.

Mature ripe bael fruits contain amber or honey coloured viscous sticky glutinous, translucent sweet fleshy aromatic pulp which can be processed into value added products. Bael pulp can be stored for six months in heat sealed containers and bael fruit powder can be stored for a year when packed in 400 gauge polypropylene pouches under dark cool place (ITDC, 2000). Bael fruit can be processed into preserve and candy from mature fruits and good quality squash, nectar, RTS beverage, jam, toffee, powder and slab (leather or bar), from ripe fruits. Bael cider is considered as one of the important products with herbal value.

Nearly 90% production of ber is consumed as fresh. Ber is dehydrated for later use and has good potential in processing industry. Mature green fruits are used for preparation of chutney, pickle and jelly and a number of products like murabba (preserve), candy, fruit powder, dehydrated fruits, RTS, squash, syrup, other beverages, jam and leather are prepared from ripe fruits. Dehydrated ber is prepared by traditional sun drying and the techniques of pre-drying treatments have been standardized.

Bilimbi is commonly consumed as fresh, and for culinary purpose as a substitute of tamarind. Several value added products viz., squash, osmodehydrated bilimbi, dried bilimbi and syrup can also be prepared.

Custard apple fruits can be processed into various value added products and its pulp is used to prepare pastries, chocolates, jam, mixed fruit jam, milk shakes, ice cream etc. Freeze drying is one of the important methods of processing custard apple pulp. Technologies have been standardised for a range of processed products viz. RTS beverage, jelly and fresh/frozen fruit pulp, jam, fruit mix and spray-dried powder (Revathy et al., 2003), dehydrated fruits (Rajarathnam et al., 2003$_a$), cereal flakes, nectar (Rajarathnam et al., 2003$_b$) and wine (Jagtap and Bapat, 2015).

Bilimbi juice

Ripe carambolas are eaten as fresh, sliced and served in salads, processed into juice, puree, preserve, candy, jelly, juice blends with pineapple juice and also cooked in puddings, tarts, stews and curries. Under ripe fruits are pickled and made into jam. The possibility of commercial utilization of fruits for the development of acceptable squash with health benefits has been proved.

When guava processing sector is analysed, major share of 62% constitutes juice, which is produced by enzymatic de- pectinisation and centrifugation to separate clear juice and preservation by freezing/pasteurization in hermetically sealed cans. Juice concentrate, jelly and nectar are the other value added products from guava.

Guava Nectar

Guava cheese   Guava fruit squash

Research to develop and popularize value added jackfruit products preserving its natural qualities are encouraging nowadays. The work done in Kerala Agricultural University in this line is worth mentioning. Minimally processed tender jackfruit, pickle, dehydrated flakes, papad, flour etc have been developed. Similarly, recipe for preparation of ripe fruit products viz. squash, nectar, candy, dehydrated sweetened pulp, payasam mix, fruit bar, wine, kesari, halwa etc have been standardized. Different value added products have been standardized by UAS, Bangalore. Blended Jackfruit R.T.S. with soya milk whey at 50:50 ratio and blended jam with avocado and kokum were also standardized. (Kushala *et al.*, 2012).

 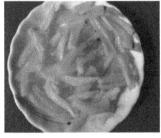 

Jackfruit chips   Osmo dehydrated jackfruit   Frozen jackfruit bulbs

Jamun is used to produce beverages, jelly, jam, squash, vinegar and pickle. Jamun squash is a refreshing drink in summer season. Trials at IARI, New Delhi for the extraction of jamun juice have shown that the maximum juice yield with high anthocyanin content and other soluble constituents can be obtained by grating the fruit, heating up to 70°C and pressing through a basket press.

Mature fig fruits with bright colour and characteristic flavour can be used for the preparation of different processed products. Processes have been standardized by CFTRI to prepare a shelf stable (6 - 12 months) beverage, syrup, jam and jelly from figs. Fully mature fig fruits can be osmodehydrated in 60°brix sucrose solution at 70°C, fruit to solution ratio being 1:5 (Kant, 2012). Fortified fig fruit toffee has been formulated using soy protein isolate, ragi powder, papaya pulp, liquid glucose, sucrose, edible fat and skim milk powder (Mhalaskar *et al.*, 2012). Fig fruit powder is utilized in the preparation of low cost value added nutritious product like burfi. Technology for drying locally available fig has been standardized by Kerala Agricultural University, which was superior to that of commercially dried ones in terms of quality and value.

Osmo dehydrated figs

Value addition of kokum fruits through processing assumes an important activity because raw/ripe fruits need to be processed before their consumption. The extract from kokum can be converted to many health beverages. In summer, the ripe rinds are ground in a blender with sugar and cardamom and consumed as a cooling drink. Juice extracted from this fruit is sweet and sour. A glass of cold kokum syrup is refreshing and it also improves the digestive system. Kokum extract is having approximately 4% sugar which can be fermented to produce high quality red wine. RTS beverage and fruit bar with 75 days of storage stability could be formulated from kokum.

Karonda fruits have great potential for processing. Usually, the fruit is pickled before it gets ripened. Ripe fruits can be processed into juice, squash, pickle, preserve and fermented beverages. Ripe fruits contain high amount of pectin and used in making jelly. Most important processed product is candy, prepared after suitable colouring, which is utilized by bakery industry in absence of cherry candy. Karonda pulp @ 20% can be used in ice cream as natural flavouring agent. Karonda juice @ 10% can be used in the manufacture of flavoured milk (Hanwate *et al.*, 2005).

Karonda sweet pickle

Karonda cherry   Mangosteen juice

Lovi-lovi fruits are dark coloured with tart flavour, rich in pectin and acids and are suitable for making jam, jelly, marmalade, syrup, squash and preserve.

Malayan apple fruit is refreshing and used in mixed fruit salads and half ripened can be pickled. Pickled or preserved slices and sauces are found in Southeast Asia and wine in Puerto Rico. Fruit is thick fleshed, juicy, fragrant and eaten in fresh forms and is mixed with other fruits of more pronounced falvour for jam and jelly making (Nath *et al.*, 2008).

Mangosteen fruits are used for fresh consumption and also for topping of ice creams, in fruit salads, canning as segments and in the preparation of squash, jelly, jam and syrup.

Papaya is consumed primarily as a dessert fruit and unripe fruits are consumed as vegetable. Both ripe as well as unripe fruits can be used for the preparation of a wide variety of processed products. Work done in Kerala Agricultural University under a NHB funded research project developed technology for extraction and preservation of papain. Recipe for preparation of an array of value added ripe papaya products viz., squash, syrup, crush, RTS, nectar, jam, sauce, fruit bar, toffee, halwa, jelly, cheese, candy, tutifruity, chutney, pickle, ready to cook papaya shreds, papaya blended products (Sheela, 2007) and fresh cuts (Amith, 2012) have been standardized

Papaya tutti frutti

Osmo dehydrated papaya

Passion fruits are most suitable for processing at maturity when the outer skin has a smooth or slightly crinkled surface. Fruit is known for its unique musky, guava-like flavour and aroma which makes its pulp an important flavouring agent in drinks, desserts, sauces and many other foods. Yellow passion fruit is known for its natural sweet taste and mainly consumed as juice, jelly and jam.

Papaya juice

Passion fruit beverage concentrate has been developed by IIHR, Bangalore from Hybrid Kaveri. Very strong and characteristic flavour of the fruit juice makes it ideal for blending with other fruits also. Passion fruit juice concentrate can be made from succulent purple and yellow types retaining freshness, wholesomeness and unique qualities of fruit. Passion fruit juice is extracted,

Passion fruit Juice   Passion fruit syrup   Passion fruit jam

centrifuged; air is removed, concentrated by removing water content at low temperature, pasteurized and canned to the desired Brix level. Passion fruit jam with good acceptability has been developed using passion fruit skin pulp and passion fruit juice in 1:1 ratio (Jena, 2013).

Pomegranate can be used for the production of sparkling juice and blended juice beverages. Its juice can be used for the preparation of wine and good quality vinegar. Fruit is used for the preparation of squash, syrup and jelly, and the juice is also used for making gelatin desserts, pudding and as sauces (Nath *et al.*, 2008). Protocol has been standardized for the fresh cut pomegranate aril preparation (Amith, 2012).

Pummelo is usually not preferred as a fresh fruit. Ideal process has been standardized by Kerala Agricultural University for easy peeling of segments. Dipping the segments in 15% brine for 15 min. and then steaming for 15 minutes helps in easy separation. Bitterness is the main problem and naringin is one of the compounds responsible for bitterness. Osmoextraction (by mixing juice vesicles with 30% sugar and incubating for 3 hrs.) is considered as the better method for juice extraction from the peeled segments. Addition of 30% sucrose masks the bitterness in juice. Treatment of juice with immobilized naringinase enzyme is also found suitable for production of de-bittered acceptable juice (Ghosh and Gangopadhyay, 2003). Jam and RTS beverage can be prepared from pure and blended pummelo pulp. Blending pummelo juice with orange juice at 70:30 has high acceptability.

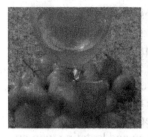  

Rose apple wine   Osmo dehydrated sapota   Sapota juice

Rose apple fruit is eaten fresh or used for making different value added products. Product and process standardization have been done at Kerala Agricultural University to prepare jam, jelly, candied fruit and syrup. It can be processed into sauce, flavoured cold drinks (Nath *et al.*, 2008), juice and wine.

Sapota fruit is highly perishable and is also sensitive to cold storage. Therefore, bulk of the produce is used for table purpose and is handled at ambient climatic conditions causing considerable postharvest losses. Commercial processing is negligible due to the sensitivity of the fruit to heat resulting in change of flavour and colour of the pulp, high labour requirement in peeling, removal of seeds etc. Nowadays dry segments and flakes of the fruit are being processed but to a limited extent. Processed food items like jam, jelly, squash and fruit drinks are produced from sapota after blending it with other fruits. Sapota powder has been prepared and incorporated @ 20% by weight, into traditional Indian recipes retaining the natural colour, aroma and flavour of sapota (Ganjyal *et al.*, 2005).

Value added vegetable products

In India, value addition of vegetables is limited to pickles, chutneys, preserves and candies at cottage level, which is highly decentralized and large units are located at small scale with unskilled labours. Vegetable processing is restricted mainly on small scale with processing capacities of 250 tons/per annum as against the same level of processing in one day by Multi National Companies (Singh & Singh, 2015). There are various methods of vegetable preservation including pickling, dehydration, freezing etc. The technology for preservation also varies with type of products and targeted market.

Vegetable dehydration techniques have been developed, well established and proven, by which vegetables in dehydrated form are preserved for a longer period and are made available during off-season. CFTRI, Mysore, has successfully developed the technical know-how for vegetable dehydration. Single layer of sliced and shredded vegetables after blanching is spread on trays in the dryer. Initial dryer temperature in cabinet or tray dryer is normally adjusted to 60-65°C for 4-5 hrs and afterwards the dryer temperature is reduced to 50-55°C for 2% final moisture content.

Most parts of the drumstick tree are edible. The leaves and flowers are eaten as salad, as cooked vegetables, or added to soups and sauces or used to make tea. Pait (2006) has standardized development of dried pulp, pulp powder, dried pieces, pickles and bottled pieces in brine from tender fruits. Fried seeds taste like groundnuts. It can be processed as a dehydrated powder. Blanched leaves are dried and blended to pass through 30-25 mesh sieve to produce leaf powder. Dried leaf powder is a good option to supplement diets of children in school feeding and community nutrition programmes and pregnant and lactating women. Drumstick leaves are often used to mutually supplement and enhance the quality of cereal based diets. Dried moringa leaves are fried in refined oil along with spices followed by addition of water for rehydration during curry preparation.

Drumstick powder   Watermelon rind candy   Pumpkin seed oil

Pumpkin has a vast scope for diversification and can be utilized in the production of processed products like jam, pickle, beverage, candy, bakery or fermented products and confectionary. Pumpkin is added as thickening agent in vegetable soups. Pumpkin dehydration results in superior quality, concentrated source of carotenoids with shelf stability. The technology involves blanching, sulphiting, followed by drying under specified conditions and packaging the powder in light/oxygen barrier type packs. Pumpkin can be processed into flour which has a longer shelf-life and can be used because of its highly-desirable flavour, sweetness and deep yellow-orange colour. It has been used to supplement cereal flours in bakery products like cakes, cookies, bread, and soups, sauces, instant noodle and spice as well as a natural colouring agent in pasta and flour mixes. Kulkarni and Joshi (2013) prepared acceptable biscuits by replacing wheat flour with pumpkin powder at 2.5% (w/w). Wholesome and nutritious pumpkin blended cakes having high β-carotene were prepared by substituting refined and whole wheat flour with pumpkin powder in 70:30 ratios (Bhat & Bhat, 2013)

Flowers of agathi are eaten as a vegetable in Southeast Asia, including Thailand, Java, Indonesia, Vietnam and Philippines. Flowers are cooked in curries. The young pods are also eaten. In Sri Lanka, agati leaves, known as Katuru murunga in Sinhala language, are widely eaten as curries. In India both leaves and flowers have culinary uses.

Snap melon fruits are highly perishable in nature and they have poor storage life. To prolong its utilization, ripe snap melon fruits are processed into value added products like syrup and jam. Milk shake of snap melon is served as refreshing drink. Unripe fruits are dehydrated for off season use. Seed kernels are extracted commercially and after removing seed coats they are extensively used in sweets and bakery products.

Dehydrated bathua leaves are added in various conventional food products in order to increase their nutritive value and to add variety in the diet. Dried bathua leaves are fried in refined oil along with spices followed by addition of water for rehydration during curry preparation. Technology for production of quality intermediate moisture vegetable has been standardized by IIVR, Varanasi where a moisture range of 15-25% is maintained.

New technologies/innovations in value addition

Food processing involves the transformation of raw plant materials into consumer-ready products, with the objective of stabilizing food products by preventing or reducing negative changes in quality. The need of food processing is to increase the shelf life of perishables, ensure safety by the inactivation of microorganisms, value addition, product diversification, increase digestibility and to meet specific consumer needs. To consumers, the most important attributes of a food product are its sensory characteristics. Goal of food manufacturers is to develop and employ processing technologies that retain or create desirable sensory qualities or reduce undesirable changes in food due to processing. The basis of traditional processing methods involves reducing microbial growth and metabolism to prevent undesirable chemical changes in food; however, they tend to reduce the product quality and freshness. One of the main concerns of the food industry is the need for high-quality fresh fruits and vegetables and value added products with good sensory quality, long shelf life, and high nutritional value. Fast growing consumer demand for natural, healthy and convenient food products lead to the innovation of novel food preservation technologies, having the potential to provide safe and nutritive foods with prolonged shelf life. Some of these techniques have already been commercialized; some are still in the research or pilot scale. These novel technologies include non-thermal methods *viz*, High Pressure Processing, Pulsed Electric Field, Ultrasound Technology, High Intensity Pulsed light, Irradiation, Ozone and Cold Plasma Technology.

The novel dehydration techniques commonly used in food industry, besides conventional hot air drying are freeze, spray, osmotic and vacuum drying. Advances in dehydration techniques and development of newer innovative drying methods have enabled the preparation of a wide range of dehydrated products and convenience foods from fruits and vegetables meeting the quality, stability and functional requirements coupled with economy. Freeze drying, spray drying and osmodehydration processes, which retain the original fruit flavour, are likely to catch on. The use of additives and pre treatments improves in retaining natural colour, nutrients and bioactive compounds in dried products. Novel dehydration process also improves the rehydration quality and thus improves the aesthetic quality in dried fruits and vegetables. Department of postharvest technology, College of Agriculture, Vellayani under Kerala Agricultural University has standardized a sugar free osmodehydrated cashew apple , a 'ready to eat' value added food, preserving natural qualities of cashew apple (Mini and Archana, 2016). Technology of osmo extraction of juice and osmodehydration of fruits of watery rose apple and malay apple was developed at College of Horticulture, Vellanikkara, Thrissur, Kerala Agricultural University. IIHR, Bangalore has standardized osmotic dehydration technology in amla, papaya, jackfruit and sapota. These can be eaten as snacks or used as dried fruit in ice-cream industry, confectionary etc. Maya (2004) standardized technology for the production of instant sapota-milk shake powder using spray drying technology containing 1:1 ratio of sapota pulp and milk solids. Process protocol has been standardized for spray dried instant cashew apple juice powder using resistant dextrin as carrier in 40:60 juice solid to carrier ratio (Rafeekher, 2017).

      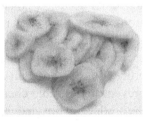

Cashew powders   Freeze dried Okra   Vacuum dried banana

Snack food industry has emerged as one of the important sectors for modern consumers with a special desire for fried snack foods. Oil uptake is one of the most important quality parameters of fried foods, posing significant health problems and is irreconcilable with consumers' awareness towards the consumption of healthier and low fat food products. Frying under vacuum condition is an alternative and novel technology that can be used to improve the quality of fried foods since it is working in low temperatures (lower than 90°C), at sub atmospheric pressures (<8 kPa) using minimum oxygen content (Govind *et al.,* 2015). It has been used for different foods, but mostly for fruits and vegetables including minor crops like jackfruit, sweet potato and purple yam. Evaluation of the effect of frying temperatures and durations on the quality of vacuum fried jackfruit chips revealed that maximum acceptable chips can be produced when fried at 90°C for 25 min. Frying under vacuum at lower temperature was found to retain bioactive compounds such as total phenolics, total flavonoids, and total carotenoids in jackfruit chips (Maity *et al.*, 2014).

Success stories in processing sector

Analysis of market potential and linkage development for underutilised fruits like aonla, tamarind, karonda, citron, jackfruit, etc. in Karnataka, Maharashtra and Gujarat revealed the availability of underutilised fruit products like citron pickle, tamarind paste and jackfruit chips in Karnataka and Maharashtra, the aonla products (pickle, squash, supari) and tamarind concentrate in Maharashtra and aonla and ber products in Gujarat. Consumers have shown willingness to accept these products, provided the details and related information are properly labelled on the products. The economics of manufacturing products from under-utilized fruits has been found profitable in all these three states (Kumar *et al.*, 2011).

Jackfruit Promotion Council has reported the success story of Anula Sirisena, a poor Sri Lankan housewife, who has seven jackfruit trees on her piece of land. After enrolling at the Horticulture Crop Research and Development Institute (HORDI) run by the Sri Lankan Government's Ministry of Agriculture, she was trained in value addition of jackfruit. Anula now runs a microenterprise in jackfruit products under the brand name Samanala (means butterfly). Her neatly packaged and labelled products are retailed at the Ministry of Agriculture's sales centres. Her family now earns 50,000 Sri Lankan rupees a year.

During the past 10 years, HORDI, funded by the International Centre for Underutilised Crops (ICUC), trained street vendors, housewives and entrepreneurs in minimal processing, dehydration, and bottling technologies. The institute's ex-students now manufacture a range of jackfruit products for the domestic and export markets. In addition, Industrial Training Institute (ITI) has in the last 20 years organized 200 workshops and trained 2,000 people in minimal processing of jackfruit. As a result, Sri Lanka has become the world leader in making jackfruit the key to food security and raising the income of the poor. Short duration training and support have empowered rural families. Every household has a few jackfruit trees, fruits of which are converted into products for sale in urban markets. Most jackfruit enterprises on the island are not high-end companies but medium scale operators and home industries. This strategy has made jackfruit products affordable for everyone. According to HORDI, the total area under jackfruit on the island is 50,000 hectares and annual production of jackfruit is 1400,000 tons. ICAR Krishi Vigyan Kendra, CARD, Pathanamthitta has reported several similar success stories of entrepreneurs in jackfruit processing.

Deccan Herald on 29th March, 2017 reported the success story of a Puttur (Karnataka) based nonagenarian agriculturist, Balliakana Narayana Rai, who has been carrying out kokum cultivation, processing and value addition and found himself successful as he has attained a level where he is exporting them to foreign countries. According to him, kokum does not need water, manure and insecticide and the plants start bearing fruits in five years. Market rate has gone up to Rs 150 per kg. As a commercial crop, kokum could fetch Rs 18 lakh per acre. Even if it is cultivated in a five-acre land, one could get not less than Rs 50,000 per month. He is producing kokum syrup, which is exported to foreign countries. Rai has now set out to start an industry to produce hair oil, syrup and juice from the fruit.

Smt. Bhagirathi Bhat, a house wife from remote Karva village in Karnataka encashed her own indigenous knowledge, skills, limited resources and with technical guidance from M/S Ashoka Kokum Industry, Gokarna in 2007 by preparing syrup and squash from Indigenous minor forest fruits through process of osmosis and juice extraction. She made her humble beginning with preparation of 20 kg kokum syrup and started selling in the local market. Today she is producing nearly 26 tons of value added products like squash, juice and syrup using kokum, brahmi, amla, ginger, pineapple, honey, *Cynadon dactylon*, pumpkin etc. She is engaging three women labourers on regular basis. Today, the venture which she started with initial investment of Rs.5000/- has reached to Rs.18,00,000/- per annum and her net income has increased from Rs. 10,000 to Rs. 39,00,000/- per annum. The products are sold throughout South Karnataka in the brand name "Swastik".

Scope of by-product utilization

Fruit and vegetable processing has increased considerably during the last 25 years. This has reflected the increase in demand for pre-processed and packaged food, particularly ready meals. During the period that many modern processes were developed and implemented, disposal of waste was not the major issue but today it is. These were often disposed by land filling, land spreading, or selling as animal feed or for its production. However, subsequent to the Kyoto agreement, the issue of waste in our modern society has become more prominent since it contributes too many problems of global environmental sustainability. For most fruits and vegetables, one can estimate approximately 30% or more of processed material or even in some processes it may be up to 75%. According to the recent research conducted by FAO, about 1.3 billion tons of food has been wasted worldwide per year, which represents one third of the total food industry production. Largest amount of loss is verified by fruits and vegetables, representing 0.5 billion tons. In developing countries, fruit and vegetable losses are severe at the agricultural stage but are mainly explained by the processing step, which accounts for 25 % of losses.

Processing and preparation of fruits and vegetables result in the production of varying degree of diverse waste materials due to the use of wide variety of fruits and vegetables, the broad range of processes and the multiplicity of the products. Vegetables and some fruits yield between 25% and 30% of non-edible products. Full utilization of horticultural produce is a requirement and a demand that needs to be met by countries wishing to implement low-waste technology in their agribusiness. Waste materials such as peels, seeds and stones produced by the fruit and vegetable processing can be successfully used as a source of phytochemicals and antioxidants.

New aspects concerning the use of these wastes as by-products for further exploitation on the production of food additives or supplements with high nutritional value have gained increasing interest because these are high value products and their recovery may be economically attractive. The by-products represent an important source of sugars, minerals, organic acids, dietary fiber and phenolics, which have antiviral, antibacterial, cardio-protective and anti-mutagenic activities (Jasna *et al.*, 2009). Because of increasing threat of infectious diseases, the need of the hour is to find natural agents with novel mechanism of action. Fruit and vegetable peels are thrown into the environment as agro waste which can be utilized as a source of anti-microbes. However, there is currently no major exploitation of these sources, due to the poor understanding of their nutritional and economic value.

Jackfruit rind is a rich source of pectin. Madhav and Pushpalatha (2002) standardized technology for extraction of pectin from jackfruit rind and utilized it for manufacture of quality jelly. Technology for extraction of quality seed flour and preparation of instant health mixes from seed flour is standardized (Ukkru and Pandey, 2011). Jacalin, the major protein from jackfruit seed is a useful tool for the evaluation of immune status of patients infected with HIV-1 and has immense anticancerous properties.

Latex of papaya fruit contains papain, a proteolytic enzyme used in industry. Papain is used to tenderize meat, clarify beer, treat digestive disorders, degum natural silk and extract fish oil. It is also used in shampoos and face lifting preparations, in leather and rayon industries, in the manufacture of rubber and chewing gum. USFDA has approved the use of chymopapain for treatment of lumbar hernia in humans.

In addition to the edible part of the fruit, the pummelo fruit peel is rich in nutrients and functional components and has high utilization value. It is used to prepare jam. It is fully utilized as adsorbents, processed foods and by extracting active components. Pomegranate peel polyphenols has a high value of application in food, medicine and other fields. Most of the peel is made into feed for animal consumption.

Utilization of sapota pulp residue was standardized for the preparation of value added products such as sapota powder and Instant sapota milkshake mix (Relekar *et al.*, 2014). White rind of watermelon, normally discarded after eating the fruit, can be converted in to value added products like candy and pickle (IIHR). Extraction of natural pigments was done at College of Horticulture, KAU Vellanikkara from water melon rind. Naturally coloured instant watermelon milk beverage powder and concentrates of watermelon and snap melon were also standardized. Extraction of anthocyanin from malay apple flowers were developed which could be used for imparting colour to different products.

Production of pectinase enzyme was standardized by KAU from pectin containing fruit wastes through solid state fermentation technology using selected micro organisms. Simple methods have been standardized for production of high value pectin from tropical fruit wastes for the production of quality jelly.

Future scope of research and limitations

In India, potential for cultivating fruits and vegetables for the domestic and export markets is high. Despite there being a demand of fruits and vegetables abroad, the country faces a serious problem in finding exportable quality fruits in sufficient quantities. Most of them still remain as underutilised, growing in unexploited areas without proper marketing strategies. Developing market potential for underutilised fruits depend on better marketing and reliable supply of the end product. All kind of fruits are rich in nutrients, vitamins and energy; and are highly used in ayurvedic and traditional medicine to treat various diseases. Research and scientific investigation are needed to explore those properties. New consumption trends observed in the current society forces the companies to develop new and healthier products to satisfy the consumer demands.

There have been many advances in technology associated with the industrial fruit and vegetable drying including pre-treatments, techniques, equipment and quality. Recent research revealed that novel dehydration approaches such as microwave- or ultrasound-assisted drying, high electric field drying and heat pump drying can now be undertaken to improve the efficiency of drying so that energy consumption can be reduced at the same time preserving the quality of the end product. Research has showed these technologies to be successful, having the greatest potential promoting sustainability in the food industry.

Many fruit and vegetable juices such as aonla, papaya, ber, watermelon, and vegetables including bottle gourd turn bitter after extraction due to conversion of chemical compounds and hence the utilization of these underexploited fruits and vegetables is very limited due to high acidity, astringency, bitterness, and some other factors. Still, all these fruits and vegetables are valued highly for their refreshing taste, nutritional value, pleasant flavour and medicinal properties. Therefore, blending of two or more fruit and vegetable juices with spice extract is thought to be a convenient and economic alternative option for utilization of these fruits and vegetables and is an area to be expanded for the preparation of novel fortified nutritive beverages, which can serve as an appetizer.

Fruit peel has great potential for extraction of bioactive components. Further research is required in the extraction of its functional and medicinal components in order to increase its added value. Research and development of fruit peel is still in the initial stage, many machining process are yet to be solved, such as fruit peel component extraction, processing technology processes etc need to be improved. Industrial research need to be studied deeply, chemical fertilizer and pesticide residues in fruit peel need to be tackled. Since, fruit peel is a rich resource, if people pay attention to, it has very high value in use and can bring higher economic efficiency, and reduce pollution to the environment.

Due to change in life styles, changing demography at work place and reduced time available for cooking, demand for high quality processed products are increasing. Demand for ready-to-eat, easy-to-cook fruits and vegetables is increasing. Processed products are very popular as it saves time, labour and having extended shelf life. In this context, low cost processing and value addition are very important to minimize huge post-harvest losses to a greater extent for nutritional security of large section of population.

Safety and quality aspects of products

Food quality includes external factors as appearance (size, shape, colour, gloss, and consistency, texture and flavour) and internal (chemical, physical and microbial) factors. Food quality is an important food manufacturing requirement, because, food consumers are susceptible to any form of contamination that may occur during the manufacturing process. Many consumers also rely on manufacturing and processing standards, particularly to know what ingredients are present, due to dietary and nutritional requirements. Besides ingredient quality, there are also sanitation requirements. It is important to ensure that the food processing environment is as clean as possible in order to produce the safest possible food for the consumer. Food quality also deals with product traceability of ingredient and packaging suppliers. It also deals with labelling issues to ensure that there is correct ingredient and nutritional information. Quality specifications of any processed product are controlled and monitored by FSSAI.

Organic fruit and vegetable processing standards generally prohibit the use of synthetic chemicals, many preservatives and other food additives that are widely used in processing of conventional foods. However, there are frequent discussions about the underlying rationales, principles and criteria to be used to allow some processing methods and additives. Consumers of low-input and organic food have specific expectations regarding quality characteristics of processed foods. Organic processed products should therefore be sustainable and fulfill consumers' expectations as much as possible. Sensory evaluation for quality assurance, inspection of raw materials and packaging materials can ensure food safety and quality aspects of underexploited fruit and vegetable products.

Demand for underexploited fruits and vegetables is growing rapidly in inland markets, but only slowly for export. Advertising through various media channels should further increase consumer awareness and may lead to higher demands. Demand for processed products can also be similarly created if the standards of the products are improved and the quality maintained.

Packaging of products

Packaging has enormous impact on food preservation. Packaging is the key to successful marketing of modern food products. Apart from extending shelf life and aesthetic appeal, packaging provides convenience of size, shape and ease of opening. Packaging materials with appropriate barrier properties are chosen to design suitable package for foods depending on their physical nature and to preserve the quality and nutritive value of foods by exclusion of oxygen and control of moisture transfer. Packaging also prevents spoilage of food by microbial or insect attack. Proper selection of the package depends on product characteristics, their nature of deteriorative changes and susceptibilities, conditions like physical hazards in handling, transportation, storage and climatic conditions of RH and temperature that brings out changes. They should not be over or under packed, as over packaging increases packaging cost and under packaging fails to offer desired protection. Hence, a functional and economic package is desirable for processed fruit and vegetable products. Situation is not different in packaging of products from underexploited fruits and vegetables too.

Traditional packaging is only suitable provided the climate does not cause an increase in moisture content of food which will result in mould growth. Boxes are used to prevent crushing of dried foods, and in humid climates, moisture-proof flexible films can be used. Some semi-moist foods such as osmotically dehydrated fruits have special needs to prevent the re-absorption of water. A wide range of flexible packaging materials are also available, but use of many of these is limited due to high cost. Low-density polyethylene is a moderately good moisture barrier, cheaper than other films and easily sealable. Flexible materials may be used as the sole component of a laminated package, but for most foods, a sturdy outer container is also needed to prevent crushing or to exclude light.

Products that are packaged inappropriately or with uninformative labelling, regardless of the quality have less selling and are against the packaging and labeling regulations of FSSAI. Jam, chutney, and ready-to-serve beverages are commonly packaged in glass bottles, while higher-volume fruit juices are packed in plastic cans as well. Market indicated a preference for standard, transparent plastic bottles. Use of attractive, internationally competitive packaging for local products would represent a successful marketing strategy. Size of the product is also important. Most consumers prefer single-serving products, or sizes that can serve 4-6 persons at the household level.

Conclusion

It is an established fact that seasonal, locally available, and cheap fruits and vegetables can also keep the population healthy and nutritionally secure rather than costly off-season ones. Also, the underutilised crops have the potential to give economic security by giving employment and by fetching good returns from their sale in raw form as well as value added products. Hence, there is a need to concentrate on research efforts in diversification and popularization of such underutilised fruit crops. To achieve this, there is a need to create demand for such fruit crops in the domestic and international markets. This, to some extent, can be achieved through developing suitable processing and marketing strategies for these underutilised fruits.

References

Amith, P. K. 2012. Protocol Development for Fresh Cut Fruits and Fruit Mix Preparations. *Thesis,* Kerala Agricultural University.

Bhat, M. A. & A. Bhat. 2013. Study on Physico-Chemical Characteristics of Pumpkin Blended Cake. *J. Food Process Technol.* 4: 262.

Chadha, K. L. 2001. New Horizons in Production and Post Harvest Management of Tropical and Subtropical Fruits. *Indian. J. Hort.* 58 (1-2): 1-6.

Ganjyal, G. M., M. A. Hanna & D. S. K. Devadattam. 2005. Processing of Sapota (sapodilla): Powdering. *J. Food Technol.* 3: 326-330.

George, S., P. J. Benny., S. Kuriakose & C. George. 2011. Antibiotic Activity of 2, 3-Dihydroxybenzoic Acid isolated from Flacourtia inermis Fruit Against Multidrug Resistant Bacteria. *Asian J. Pharmaceutical and Clinical Research* 4 (1): 126-130.

Ghosh, U. & H. Gangopadhyay. 2003. Reduction of Bitter Component of Pomelo Juice by Chemical Treatment and Immobilized Enzyme. *Indian J. Chemical Technol.* 10:701-704.

Govind, T., S. Nehul & S. Nilesh. 2015. Vacuum Frying Technology- New Technique for Healthy Fried Food. *National Seminar* on *Indian Dairy Industry - Opportunities and Challenges* pp.194-197.

Hanwate, B. D., R. M. Kadam., S. V. Joshi & D. N. Yadav. 2005. Utilization of Karonda (*Carissa carandas* L.) Juice in the Manufacture of Flavoured Milk, *Souvenir - National Seminar on Value Added Dairy Products*, National Dairy Research Institute, Karnal, December 21 to 25.

ITDC, 2000. Processing of Wild Bael Fruit for Rural Employment and Income Generation. *ITDC Food Chain* 27: 15-17.

Jagtap, U. B. & V. A. Bapat. 2015. Phenolic Composition and Antioxidant Capacity of Wine Prepared from Custard Apple (*Annona squamosa* L.) Fruits. *J. Fd Processing and Preservation* 39: 175- 182.

Jasna, S. D., C. Brunet & G. Aetkovae. 2009. By-products of Fruits Processing as a Source of Phytochemicals. *Chemical Industry and Chemical Engineering Quarterly* 15: 191-202.

Jawanda, J. S., J. S. Bal., J. S. Josan & S. S. Mann. 1981. Ber Cultivation in Punjab. *Punjab Horticultural J.* 21: pp.17-22.

Jena, S. 2013. Development of a Preserved Product from Underutilised Passion Fruit and Evaluation of Consumer Acceptance. *J. Food Research and Technol.*1(1): 11-20.

Kahane, R., T. Hodgkin., H. Jaenicke., C. Hoogendoorn., M. Hermann., J. D. H. Keatinge., J.D.A. Hughes., S. Padulosi and N. Looney. 2013. Agrobiodiversity for Food Security, Health and Income. *Agron. Sustain. Dev.* 33: 671–693.

Kant, R. 2012. Osmomechanical Dehydration of Fig (*Ficus carica*) and its Value Addition. *Thesis,* Punjab Agricultural University, Ludhiana.

Kevilli, K. 2005. Herb Profile: Garcinia. American Herb Association, *Quarterly Newsletter, Winter.* 20 (3): 3.

Kulkarni, A. S. & D.C. Joshi. 2013. Effect of Replacement of Wheat Flour with Pumpkin Powder on Textural and Sensory Qualities of Biscuit. *Int. Food Res. J.* 20 (2): 587-591.

Kumar, A., H. Singh., S. Kumar & S. Mittal. 2011. In: XVII AERA Annual Conference, *Agricultural Economics Research Review* 24: 169-181.

Kushala, G., K. N. Sreenivas & R. Siddappa. 2012. Product Development Acceptability and Cost Effectiveness of Jackfruit Jam Blended with Avocado and Kokum. *Food Sci. Res. J.* 3 (1): 78-80.

Madhav, A. & P. B. Pushpalatha. 2002. Quality Degradation of Jellies Prepared Using Pectin Extracted from Fruit Wastes. *J. Tropical Agriculture.* 40: 31-34.

Maity, T., A. S. Bawa & P. S. Raju. 2014. Effect of Vacuum Frying on Changes in Quality Attributes of Jackfruit (*Artocarpus heterophyllus*) Bulb Slices. *Int. J. Food Science* P. 8.

Maya, T. 2004. Value Addition in Sapota [*Manilkara achras* (Mill.) Fosberg]. Ph.D. thesis, Kerala Agricultural University, Thrissur, India. 112p.

Mbikay, M. 2012. Therapeutic Potential of *Moringa oleifera* Leaves in Chronic Hyperglycemia and Dyslipidemia: A Review. *Front Pharmacol.*3: P.24.

Mhalaskar, S. R., S. B. Lande., P. N. Satwadhar., H.W. Deshpande & K. P. Babar. 2012. Development of Technology for Fortification of Fig (*Ficus carica* L.) Fruit into its Value Added Product-Fig Toffee. *Internat. J. Proc. & Post Harvest Technol.* 3(2): 176-179.

Mini, C. & S. S. Archana. 2016. Formulation of Osmo-Dehydrated Cashew Apple (*Anacardium occidentale* L.) *Asian J. Dairy & Food Res.* 35(2): 172-174.

Moghadamtousi, S .Z., M. Fadaeinasab., S. Nikzad., G. Mohan., H .M. Ali & H. A. Kadir. 2015. *Annona muricata* (annonaceae): A Review of its Traditional Uses, Isolated Acetogenins and Biological Activities. *International J. Molecular Sciences* 16: 15625-15658.

Nandal, U. & R. L. Bhardwaj. 2015. Medicinal, Nutritional and Economic Security of Tribals in Aravali Region Sirohi, Rajasthan, India.

Nath, V., D. Kumar & V. Pandey. 2008. *Fruits for the Future: Well Versed Arid and Semi Arid Fruits* Vol.1: Satish Serial Publishing House , Delhi, India.

Pait, M. 2006. Product Development in Drumstick (*Moringa oleifera* Lam.). *Thesis,* Kerala Agricultural Univesity. 88p.

Radha, T. & L. Mathew. 2007. *Fruit Crops.* Horticulture Science Series. Vol.3. New India Publishing Agency, New Delhi.

Rafeekher, M. 2017. Instant Juice Powders of Cashew Apple (*Anacardium occidentale* L.) and Pineapple (*Ananas comosus* (L.) Merr.) *Ph.D. thesis,* Kerala Agricultural University. P.396.

Rajarathnam, S., M .N .Shashirekha., M. R. Vijayalakshmi & B. Revathy. 2003ₐ. A Process for the Preparation of Dehydrated Product from Custard Apple. Indian Patent Application No. 382/DEL/03.

Rajarathnam, S., M. N. Shashirekha., M. R. Vijayalakshmi & B. Revathy. 2003_b. *A Process for the Preparation of Nectar from Custard Apple. Indian Patent Application: No. 402/DEL/03.*

Ravani, A. & D. C. Joshi. 2014. Processing for Value Addition of Underutilised Fruit Crops. *Trends in Post Harvest Technology.* 2(2): 15-21.

Relekar, P.P., A. G. Naik & B. V. Padhia. 2014. Waste Utilization of Sapota for Value Addition. *Ecr Ecology, Environment and Conservation.* 20(1):111-114.

Revathy, B., S. Rajarathnam., M. N. Shashirekha & M.R. Vijayalakshmi. 2003. A Process for the Preparation of Fruit Mix from Custard Apple. *Indian Patent Application No. 384/DEL/03.*

Sheela, K. B. 2007. Standardization of Minimal Processing of Fruits and Vegetables. *Ann. Rep.* 2006- 2007, KAU, Thrissur. 59p.

Singh, R., J. Sharma & P. K. Goyal. 2014. Prophylactic Role of *Averrhoa Carambola* (Star Fruit) Extract against Chemically Induced Hepatocellular Carcinoma in Swiss Albino Mice.

10

Scope of Entrepreneurial Developments in Cocoa Processing

Suma B & Minimol J S

Introduction

Cocoa, *Theobroma cacao* L. popularly known as 'Food of Gods', the original source of chocolate, is indigenous to tropical humid forests on the lower eastern equatorial slopes of the Andes in South America. Cocoa spread to all over the tropical regions of the world from 18th century onwards and is now grown in 58 countries covering 6.9 million ha worldwide, producing 4 million tons of cocoa. Cocoa is an important agricultural commodity and the key raw material in chocolate manufacture. Unlike large, industrialized crops, 80 to 90% of cocoa comes from small, family-run farms, with approximately five to six million cocoa farmers worldwide. Total production has increased by 13% from 4.3 million metric tons in 2008 to 4.8 million metric tons in 2012. This represents an average year-over-year production increase of 3.1%. This rate of increase may slow in the coming years, as cocoa trees are sensitive to changing weather patterns. Periods of drought and of excessive rain or wind can negatively impact the yield, and will continue to fluctuate as climate change intensifies. Most of the world's cocoa is produced in West Africa (70%) followed by Asia and Oceania (15.6%) and Latin America (14.1%). World leaders in cocoa bean production are Ivory Coast, Ghana, Indonesia, Nigeria, Cameroon, Brazil, Ecuador, The Dominican Republic, and Malaysia, supplying 90% of the world production.

Though, cocoa was introduced in India in the early 20th century; its exploitation as a crop of significant economic value is just five decades old and confined as an inter crop in Kerala, Karnataka, Tamil Nadu and Andhra Pradesh in small/middle holder sector. As a cash crop, a cocoa plantation can last between 15 and 40 years. Cocoa constitutes a significant source of income for the small scale operators who are responsible for majority of worldwide production. Cocoa flavour is unique, complex, and fascinating. Cocoa and cocoa based products such as chocolates are one of the major natural sources of flavonoids because of its high content of polyphenols, epicatechin, catechin and their oligomers, and the procyanidins. Both non-volatile and volatile chemical components contribute to the cocoa aroma. When fermented and dried, it contains 50 - 57% lipids, 10% proteins, 12% fibre, 8% carbohydrates and 5% minerals.

History of chocolate

Cocoa was domesticated by the natives of Central America and the produce was used for consumption for the first time by Mayas and Aztecs. Spanish were the first Europeans to drink cocoa when they invaded and conquered the empire of Mexico in 16th century. The Spanish learnt from the Aztecs the art of making 'xocoatl', a drink made from roasted beans after roasting and grinding. The word 'chocolate was originated from 'xocoatl'. The recipe of xocoatl was later modified by a Spanish explorer, Herman Cortz in 1517. Spanish people kept this as a trade secret for more than 100 years. In 1580, world's first chocolate factory was established in Spain. The mesmerizing taste of cocoa spread throughout the European countries and first cocoa product in solid form was made by Joseph Fry and Sons in 1847. Within two years, Cadbury entered the field and it was William Peter in 1876, who made world's first milk chocolate.

Post harvest handling of cocoa

Quality of finished product depends upon a number of factors like the variety, agro techniques adopted, environmental conditions during the development of pod and processing technology adopted. Hence, it is necessary to maintain the quality standards throughout the processing period. Cocoa processing is mainly divided in to two; primary and secondary processing.

Primary processing

Primary processing is actually curing of cocoa. Curing is a process by which cocoa beans are prepared for the market. It involves fermentation and drying and final quality of the product is solely dependent on these two processes. Any defect occurring during this process will ultimately affect the quality of the end product irrespective of the superiority of the genotype selected for processing.

Harvesting

Harvesting is the process by which ripe pods are removed from the tree. Stage of maturity of pods is best judged by change of colour of pods. Pods, which are green when immature, turn yellow when, mature (forastero) and the reddish pods (criollo) turn yellow or orange (Prasannakumari *et al.*, 2009). Only healthy ripe pods should be harvested. Use of unripe and over ripe pods may be avoided which may affect fermentation process. Time taken for ripening varies with agro climatic conditions. There is considerable correlation of maturation with temperature, ie during high temperature, time required to attain maturity is less (Minimol *et al.*, 2015). Harvested pods when heaped under shade condition for one or two days enhance fermentation process (Anon, 1981). Pods are opened by hitting with each other or on a hard surface. Iron knife is not recommended because, the iron cause blanching effect on seeds. Seeds are extracted from broken pods and collected for fermentation without placenta.

Fermentation

Fermentation is the most important step in processing cocoa. The process involves keeping together a mass of wet beans. Numerous microorganisms act on sugary pulp covering the beans and result in temperature build up with in the mass. Most of the pulp will drain out within 24 hours. During fermentation, a series of biochemical reactions occur in the beans. Most important one is oxidation reaction which will continue even during drying (Wood and Lass, 1985).

Types and duration of fermentation

Types of fermentation vary from country to country or even from farmer to farmer. Duration of fermentation depends upon variety and season. Criollo cocoa will ferment within 2 – 3 days, while forestero will take 3 – 7 days and sometimes even more (Wood & Lass, 1985). Recovery per cent is highly influenced by the environmental conditions. Recovery percentage during dry season is about 38% compared to 34 % in wet season (Prasannakumari *et al.*, 2002).

Methods for large scale fermentation

Among various methods adopted for fermentation in different cocoa producing countries, heap, tray and box methods are considered as the standard methods.

i) *Heap method*: Cocoa beans of 50-500 kg are heaped over a layer of banana leaves in a slopy floor. The sweating will flow off within 24 hours. Then, it is covered with gunny bags keeping weight above the beans. Dismantle the heaps and mix the beans on the third and fifth days. Fermentation will be usually completed by sixth day and beans are taken out for drying on seventh day. This method is practised in West Africa and large fermentaries of India.

ii) *Tray method*: Wooden trays of size 90x60x10 cm are made by fixing reapers at the bottom, with gaps in between, such that beans do not fall through; at the same time allow free flow of sweating. Each tray can hold 45 kg of wet beans. Stack the trays one over the other and minimum six trays are required. An empty tray will be kept at the bottom to allow drain of the sweating. After stacking, the top most tray is covered with banana leaves and after 24 hours it is covered with gunny bags. Beans up to a depth of 10 cm will ferment in trays stacked one above the other. Normally, fermentation will be completed by four days and on fifth day the beans can be taken out for drying.

iii) *Box method*: This method is commonly used in cocoa estates of Malaysia. Boxes made of wood of standard dimension of 1.2x0.95x0.75 m which can hold one ton of beans are used. Holes are provided at the bottom and sides to allow free flow of sweating and to facilitate aeration. This will necessitate having a minimum of three boxes. Beans are mixed by transferring from one box to another. Though, box method of fermentation is convenient to handle large number of beans, the quality of beans is often rated as inferior.

Methods for small scale fermentation

All the standard methods of fermentation need relatively minimum quantities of wet beans. In heap method, the smallest batch size is 50 kg. Hence, in areas where cocoa is grown in small holdings, alternate method of fermentation involving small quantities of beans is necessary. It is not easy to develop small scale fermentation methods since small quantity of beans will make it difficult to develop adequate temperature of fermenting mass. Attempts were made to find out suitable small scale methods of fermentation using bean lots substantially smaller than those required for standard methods. Some of these methods are identified as heap, basket and plastic bag methods, as judged from temperature development of the ferment, pH of the beans and cut test.

i) *Heap method*: It is same as that of large scale fermentation process but the quantity of beans kept for fermentation is not less than 50 kg.

ii) *Basket method*: In this method, bean lots ranging from 2-6 kg can be fermented successfully. Mini baskets made of bamboo matting with a size of 15 cm height and 20 cm diameter can hold 2 kg wet beans. Baskets are lined with banana leaves to facilitate drainage of sweating. Wet beans are then filled in these baskets and covered with banana leaves. Baskets are placed on a raised platform to allow the flow of dripping. After 24 hours, it will be covered with gunny sackings. Beans are to be taken out and stirred well on third and fifth days after initial sweating. Fermentation will be completed in six days and beans can be taken for drying on seventh day.

Heap   Basket

Box   Sack

iii) *Plastic gunny bag method*: Cocoa can be fermented satisfactorily in clean plastic gunny bags without lining. Beans are filled ¾ ᵗʰcapacity and tie loosely. Keep the bags on an elevated position to facilitate drainage of sweating. On the second day, plastic bags are heaped one over the other and insulated properly to conserve heat by covering with tarpaulin or polythene sheet. Beans are stacked and mixed thoroughly without opening the bags on third and fifth days and restacked. Beans are taken out for drying on seventh day.

Factors affecting fermentation

In addition to the method of fermentation, large number of factors affects the final quality of the beans. The factors are explained below.

Large scale fermentation- Box method

i) *Ripeness of pod*: Pods should be at a fairly uniform state of ripeness. Under ripe or over ripe pods will affect the whole process of fermentation. Cruz *et al.*, (2013) found that under ripe pods do not ferment properly and the temperature of fermenting mass will initially rise to 40°C and then continues to remain at 35°C till the fermentation is over.

ii) *Types of cocoa*: Criollo cocoa gets fermented in 2-3 days while forastero takes 3-7 days. Hence, it is recommended not to mix criollo and forastero types in a lot for fermentation.

iii) *Diseases*: Even if the beans are not destroyed completely due to pod diseases it is undesirable to use such type of beans in a fermentation lot. An increase in free fatty acid content will occur if disease affected pods are used which adversely affect the quality of chocolate.

iv) *Seasonal variation*: The per cent pulp recovery will be less during wet season when compared to dry season (Prasannakumari *et al.*, 2002).

v) *Storage of pods*: Pods stored in shades for 2 or 3 days before fermentation will accelerate fermentation process (Mossu, 1992).

vi) *Quantity of beans*: Minimum quantity of beans to be kept for fermentation is fixed to increase the temperature inside the lot. This is very important to get good quality fermented beans.

vii) *Duration*: Duration of fermentation varies depending on the genetic structure of cocoa mass, the climate, volume and method adopted which varies from 1.5 to 10 days.

viii) *Turning*: Frequency of turning the pods varies depending on the method adopted and the country in which it is processed. Usual practice is to turn on alternate days so as to ensure uniformity.

The end point of fermentation

The end point of fermentation can be judged by observing certain parameters.

i) *Color of beans*: Well fermented forastero beans will show a bleached centre with a brownish ring at the periphery.

ii) *External shell colour*: The pulp which is whitish initially turns to pinkish white after sweating. At the end of fermentation, the shell surface will attain reddish brown colour.

iii) *Smell of fermenting mass*: Faint sweet smell of fresh pulp will change to a characteristic acid smell as fermentation proceeds and persists till the end of fermentation. Over fermented beans will produce an ammonia smell.

iv) *Development of heat*: After setting for fermentation, temperature of the mass increases steadily and reaches a peak of 47 to 49°C by the third day. Then, temperature decreases slowly till the end of fermentation to a range of 45 to 46°C. However, mixing of beans during fermentation will raise the temperature.

v) *Plumpy nature of beans and colour of exudates*: Well fermented beans will be plump and full, and reddish brown exudates flow out on squeezing. Usually, beans in the fermenting lot will not be in same stage of fermentation. Hence, all the beans in a sample drawn from the lot will not show the above mentioned indices. When 50% of the bean is fully fermented it can be presumed that fermentation is complete.

Drying

Fermented beans will have a moisture content of 55%. Main objective of drying is to bring this moisture content to 6-7% for safe storage and transportation. Moreover, drying process is a continuation of the oxidative stage of fermentation and this plays an important role in introducing bitterness and astringency and to develop chocolate brown colour in the final product. Thus, a very quick drying or excessive heating of bean will not be suitable. A very slow drying also will not suit as the beans get mouldy if they continue to remain moist too long. Cocoa can be dried by sun drying or artificial drying.

Sun drying is the simplest and most popular method in most of the cocoa producing countries. Depending up on the climatic conditions, beans are exposed to sun for about 4-12 days. This method generally gives good quality beans in traditional areas of cocoa production where the weather is sufficiently sunny. In West Africa and in India, beans are dried on raised ground or on concrete floors. In West Indies and South America, wooden drying floors are constructed with movable roof structure, which is referred as 'Boucans' in Trinidad and 'Barcacas' in Brazil (Wood and Lass, 1985). Attempts to improve the efficiency of sun drying using solar cabinets have been made (Mossu, 1992). Whatever be the method adopted, beans has to be spread in very thin layer and with frequent mixing.

During rainy or wet conditions artificial drying is recommended. Heated air is commonly used to dry cocoa beans artificially. Temperature, rate of air flow, depth of the beans and extent of stirring are important factors affecting quality of artificial dried beans. Maximum permissible temperature for drying is 60°C. A convenient thickness could be about 12 to 15 cm, when mixing is done manually. To obtain optimum quality, cocoa beans are dried at 45°C for 9 hours.

Sun drying   Artificial drying

Secondary processing

Secondary processing involves the conversion of cured beans in to different products. Most important product obtained from cocoa is chocolate. The essence of cocoa and chocolate manufacture lies in the development of flavour obtained by roasting the beans followed by extraction of cocoa butter from the nib to produce cocoa powder. The steps involved are

Cleaning

Beans coming for secondary processing are to be cleaned thoroughly so as to remove foreign matters like broken beans, metallic foreign matter and dust.

Alkalization

Cocoa beans are treated with alkali to improve the colour and to develop the flavour. This process is generally known as ditching. Alkalization helps to reduce astringency by complex polymerization of polyphenols and decrease the bitterness. Saturated solution of sodium or potassium carbonates or bicarbonates are mostly used while ammonium carbonate, magnesium oxide or carbonate or bicarbonate or mixtures of certain of the above chemicals are favoured by some manufacturers. Alkalized cocoa is commercially known as 'soluble cocoa'.

Roasting

Roasting is the most important step in cocoa processing. Cocoa beans are roasted in hot air and at this step characteristic chocolate flavour and colour are developed. During roasting, undesirable volatiles (acetic acid) are eliminated and moisture content will be reduced to 1-2%. It also decreases astringency and results in the development of aldehydes and ketones essential for cocoa flavour. Biogenic amines which are non volatile formed during roasting do not contribute to flavour of cocoa and cocoa products. Roasting parameters depend upon the raw material, the variety of cocoa and the type of desired flavour. Roasting time ranges from 10-35 minutes at a temperature between
110°C- 160°C. Over roasting causes the development of burnt taste as well as off flavour.

Three types of roasting process namely bean roasting, nib roasting and mass roasting are used commercially. In bean roasting, whole cocoa beans are roasted at a higher temperature and shells are loosened for easy separation from the nibs. The problems like development of off flavour and fat migration to the shell during roasting are overcome by nib roast. In mass roast, the nibs are converted to mass or liquid and then roasting is carried out. This will help to overcome the problem of uniformity in heating of beans.

Kibbling and winnowing

Kibbling is a process by which the shells are separated from the beans. The purpose of winnowing is to separate the shells from cocoa nibs. Roasted cocoa bean will contain 10-15% shell depending on the source (Prasannakumari *et al.*, 2009).

Grinding

The cotyledons are ground to get mass or liquor.

Kibbling process

Cocoa mass contains about 55-58% fat, which is also known as cocoa butter. This butter has the characteristic of melting at body temperature. Cocoa nibs are ground at relatively high temperature which will result in the production of cocoa mass.

Extraction of cocoa butter from cocoa mass

Cocoa butter is extracted from the cocoa mass or liquor with the help of a hydraulic press. Another method of fat removal is solvent extraction by using butane as the solvent. Powder and butter obtained by solvent extraction will contain solvents which may cause undesirable flavour. Therefore, this has been filtered, neutralized and tempered. Butter under room temperature is hard in consistency, waxy, slightly shiny, pale yellow in colour and oily to touch.

The extraction methods are crucial in cocoa industry so as to minimize the cost of processing, to get maximum extraction yield and to preserve bioactive compounds.

Shelling machine   Colloid mill for Grinding

Hydraulic butter press   Conching and tempering machine

Making of cocoa powder

The cake left behind at the bottom of the press after extraction contains 12-15% butter. This cake is milled and sieved to obtain cocoa powder. There are two types of cocoa powder. High fat powder with 20-25% butter and low fat powder with 10-13% butter. High fat powder is used in drinks while low fat powder is used in cakes, biscuits, ice creams and other products.

Production of chocolate

Chocolate is derived from sugar, milk and cocoa solids which in turn contribute to the flavour, aroma, colour and form of the final product. The proportion of mass, sugar and cocoa butter varies with manufacturers and remains to be a trade secret. Different types of chocolates are

Dark chocolate

This contains chocolate liquor, sugar and cocoa butter. Milk solids are not added in dark chocolate. The cocoa content of commercial dark chocolate varies from 30% (sweet dark) to 70-80% (extremely dark).

Milk chocolate

This is solid chocolate made with milk powder, liquid milk or condensed milk in addition to cocoa powder, cocoa butter and sugar. The percentage of cocoa powder in milk chocolate ranges from 25-29%.

White chocolate

This is made from sugar, milk and cocoa butter.

In chocolate manufacturing, the mixture of mass, sugar and milk powder are combined and it is refined using 2-5 roll refiner. This gives an absolutely homogenous mixture with fine grain size. The mass then become dry and flaky. It is kneaded again in a blender and then cocoa butter is added along with flavouring agent if necessary. Next step is conching which is actually agitating chocolate at a temperature above 50°C for few hours. This stage is very important because it contributes the development of final flavour and smooth texture of chocolate. This process will improve the flavour profile and reduces the concentration of free fatty acids and other volatile by-products of cocoa beans (Giacometti *et al.,* 2015). Temperature during conching depends upon the type of chocolate. For dark chocolate it is between 70-82°C.

Next step is tempering which is done in automatic tempering machines with a reduced temperature of 28-30°C. Objective of tempering is to promote crystallization of fat into a polymeric form. This will help to generate a more stable and acceptable final product. Next step is pressing and includes molding the chocolate paste into different molds. The molds are shaken continuously in order to distribute the mass evenly without air bubbles. It is refrigerated at 7°C and chocolates are removed by turning out the molds. Finally, wrap chocolates in attractive packages.

Establishment of a small scale chocolate processing unit

The list of equipment required to establish a small scale cocoa processing unit to produce cocoa butter, cocoa powder and milk chocolate are given in Table 1. The cost of equipment varies depending upon manufacturer and capacity.

Table 1. Equipment required to establish a small scale cocoa processing unit

Sl. No.	Equipment	Amount(Lakhs)
1.	Pod breaker	0.50
2.	Roaster (10 kg)	2.00
3.	Grader cum bean breaker cum winnower	2.50
4.	Colloid mill (10 kg)	1.50
5.	Butter extractor (2 kg)	4.00
6.	Conching machine (10 kg)	6.00
7.	Tempering machine (30 kg)	10.00
8.	Packing machine (fully automatic)	8.50
9.	Deep freezer (100 L)	0.15
10.	Refrigerator (230 L)	0.35
11.	Miscellaneous (molds, table top grinder, oven etc)	0.50
	Total	**36.00**

Chemistry of chocolate

Cocoa contains more than 300 volatile compounds; the most important components are aliphatic esters, polyphenols, aromatic carbonyls and theobromine, which also prevent rancidity of fat. Pharmacologically active ingredients of cocoa seeds include amines, alkaloids, theobromine, caffeine, theophylline, fatty acids, polyphenols, tyramine, trigonelline, magnesium, phenylethylamine and N-acylethanolamines. A standard chocolate bar (40 to 50 g) contains theobromine (86 to 240 mg) and caffeine (9 to 31 mg) (Matissek, 1997). Characteristic bitter taste of cocoa is generated by the reaction of diketopiperazines with theobromine during roasting. Theobromine is produced commercially from cocoa husks. Cocoa butter contains fatty acids consisting mainly of oleic, stearic and palmitic acids. It also contains myristic, arachidic, lauric, palmitic, linoleic and α- linolenic acids. In cocoa, the polyphenols of particular interest are flavanols, a subclass of flavonoids. Cocoa contains more than 10% flavanol by weight. Chemical composition of beans after fermentation and drying is given in Table 2.

Table 2: Chemical composition (%) of beans after fermentation and drying

Composition (%)	Nib (Max)	Shell (Max)
Water	3.2	6.6
Fat(cocoa butter, shell fat)	57	5.9
Ash	4.2	20.7
Total nitrogen	2.5	3.2
Theobromine	1.3	0.9
Caffeine	0.7	0.3
Starch	9	5.2
Crude fibre	3.2	19.2

Composition of cocoa powder varies depending on roasting, alkalisation and pressing processes undertaken. If standard procedures are followed, normally, cocoa powder contains the components given in Table 3.

Table 3: Composition of cocoa powder

Constituents	Quantity
Moisture (%)	3.0
Cocoa butter (%)	11.0
pH (10% suspension)	5.7
Ash (%)	5.5
Water soluble ash (%)	2.2
Alkalinity of water soluble ash as K_2O in original cocoa (%)	0.8
Phosphate (as P_2O_5) (%)	1.9
Chloride (as NaCl) (%)	0.04
Ash (insoluble in 50% HCl)	0.08
Shell (%) (calculated to unalkalised nib)	1.4
Total nitrogen (%)	4.3
Nitrogen (corrected for alkaloids) (%)	3.4
Protein(%)	20.0
Nitrogen corrected for alkaloids x 6.25(%)	21.2
Theobromine (%)	2.8

Health benefits of chocolate

The chocolates based products have high energy value in relation with its volume. They contain a proportion of carbohydrates and proteins together with B complex vitamins and minerals. Plain chocolates contain 64.8 g carbohydrates, 29.2 g fat, 4.7 g protein, 11 mg sodium, 300 mg potassium, 38 mg calcium, 100 mg magnesium, 140 mg phosphorus per 100 g.

The health benefits of cocoa include relief from high blood pressure, cholesterol, obesity, constipation, diabetes, bronchial asthma, chronic fatigue syndrome and various neurodegenerative diseases (Andujar et al., 2012). Cocoa offers anti-inflammatory, anti-allergic, anti-carcinogenic and antioxidant qualities and has demonstrated positive effects by imparting numerous health benefits. It is beneficial for quick wound healing, skin care and it helps to improve cardiovascular health and brain health. It also helps in treating copper deficiency. It possesses mood-enhancing properties and exerts protective effects against neurotoxicity. Flavonoid-rich cocoa aids in lowering blood pressure and improving the elasticity of blood vessels. Furthermore, this helps in maintaining a healthy circulatory system (Serafini *et al.*, 2003). Cocoa polyphenols have strong antioxidant properties. The effect of cocoa is more predominant and pronounced in controlling oxidative stress ageing, nutritional deficiency and pathological stress (Allgrove and Davison, 2014). A diet containing cocoa has shown to reduce triglycerides, LDL cholesterol and glucose level. Consumption of cocoa has been shown to be effective in improving insulin resistance and glucose metabolism. Studies have confirmed the protective antioxidant activity of cocoa in the treatment of long term diabetic complications such as diabetic nephrotoxicity. Cocoa extracts, trusted for their therapeutic and wound-healing properties are used to manufacture natural medicinal products. The extracts help in preventing the development of various kinds of infections in the body. Potential benefits of cocoa in preventing high-fat diet-induced obesity have been reported. It is also essential in modulating lipid metabolism and reducing the synthesis and transport of fatty acids. Polyphenols present in cocoa has a capacity to reduce the occurrence of stroke artery diseases, heart failure and cardiovascular disease related mortality (Crozier and Hurst, 2014). Cardio protective properties of cocoa are related to improvement in antioxidant status, metabolic and anti-inflammatory process, regulation of blood pressure etc. Cocoa flavanols help to enhance mood, combat depression, and promote improved cognitive activities during persistent mental exertion. Beneficial effect of cocoa in inhibiting the growth of cancer cells without effecting the growth of normal healthy cells has been reported. Healing effects of cocoa have proven extremely valuable in the treatment of various types of cancers including colon and prostate cancers (Yamagishi *et al.*, 2002). Cocoa has also been found to be effective in maintaining good skin health. Research findings suggest that the consumption of flavonol-rich cocoa helps in decreasing the effects of UV- induced erythema and reducing skin roughness and scaling. Presence of major flavonols like epicatechin and catechin in cocoa has shown beneficial effects in treating neurodegenerative diseases like Alzheimer's.

Home level chocolates-from Kerala Agricultural University

Chocolates are usually manufactured in large factories with huge investment. Production of chocolates of acceptable quality was thought to be impossible till 2000. Studies on secondary processing at Kerala Agricultural University on small scale led to the technology of converting beans to bar and production of a number of cocoa based products without any chemicals or preservatives. This opened up way to utilize cocoa in homesteads, ensure women empowerment, improve income from unit area of land and make available farm fresh and natural chocolates to large sections of the society at comparatively affordable price. One product 'Chocolate 4 U', a milk chocolate was launched in February 2008.

Milk chocolate (Chocolate 4 U) and White chocolate

Chocolate 4 U

Ingredients: roasted nibs of cocoa- 400 g; cocoa butter- 500 g; powdered sugar- 1 kg; milk powder- 500 g; and vanilla powder (natural) -30 g.

White chocolate

Ingredients: cocoa butter- 500 g; powdered sugar- 400 g; milk powder- 360 g; vanilla powder (natural)-20 g.

Machinery required: wet grinder/small scale tempering cum conching machine and refrigerator

Chocolates are prepared by grinding the different ingredients and adding vanilla essence in a phased manner. Continue grinding for about 7 hours until the mixture reaches moldable consistency.

Pour chocolate on to the molds, shake slightly and refrigerate at 4°C for 2 hours. Pack the chocolates and store in a refrigerator.

Drinking chocolate

Drinking chocolate is prepared by mixing the cocoa powder with sugar in 1:4 ratio and powdering in a mixer. This can be mixed with hot milk (1 teaspoon/ cup) and consumed.

Machinery required: mixer grinder

Cocoa delite (black and white)

Black delite

Ingredients: cocoa powder- 40 g; milk powder-150 g; sugar-150 g; butter-40 g; water-75 ml; vanilla essence- 1-2 drops.

White delite

Ingredients: cocoa butter-40 g; milk powder-150 g; sugar-150 g; butter-40 g; water-75 ml; vanilla essence- 1-2 drops.

Machinery/ equipment: uruli, gas burner, tray.

Dissolve sugar in water by heating. To this, add vanilla essence and butter. When butter gets fully dissolved, add rest of the ingredients, mix the contents thoroughly and cool.

Milk chocolate candy

Ingredients: cocoa powder-800 g; cocoa butter- 350 g; milk- 15 liters; sugar- 4 kg; sugar candy-500 kg; liquid glucose-500 g; ghee- 1 teaspoon.

Machinery/ equipment: milk cooker, uruli, gas burner, tray

Mix sugar, sugar candy and liquid glucose with milk, boil to condense the mixture. After cooling, transfer the contents to a hard bottomed vessel. To this mixture, add cocoa powder, heat and then add cocoa butter. When it comes to bubbling stage, stir the mixture continuously till the mixture starts leaving the sides of the vessel. Transfer this chocolate mass to a tray smeared with ghee and level it. After cooling, mold to different shapes.

Cocoa bite

Machinery/equipment: milk cooker, uruli, gas burner, tray

Cocoa bite can be prepared from the chocolate mass used for making milk chocolate. Make small balls of about 10 g and decorate with cashew nut pieces roasted in ghee. For 7 kg mixture 1 kg roasted cashew nuts can be used.

White chocolate candy

Ingredients: cocoa butter-175 g; milk- 7.5 liters; sugar-750 g; sugar candy-750 g.

Machinery/equipment: milk cooker, uruli, gas burner, tray

Mix sugar and sugar candy with milk and boil. When the mixture is reduced to about 1/3rd add cocoa butter. Stir the mixture continuously. When the mixture starts leaving the sides of the vessel, transfer the content on to a greasy tray. This mass can be molded into chocolates of different shapes. These are packed and stored in a refrigerator.

Cocoa burfi

Ingredients: sugar- 250 g; milk-250 g; coconut-250 g; cocoa powder- 50 g; ghee- 25 g.

Machinery/equipment: milk cooker, uruli, gas burner, tray

Roast grated coconut in ghee until it becomes golden brown and powder it. Boil milk and add sugar, stir continuously till the milk- sugar mixture becomes 3/4 th of the original volume. Add cocoa powder and stir continuously. Add powdered roasted coconut to the mixture and stir well. Stop the cooking when mass become sticky to touch. Grease a tray with ghee and pour the mass, spread it with spatula and cut it in to uniform pieces.

Cocoa cookies

Ingredients: maida- 500 g; dalda- 400 g; sugar- 300 g; cashew nut- 100 g; cocoa powder- 50 g; salt- a pinch; vanilla essence-1 teaspoon; baking powder-1 teaspoon.

Machinery/ equipment: hot air oven, roaster, blender

Mix maida, salt and baking powder in a bowl and sieve twice. Soften dalda with the blender and add vanilla essence. Add powdered sugar to the dalda and mix. After this, add sieved maida and knead well. Add powered roasted cashew nuts, cocoa powder and knead thoroughly till no cracks are formed in the mass. Flatten the mass on a tray with 1 cm thickness and cut with the mold and put the cookies in a preheated hot air oven, bake at a temperature of 185°C for 20 minutes.

Cocoa nutribar

Ingredients: puffed rice- 300 g; corn flakes- 300 g; rice flakes- 200 g; oats-200 g; sugar- 400 g; cashewnut- 100 g; jaggery- 200 g; liquid glucose- 300 g; raisins- 100 g; cocoa powder- 100 g.

Machinery/ equipment: roaster, mixer grinder

Roast puffed rice, rice flakes, corn flakes, oats and cashew nuts separately. Crush all the items. Roast raisins in ghee. Make syrup of jaggery, sugar and liquid glucose in to a thread consistency and mix all the roasted ingredients, roasted raisins and cocoa powder in to the syrup. Pour the mass in to a greasy tray and spread uniformly with a spatula. Cut the nutribar into pieces of 1.5 cm thickness, 11 cm length and 3 cm width.

Cocoa pudding

Ingredients: sugar-600 g; china grass- 3 packet; milk- 6 packet; milkmaid- 1 tin; cocoa powder- 30 g; cherries- 100g.

Machinery/equipment: milk cooker, refrigerator

Boil milk and sugar. Boil and melt china grass in water. Add melted china grass in boiled milk and stir. Then, add cocoa powder and milkmaid and mix well. After cooling, pour the mixture in to pudding plates and decorate with cherries.

Cocoa sip up

Ingredients: milk- 4 packet; water- 4 glass; sugar- 1 kg; cocoa powder- 10 g; vanilla essence- 2 teaspoon.

Machinery/ equipment: milk cooker, refrigerator, packing machine

Boil water, milk and sugar until it becomes ¾ᵗʰ of the original volume. Add cocoa powder in to half a cup of hot milk and pour it in to the whole mixture. Add vanilla essence to the mixture. After cooling, fill in the sip up covers and seal.

Cocoa ice cream

Ingredients: milk- 2 liter; sugar- 500 g; corn flour- 6 teaspoon; chocolate 4 U-200 g; whipping cream- 1 cup; milkmaid- 200 g.

Machinery/ equipment: milk cooker, deep freezer

Boil milk with sugar. Add corn flour- milk paste to the mixture and make custard. After cooling, refrigerate for 4-5 hours. Blend the custard and add chocolate mass and vanilla essence. Whip the whipping cream in another bowl till it becomes a peak. Add whipped cream and milkmaid to the custard mix and blend together. Keep it in the freezer for 6 hours. After 6 hours again blend the mixture thoroughly, cover the tray with aluminium foil and keep for another 6 hours in the freezer.

Chocos

Ingredients: sugar- 500 g ; milk powder- 500 g; cocoa powder- 6 teaspoon; cocoa butter- 200 g; cashew nut- 1 cup; vanilla essence- 200 g.

Machinery/ equipment: uruli, gas burner, tray

Mix sugar and vanilla essence in water and boil. Add cocoa butter to the boiling mixture. Keep it for cooling. After cooling add cocoa powder and milk powder along with roasted cashew bits. Make the mixture into small balls.

Success stories in cocoa processing sector

Cost of production of milk chocolate is approximately Rs.300/kg when materials are purchased from retail market. Cost of production is expected to come down significantly when the activity is initiated on regular basis. Cost of branded chocolates varies from Rs. 1000-1500/kg. In recent times in tourist hot spots a number of homemade chocolate units have come up and is spreading like a cottage industry in India. Training for secondary processing is given at regular basis at Cocoa Research Centre under Kerala Agricultural University which resulted in the establishment of processing units by self help groups. List of chocolate units started by women who were trained in the centre are listed below.

1. Sargasree chocolates, Karalam, Thrissur

2. Kukkoos chocolates, Poonkunnam, Thrissur

3. Real dreams, Guruvayoor, Thrissur

4. Pentas cocoas, Thrissur

5. Bliss chocolates, Thodupuzha, Idukki

6. Mariyan chocolates, Idukki

7. Leela foods, Kottayam

8. Roshni organic farming club, Kozhikode

9. Devamrutham herbal products, Kollam

10. CARD (Christian Agency for Rural Development) Krishi Vigyan Kendra, Pathanamthitta

11. Kids chocolates, Krishnapuram, Thrissur

12. Tasty chocolates, Puranattukara, Thrissur

13. Madhurima, Madakkathara, Thrissur

In addition to this, certain entrepreneurs have established sophisticated cocoa processing units. One example is Mr. Boby of Chalakudy who have a chocolate unit with a capacity to manufacture 100 kg chocolate per day.

Scope of by-product utilization

Processing of cocoa both at primary and secondary levels gives a large quantity of waste materials. Disposal of waste is a major problem of cocoa growing countries. Research on utilization of these indicates that by-products can be made from cocoa waste.

Pod husk

About 70-78% of pod weight is constituted by pod husk. These are generally discarded after collection of beans. Pod husk contain crude protein (5.69-9.69%), fatty substances (0.03-0.15%), glucose (1.66-3.92%), nitrogen free extract (44.21-51.27%), crude fibre (33.19-39.45%), theobromine (0.20-0.21%) and ash (8.83-10.18%). Cocoa pods can be used as potential fiber sources for pulp and paper production to promote the concept 'from waste to well' and 'recyclable material to available product' for reducing the environmental problem (Daud *et al.*, 2013). Dry pod contains 5.3-7.08% pectin and is higher than the pectin content of orange pulp, lemon pulp and apple pomace, the established raw materials for pectin (Marsiglia *et al.*, 2016). Nitrogen and phosphorus content of cocoa pod husk is comparable to farmyard manure from animals (Simpson and Oldham, 1985).

Pod husk contains less theobromine than cocoa shell and makes it a less dangerous feed stuff. Incorporation of 20% pod husk in cattle feed had shown beneficiary effect (Ashade and Osineye, 2013).

Mucilage consistency

Concentration of alcohol in the sweating is about 2-3% and of acetic acid is 2.5%. Sweating contains water (79.2-84.2%), dry substances (15.2-20.8%), citric acid (0.77-1.52%), glucose (11.60-15.32%), sucrose (0.11-0.92%), pectin (0.90-1.19%), proteins(0.56-0.69%) and salts (K, Na, Ca, Mg) (0.41-0.54%) with a P^H of 3.2-3.5. Sweating can be used to make jelly or jam. Cocoa sweating is also used for making alcohol and vinegar.

Dias *et al*, (2007) successfully made fruit wine by fermenting fresh cocoa pulp using *Saccharomyces cerevisiae* strain. Since, high amount of pectin and sugar are present in cocoa pulp, it can be used for making juice and jelly (Malaysian cocoa board, 2004). In Ghana, unfermentable pulp is used for making jam (Cocoa Research Institute of Ghana, 2010$_a$). Gin and brandy from cocoa pulp are commercially available in Ghana (Cocoa Research Institute of Ghana, 2010$_b$).

Shell or testa

Availability of bean shell is of the order of 11-12% of dry beans. It contains 2.8% starch, 6.0% pectin, 18.6% fibre, 1.3% theobromine, 0.1% caffeine, 2.8% total nitrogen, 3.4% fat, 8.1% total ash, 3.3% tannins and 300 IU vitamin D. The yield of furfural is about 5-6%. The fat present in cocoa shell with a lipid profile similar to cocoa butter can be extracted and used (Okiyama, 2015) for different purposes. Even though, chocolate aroma was found to be less pronounced, it will not significantly affect the overall acceptability of the products. Cocoa shell ash is also used as an alkalizing agent for cocoa nibs. Shell is found to be very good mulch and it is used in anthurium, orchid and foliage plants.

Future scope of research

Quality of primary processed cocoa beans depends upon the free fatty acid content. High amount of free fatty acid will reduce the quality and economic value of cocoa beans (Guehi *et al.*, 2008). The reason for accumulation of free fatty acid is due to unscientific practises followed during fermentation and storing (Hiol, 1999). Therefore, a thorough research has to be done in order to standardise different steps of primary processing in such a way that the amount of free fatty acid falls within the prescribed limit.

Scope of entrepreneurship development in cocoa processing sector in India

India produces a large range of cocoa and non-cocoa based confectionery items, besides other cocoa-based products. Production of confectioneries, except chocolates, is reserved for the small-scale sector. However, there are several large companies with an established markets and brands in cocoa and non-cocoa confectionery items. Confectionery output grew at a compound rate of 6 to 7% in recent years. Chocolate production is growing at the rate of 10 to 15% a year.

Total area under cocoa in the country is presently 81,274 ha, which comes to only about 1.5% of the coconut and arecanut area in India. In terms of area, Kerala was the leading cocoa growing state in the country till recently but there has been significant enhancement in area in Andhra Pradesh as an intercrop both in coconut and oil palm plantations. In terms of production, Kerala leads with 8,500 MT, followed by Andhra Pradesh (5500 MT), Karnataka (2500 MT) and Tamil Nadu (1500 MT). The annual production of cocoa in India is about 16329 MT; whereas the annual requirement to run the chocolate factories throughout the year is estimated to be 27216 MT.

India plans to increase its cocoa production by 60% in next four years to meet rising demand from the chocolate industry and to cut dependency on costlier imports. Chocolate consumption is gaining popularity in the country due to increasing prosperity coupled with a shift in the food habits, pushing up the country's cocoa imports. Experts opine that there is huge scope for expanding area under cocoa considering the rising demand and firm global price. Cocoa requirement is growing, around 15% annually. Per capita consumption of chocolates in India has increased from 40 g per person per year in 2005 to 110-120 g. Though, this is a very significant jump in consumption, it is still very nascent, leaving enough room for growth. A delectable combination of rising income, changing lifestyles and a young population's growing penchant for indulgence has transformed India into one of the world's fastest growing chocolate markets. Moreover, adults also have developed a culture of gifting chocolates in special occasions instead of traditional sweets. A report from French investment bank Societe Generale predicts that in the next five years, global confectioners will see highest growth in four markets which are India, Mexico, China and Brazil.

Safety and quality aspects of cocoa products

Quality requirements

The word 'quality' includes all the important factors of flavour and purity. It also covers the physical characteristics, which have a direct influence on value and accessibility of a lot of cocoa beans. The quality of a sample is primarily judged by the flavour of chocolate made from it. It also depends on factors such as bean size, shell percentage, fat content and number of defective beans. Cocoa of good quality will have the inherent flavour of the type of cocoa together with the relevant physical characters and freedom from defects.

Different aspects of quality are discussed here under:

Flavour

Flavour is developed during fermentation and roasting (Niemenak *et al.*, 2014). Flavour is assessed by tasting the chocolate made from a sample by a panel of experienced tasters. Flavour varies with the type of cocoa. Criollo and Trinitario give the finest of them all. Forestero types like Amelonado, Amazon and hybrids give 'bulk' cocoa, which constitute about 90- 95% of the worlds' supplies. Flavour of bulk cocoa varies from country to country. Common off-flavours detected in both fine and bulk cocoa are:

Mouldy: This arises due to the presence of moulds inside the beans. Sample with as low as 4% of mouldy beans impart an off- flavour to the chocolate. This off- flavour cannot be removed by processing.

Smoky: Contamination by smoke during drying or during storage can cause off- flavour. This off- flavour can be removed during chocolate manufacture.

Acidic: This is due to the presence of excessive amounts of volatile (acetic acid) and non volatile (lactic acid) acids. During manufacture, acetic acid is reduced to an acceptable low level, but lactic acid is not removed.

Purity or wholesomeness

It is essential that the cocoa beans delivered to the market are pure. It should not contain any impurities. In recent years, more importance is given to hygiene and safety at all levels of production and manufacture of food products. Important sources of impurities in cocoa beans are pesticides, bacteria and foreign matter.

Consistency

Quality of cocoa cured in a particular site and send to different places should be consistent. This is essential because the chocolate manufactures aims to produce chocolate of consistent quality. To some extent, consistency of bulk cocoa can be achieved by blending cocoa of the same grade standard.

Yield of edible material

A number of factors affect the yield of edible material or nib and cocoa butter.

Bean size and uniformity: Average weight of cured dry bean should be at least 1.0 g. The traditional criterion is that not more than 12% of the beans should be outside the range ± one third of the average weight.

Shell percentage: Shell should be loose, but strong enough to remain unbroken during normal handling. It should be free from lumps of dried pulp. Main crop of West Africa usually has a shell content of 11-12%.

Fat content: Cocoa grown under optimum conditions produce beans with 56-58% butter in the dry nib.

Moisture content: For safe storage the moisture content of cocoa beans should be around 6-7%.

Foreign matter, flat beans and insect damaged beans: Presence of these will reduce the quality of usable beans.

Cocoa butter characteristics

- Good quality cured cocoa beans will contain about 0.5% free fatty acid.
- Cocoa butter consists of triglycerides i. e. fats, which are made up of glycerol and three fatty acids, of which one is unsaturated. Manufacturers prefer cocoa butter that is relatively hard and consistent. Cocoa butter from West African countries gives the desired properties.

The cut test to assess quality

It is the standard method for assessing quality as defined in quality standards (Shamsuddin and Dimmick, 1986). It defines the major off- flavours- mouldy and unfermented beans based on visual examination of a cross section of the nib. Other defects, which can affect the keeping quality of cured cocoa are also detected. It is a guide to the degree of fermentation.

Cut test involves cutting 300 beans lengthwise taken from a random sample of cocoa. Based on the colour, beans are categorized as fully fermented, partly brown; partly purple, fully purple and slaty. Fully fermented beans are those which are almost brown. The beans, which are showing blue or violet colour on the exposed surface come under the second category. The fully purple category includes beans showing complete blue, purple or violet colour, over the whole exposed surface. The slaty beans are unfermented beans.

It is not possible to prepare a sample with 100% fully brown beans and it is not desirable to attempt to do so. Partly brown and partly purple are not defective ones and should be present at least to the extent of 20%. A proportion of 30-40% is acceptable, but sample with more than 50% is objectionable as these may give rise to astringent flavour to the finished product.

Safety of cocoa beans

Adulteration is the main problem in cocoa commodity. Empty shells of cocoa, fine grains of soil, rocks are often found as adulterants (Anton, 2017).

Chemical safety

Pesticide and fungicide residues in cocoa products are another major problem even though cocoa pods are not directly exposed to these chemicals. Systemic fungicide or pesticide in plants will affect the fermentation process. There is a potential danger of heavy metal content if fermentation is carried out in polluted environment.

Microbial safety

Salmonella and other enteropathogenic bacteria are the main threat for microbiological safety in cocoa beans, products, powder or chocolates. *Salmonella* posse's acid resistant strains so they can survive in cocoa beans even after fermentation and drying. To overcome these problems good agricultural practices can be adopted during cocoa processing.

International cocoa standards

The International cocoa standards were agreed at a meeting of producing and consuming countries held in Paris in 1969 (Wood and Lass, 1985). These standards form the basis of the grading regulations of several cocoa producing countries. Cocoa of merchantable quality is defined as

- Fermented, thoroughly dry, free from smoky beans, free from abnormal or foreign odours and free from any evidence of adulteration.
- Reasonably uniform size, reasonably free from broken beans, fragments and pieces of shell, and be virtually free from foreign matter.

Maximum percentages of mouldy, slaty and insect damaged, germinated or flat beans in Grade 1 will be 3% each and in Grade 2, maximum limits are 4, 8 and 6% respectively.

Conclusion

Cocoa originating from beans of the cocoa tree is an important commodity in the world and the main ingredient in chocolate manufacture. Its value and quality are related to unique and complex flavours. There is no single key component that determines the final flavour character. Both nonvolatile and volatile chemical components contribute to the cocoa aroma. Specific cocoa aroma arises from complex biochemical and chemical reactions during the post harvest processing of raw beans, and from many influences of the cocoa genotype, chemical make-up of raw seeds, environmental conditions, farming practices, processing, and manufacturing stages.

Cocoa and cocoa-based products such as chocolate are one of the major natural sources of dietary flavonoids because of its high content of polyphenols, epicatechin, catechin and their oligomers, and the procyanidins. Factors such as rising cocoa prices and lack of supply-chain infrastructure in India have not exactly dampened the enthusiasm of chocolate manufacturers. Chocolate industry in India is growing at nearly 20% every year. Still, there is a huge opportunity to expand our chocolate portfolio in the country in the coming years.

References

Allgrove, J. & G. Davison. 2014. Dark Chocolate/Cocoa Polyphenols and Oxidative Stress. In: Ross Watson., V. R. Preedy & S. Zibadi. (eds.). *Polyphenols in Human Health and Disease*. Waltham, Academic Press.

Andujar, I., M. C. Recio., R. M. Giner & J. L. Rios. 2012. Cocoa Polyphenols and their Potential Benefits for Human Health. *Oxid. Med. Cell.* P. 23.

Anon. 1981. The Relationship between Oxygen, Temperature, Acetic acid and Lactic acid during Cocoa Fermentation. *A Record of Fermentation Trials Performed at CRIG, Tafo, Ghana 1980. Unpublished Report.* Cocoa. Chocolate and Confectionary Alliance, London.

Anton, R. 2017. Cocoa Commodity Safety Reviews. *Food Review* 1-4.

Ashade, O. O. & O. M. Osineye. 2013. Effect of Replacing Maize with Cocoa Pod Husk in the Nutrition of *Oreochromis niloticus. J. Fish. Aquat. Sci.* 8(1): 73-79.

Cocoa Research Institute of Ghana.2010$_a$. *Cocoa alcohol.* http://www.crig.org/ by_products_subcat_ list.php?id=18. (accessed on May 10 2017).

Cocoa Research Institute of Ghana.2010$_b$. *Cocoa pulp juice (sweatings) products.* http://www.crig.org / by_products_subcat_list.php? id=2. (accessed May 10, 2017).

Crozier, S. J. & W. J. Hurst. 2014. Cocoa Polyhenols and Cardiovascular Health. In: Ross Watson., V. R. Preedy & S. Zibadi. (eds.). *Polyphenols in Human Health and Disease.*Waltham, Academic Press.

Cruz, J. F. M., P.B. Leite., S. E. Soares & E. S. Bispo. 2013. Assessment of the Fermentative Process from Different Cocoa Cultivars Produced in Southern Bahia, Brazil. *Afr. J. Biotechnol.* 12 (33): 5218-5225.

Daud, Z., A. S. M. Kassim., M. A. Ashuvila., H. Aripin., H. Awang & M. Z. M. Hatta. 2013. Chemical Composition and Morphological of Cocoa Pod Husks and Cassava Peels for Pulp and Paper Production. *Aust. J. Basic & Appl. Sci.* 7(9):406-411.

Dias, D. R., R. F. Schwan., E. S. Freire & R. Santos Serodio. 2007. Elaboration of a Fruit Wine from Cocoa (*Theobroma cacao* L.) Pulp. *Int. J. Food Sci. & Tech.* 42:.319-329.

Giacometti, J., S. M. Jolic & D. Josic. 2015. *Cocoa Processing and Impact on Composition.* In: Preedy, V. R. (ed.). *Processing and Impact on Active Components in Food.* Academic Press, London/Waltham/San Diego.

Guehi, S. T., M. Dingkuhn., E. Cros., G. Fourny., R. Ratomahenia., G. Moulin & A. C. Vidal. 2008. Impact of Cocoa Processing Technologies in Free Fatty Acids Formation in Stored Raw Cocoa Beans. *Afr. J. Agric. Res.* 3(3): 174-179.

Hiol, A. 1999. Contribution a Letude de deaux Lipases Extracellulaires Issues de souches fongiques isolees a partir du fruit de palme. *PhD dissertation,* Universite d,Aix, Marseille, France.

Malaysian Cocoa Board. 2004. *Products: Cocoa Pulp Jelly and Cocoa Pulp Juice.* http:// www.koko.gov.my/lkm/ (accessed Oct 16, 2010).

Marsiglia, D. E., K. A. Ojeda., M. C. Ramírez & E. Sánchez. 2016. Pectin Extraction from Cocoa Pod Husk (*Theobroma cacao* L.) by Hydrolysis with Citric and Acetic acid. *Int. J. Chem.Tech. Res.* 9(7): 497-507.

Matissek, R. 1997. Evaluation of Xanthine Derivatives in Chocolate-Nutritional and Chemical Aspects. *Z. Lebensm. Unters.Forsch. A.* 205: 175–84.

Minimol, J. S., T. K. Shija., V. Nanthitha., K. M. Sunil., B. Suma & S. Krishnan 2015. Seasonality in Cocoa: Weather Influence on Pod Characters of Cocoa Clones. *Int. J. Plant Sci.* 10(2): 102-107.

Mossu, G. 1992. Cocoa. *Harvesting and preparation of commercial cocoa.* The Macmillan Press Ltd. London, United Kingdom.

Niemenak, N., J. A. Eyamo., P.E. Onomo & E. Youmbi. 2014. Physical and Chemical Assessment Quality of Cocoa Beans in South and Center Regions of Cameroon. *Syllabus Review, Sci. Ser.* 5: 27 – 33.

Okiyama, D. C. G. 2015. Reuse of Cocoa Shell for Fat Extraction with Alcoholic Solvents. Universidade de São Paulo (USP), *Phd Thesis.*

Prasannakumari, A. S., R. V. Nair., E. K. Lalithabai., V. K. Mallika., J. S. Minimol., K. Abraham & K. E. Savithri. 2009. *Cocoa in India.* Kerala Agricultural University, Kerala.

Prasannakumari, A. S., V. K. Mallika & R.V. Nair. 2002. Harvest and Post Harvest Technology. *National Seminar on Technologies for Enhancing Productivity in Cocoa,* 29-30 November, 2002, 111-119.

Serafini, M., R. Bugianesi., G. Maiani., S. Valtuena., S. De Santis & A. Crozier. 2003. Plasma Antioxidants from Chocolate. *Nature* 424: p. 1013.

Shamsuddin, S. B. & P. S. Dimmick. 1986. Qualitative and Quantitative Measurement of Cacao Beans Fermentation. *University Proceeding of the Symposium of Cacao Biotechnology,* Pennsylvania State, pp. 55-78.

Simpson, B. K. & J. H. Oldham. 1985. Extraction of Potash from Cocoa Pod Husks. *Agricultural Wastes* 13 (1): 69-73.

Wood, G. A. R. & R. A. Lass. 1985. *Cocoa.* Tropical agriculture series, Longman House, London and New York.

Yamagishi, M., M. Natsume., N. Osakabe., H. Nakamura., F. Furukawa., T. Imazawa., A. Nishikawa & M. Hirose. 2002. Effects of Cacao Liquor Proanthocyanidins on PhIP-Induced Mutagenesis *in vitro,* and *in vivo* Mammary and Pancreatic Tumorigenesis in Female Sprague-Dawley Rats. *Cancer Lett.* 185: 123–30.

11

Spice Processing and Value Addition Opportunities for Start-up Entrepreneurs

Jayashree E

Introduction

Spices are high value export oriented crops extensively used for flavouring food and beverages, medicines, cosmetics, perfumery etc. Spices constitute a significant and indispensable segment of culinary art and essentially add flavour, colour and taste to food preparations. Farm level processing operations are important for value addition and product diversification of spices. It is essential that various operations like washing, threshing, blanching, drying, cleaning, grading, storing and packaging ensure proper conservation of the basic qualities like aroma, flavour, pungency, colour etc. Each of these operations enhances the quality of the produce and the value of the spice.

India is the largest producer, consumer and exporter of spices in the world. India produces more than 65 spices in different varieties out of the 109 spices listed by International Standards Organisation (ISO). India produces around 5.8 million tons of spices annually (2012-13), of this about 10% of the total produce is exported to over 150 countries. The USA, Europe, Australia, Japan, the Middle East and Oceanic countries are the major importers of Indian spices. Estimated world trade in spices is 1.05 million tons valued at 2750 million US $, out of which India has a significant share of 48% in quantity and 43% in value.

Health benefits of spice products

Spices are known to possess several medicinal and pharmacological properties and hence find position in the preparation of a number of medicines. Today, there is scientifically validated knowledge on spice phytochemistry, therapeutic effects of their bioactive principles and mechanism of action. Health benefits include carminative action, hypolipidemic effect, antidiabetic property, antilithogenic property, antioxidant potential, anti-inflammatory property, antimutagenic and anticarcinogenic potential. Of these, the hypocholesterolemic and antioxidant properties have far-reaching nutraceutical and therapeutic values. Most of the medicinal properties are attributed to the secondary metabolites – the essential oils and oleoresins, present in spices. Spices which possess some of the important medicinal properties are listed in Table 1.

Table 1: Major spices and their medicinal properties

Medicinal property	Spices
Cancer Preventive	Ginger, Black pepper, Nutmeg, Cinnamon, Clove, Turmeric, Cardamom, Vanilla, Allspice, Mace
Antimicrobial	Ginger, Nutmeg, Black pepper, Cinnamon, Vanilla, Turmeric, Clove, Allspice, Cardamom, Mace
Anti-Inflammatory	Black pepper, Cinnamon, Clove, Turmeric, Allspice, Cardamom
Spasmolytic	Cinnamon, Black pepper, Clove, Ginger, Nutmeg, Turmeric
Antioxidant	Vanilla, Ginger, Black pepper, Clove, Turmeric
Antiulcer	Ginger, Black pepper, Turmeric, Cinnamon, Clove, Nutmeg, Vanilla, Allspice, Mace
Hypoglycaemic	Cardamom
Antihepatotoxic	Vanilla
Antiallergic	Allspice
Antimigraine	Turmeric, Allspice, Cardamom, Mace
Antiosteoporotic	Black pepper, Allspice, Clove, Cardamom, Mace
Estrogenic/Androgenic	Cardamom
Immunostimulant	Turmeric, Mace
Antilithic	Allspice
Anti-insomniac	Allspice, Clove, Mace
Antiedemic	Vanilla

Scope of entrepreneurship development in spice processing sector

India is known as the home of spices and produces a wide variety of spices like black pepper, cardamom (small and large), ginger, garlic, turmeric, chilli, etc. Almost all the States and Union Territories of the country grow one or the other spices. It exports more than 0.40 million tons of spices annually. Due to liberalization of Indian economy, the spice industry of India has grown very rapidly. It is a source of livelihood and employment for large number of people in the country, especially for rural population. All this shows that spice production in India holds a prominent position in the world spice production. Entrepreneurs from all over the world are exploring the opportunities in this area. The Government, both at the Centre and the State level, has undertaken several measures and initiatives for the sound development of the spice industry. The 'Department of Agriculture and Cooperation' and 'Spices Board of India' are the main organizations for promoting research in this sector and scale up activities of exports.

Value added products of spices

Value addition in spices is yet another area of activity in which India is moving forward. Consistent effort by various agencies during the last one decade has improved the share of value added products in the export basket to more than 53%. India can now boast as the monopoly supplier of spice oils and oleoresins the world over. During the year 2014-15, India exported 24,650 tons of curry powder/paste worth Rs.476.26 crores and 11,475 tones of spice oils and oleoresins worth 1910.10 crores. In the case of curry powders, spice powders, spice mixtures and spices in consumer packs, India is in a formidable position.

Spices thus open ample opportunity for entrepreneurship, and to achieve this, one of the key requirements is to diversify the products from spices. Secondary agriculture is the watch word for development for both farmers and primary processors of spices. Even though, India produces a good quantity of black pepper, ginger, turmeric and cardamom, more than 85% of it is consumed within the country itself. Value addition throw ample opportunity in export (Jayashree, 2005). The details of value added products obtained from important spices are discussed.

Black pepper

Variety of value added products has been made from pepper and are classified as

- Green pepper based products
- Black pepper and white pepper based products

Green pepper based products

i). *Canned green pepper:* The despiked and cleaned berries are immersed in 2% hot brine containing 0.2% citric acid, exhausted at 80°C, sealed properly and processed in boiling water for 20 minutes. Canned pepper is cooled immediately in a stream of running cold water. Pepper harvested one month before maturity is ideal for the manufacture of canned green pepper

ii). *Green pepper in brine:* Freshly harvested green berries or spikes as such are used for preparing pepper in brine. The berries are washed and the cleaned berries are stored in brine solution having a concentration of $17 \pm 2\%$ salt with added vinegar of around $0.6 \pm 2\%$. Pepper is washed three times in a period of 45 days at an interval of 20, 20 and 15 days, respectively and each time the brine solution is changed. Pepper is then packed in high density poly ethylene (HDPE) food grade cans with sufficient quantity of freshly prepared brine solution of the same concentration just sufficient to immerse the pepper. Major applications of green pepper in brine are in making sauces, meat processing industries and in the food service sector.

iii) *Dehydrated green pepper*: Slightly immature green pepper is preferred for producing dehydrated green pepper. Freshly harvested cleaned pepper berries are subjected to blanching in boiling water for 15 minutes till the enzymes responsible for blackening the pepper are inactivated and polyphenols washed out of the berries. Berries are cooled immediately and dried in a cabinet drier at 70°C.

iv) *Frozen green pepper*: Frozen green pepper is considered far superior to green pepper in brine or dehydrated green pepper because it has better flavour, colour, texture and natural appearance. It is packed in poly pouches and hence the cost is much less compared to cans and containers. Though freezing is expensive, it is gaining popularity because of its superiority in every respect.

v) *Freeze dried green pepper*: Moisture content of fresh tender green pepper is removed by freeze drying the berries at -30°C to -40°C under high vacuum. As a result of this, a product with its natural colour, texture and of far superior quality to those of sun dried, solar dried or mechanically dehydrated green pepper is obtained. It is much lighter than frozen green pepper, since its moisture is reduced to 2-4%. Demand for freeze dried green pepper is growing and is likely to go up in due course.

vi) *Green pepper pickle*: Green pepper pickle is popular in many states notably in Kerala, Karnataka, Tamil Nadu, Gujarat and Maharashtra etc. People relish it with rice as an appetizer. When mixed with shredded fresh ginger, it becomes more tasty and piquant.

vii) *Mixed green pepper pickle*: Green pepper berries are mixed with lime, mango, brinjal, bitter gourd or mixed cauliflower and carrot pickles with or without green chillies and sliced fresh ginger. They are quite popular but their preparation is mostly limited to domestic scale.

Black pepper and white pepper based products

i) *Whole black pepper*: Fully mature green pepper is dried under sun for 5 days to obtain whole black pepper. In the modern spice processing unit, black pepper is first passed through a cleaning cum grading unit which consists of a specific gravity separator/destoner for removal of stones, an aspirator for sucking and removal of light impurities like the pin heads, husk, light berries, dust etc. and a multiple sieve grader for grading black pepper. Pepper is then passed through a spiral separator to remove flat impurities like broken spikes etc. from pepper and passed through a magnetic separator for removal of metallic impurities. Cleaned pepper is graded in to different sizes as 4.75 mm, 4.25 mm, 4.0 mm, 3.25 mm etc., and packaged in bulk or consumer packages for domestic or outside market.

ii)   *Sterilized black pepper*: The cleaned black pepper is subjected to sterilization to ensure high quality, clean and dried product free from microbial contamination. In continuous steam sterilization method, the spice is subjected to a rapid flow of superheated steam for a predetermined period of time followed by drying, rehumidification and packaging. Microbial levels as well as the enzyme activity are considerably reduced to low levels. In countries where the sterilization by chemical method is not permitted, steam sterilization is the best alternative.

Chemical sterilization involves the use of permitted chemicals like ethylene oxide for destroying microbes. Effectiveness of sterilization depends on the moisture content of pepper, concentration of gas, temperature and time of contact.

iii) *Ground pepper*: Ground pepper is obtained by grinding cleaned black pepper without adding any foreign matter. Grinding is accomplished by employing equipment like hammer mill, pin mill or plate mill. Ground product is further sieved and materials possessing the required size are packed. Overflow is sent back to the grinding zone for further size reduction.

iv) *Pepper oil*: Characteristic aroma of black pepper is due to the presence of volatile oil which ranges from 2-5% and can be recovered by steam or hot water distillation. Industrial process for the recovery of essential oil involves flaking of the black pepper using roller mills or grinding into coarse powder and distilling it in a stainless steel extractor. The steam comes in contact with ground pepper particles and vaporizes the oil present in the oil cells. On cooling, oil is separated from water. It is observed that slightly immature pepper will have more oil.

v)   *Oleoresin*: Oleoresin is the concentrated product of all the flavour components (aroma, taste, pungency and related sensory factors) obtained by cold extraction of ground pepper using solvents like hexane, ethanol, acetone, ethyl acetate etc. Pepper is flaked to a thickness of 1 to 1.5 mm and packed in stainless steel extractors for extraction with the organic solvent. Normally, solid to solvent ratio of 1:3 is employed and the recovery is 10-13%.

vi) *White pepper*: White pepper is the white inner corn obtained after removing the outer skin or pericarp of pepper berries. Traditional method of preparation of white pepper is by retting. If running water is not available, the alternative is to use fermentation tanks wherein the water is changed every day for 7-10 days. Retting converts only ripe and fully mature green berries to white pepper. Conversion of harvested berries to white pepper gives a recovery of 22 to 27%.

White pepper is preferred over black pepper in light coloured preparations such as sauces, cream soups etc. where dark coloured particles are undesirable. It imparts modified natural flavour to food stuff.

vii)   *White pepper powder*: White pepper powder is processed in the same way as black pepper powder, except the starting material is white pepper. White pepper powder can also be produced from black pepper by selective grinding followed by sieving. Before pepper is subjected to grinding, it is conditioned by adjusting the moisture content.

Cardamom

The harvested capsules of cardamom *(Elettaria cardamomum)* are washed in water to remove dust and soil particles. The capsules are then treated with 2% sodium carbonate solution for 10 minutes and spread on wire net trays of the flue type kiln drier. Heat produced by burning the firewood in iron kiln is passed through pipes into the drying chamber. They are initially dried at 50°C for the first 4h and heat is then reduced to 45°C by opening ventilators and operating exhaust fans till the capsules are properly dried (Patil,1987) . Finally, the temperature is raised to 60°C for an hour. The process of drying takes about 18- 24 h for reducing the moisture content from 80% to 10% (Korikanthimath, 1993). Dried capsules are rubbed on wire mesh to remove the stalk, dried portion of flower from the capsules and then graded according to size by passing through sieves of sizes of 7, 6.5, 6 mm etc. Graded produce is stored in polythene lined gunny bags to retain the green colour during storage.

Value added products of cardamom

i)   *Oleoresin*: Solvent extraction of ground spice yields 10% oleoresin. Cardamom oleoresin is used for flavouring food after being dispersed in salt, flour etc.

ii)   *Essential oil*: Essential oil of cardamom is extracted by steam distillation from the seeds of the fruit gathered just before they are ripe and the yield is 1-5%. Main chemical components of cardamom oil are á-pinene, â-pinene, sabinene, myrcene, a-phellandrene, limonene, 1,8-cineole, y-terpinene, p-cymene, terpinolene, linalool, linalyl acetate, terpinen-4-oil, a-terpineol, a-terpineol acetate, citronellol, nerol, geraniol, methyl eugenol and trans-nerolidol. Cardamom oil is a precious ingredient in food preparations, perfumery, health foods, medicine and beverages. A good portion is consumed for chewing or as a masticatory item.

iii) *Decorticated seeds/seed powder*: Decorticated seeds command a lower price due to rapid loss of volatile oil during storage and transportation. Seed powder is marketed to a limited extent.

iv) *Bleached cardamom*: A proportion of the crop is bleached after sun drying by exposing the capsules to fumes obtained from burning sulphur to get uniform colour and appearance. Steeping capsules in a dilute solution of potassium metabisulphite solution induces a slight improvement in keeping quality.

Turmeric

Cured fingers of turmeric (*Curcuma longa*) are sun dried by spreading the materials on clean drying floor which takes 10 to 12 days for reduction in moisture content from about 82% to 10%. Dried turmeric rhizomes are polished in a mechanical polisher with capacities varying from 100 to 500 kg/batch for about 45-60 min to improve the colour and to make the rough hard outer surface smooth (Kachru and Srivastava, 1991). The loss due to polishing is about 5-10%. Cleaned and graded turmeric rhizomes are packaged in clean gunny bags.

Value added products from turmeric

i) *Turmeric powder*: Dried turmeric is powdered to a fine mesh-60 (250 microns) to be used in various end products. Turmeric rhizomes contain 4-6% of volatile oil and there is a great chance of losing the oil when powdered. Hence, it is to be properly packed immediately after powdering.

ii) *Turmeric oil*: Dried rhizomes and leaves are used industrially to extract the volatile oil. Dried rhizomes contain 5-6% and leaves contain about 1-1.5% oil. It is generally extracted by steam distillation.

iii) *Turmeric oleoresin*: Turmeric rhizomes contain about 7-14% oleoresin. This can be extracted using organic solvents such as acetone, hexane, ethyl acetate etc. Major compound in the oleoresin is the colouring principle curcumin. It is used in food preparation and pharmaceutical products.

iv) *Curcumin*: The major colouring principle of turmeric is curcumin. Curcumin content in turmeric varieties vary from 3-9%. It is a mixture of three pigments, curcumin, demethoxy curcumin and bis-de methoxy curcumin. It is preferred in the food and pharmaceutical industries as a natural colourant.

Ginger

Ginger *(Zingiber officianale)* is used both as a fresh vegetable and as a dried spice. The crop is harvested at full maturity by lifting the clumps carefully with spade or digging fork (Kachru and Srivastava, 1988). Rhizomes are then separated from the dried up leaves, roots and adhering soil. Clumps are broken to rhizomes of sufficient length followed by partial peeling to remove the outer skin (Agarwal *et al.*, 1987). Peeling hastens the process of drying and maintains the epidermal cells of the rhizomes, which contains essential oil responsible for aroma of ginger. Peeled ginger with moisture content of about 82-84% is spread thinly on concrete floor and dried under sun for 10-12 days until the moisture content is reduced to 10%. The dry ginger so obtained is known as rough or unbleached ginger. Yield of dry ginger is 20% of fresh ginger depending on the variety and the location where it is grown. Dried ginger rhizomes are manually graded based on the external appearance and bulk packaged separately in jute or woven poly propylene bags.

Value added products from ginger

i)   *Ginger powder*: Dried ginger is powdered to a fine mesh-60 (250 microns) to be used in various end products.

ii) *Salted ginger*: Fresh ginger (with relatively low fibre) harvested at 170 -180 days after planting can be used for preparing salted ginger. Tender rhizomes with portion of the pseudo stem is washed thoroughly and soaked in 30% salt solution containing 1% citric acid. After 14 days, it is ready for use and can be stored under refrigeration.

iii) *Crude fibre*: In fully matured ginger, crude fibre varies from 3-8%. It is estimated by acid and alkali digestion of ginger powder and whatever remains is considered as fibre.

iv) *Ginger oil*: Dry ginger on distillation yields 1.5 to 2.5% volatile oil. The main constituent in the oil is zingiberene and contributes to the aroma of the oil.

v)   *Ginger oleoresin*: Dry ginger powder on treating with organic solvents like acetone, alcohol, ethyl acetate etc. yield a viscous mass that attribute the total taste and smell of the spice. Major non volatile component of oleoresin is gingerol. The oleoresin content varies from 4 -10%.

vi) *Others*: Sweet and salty products like ginger candy, ginger paste, salted ginger, crystallized ginger etc. can be prepared from fresh ginger.

Nutmeg

Nutmeg and mace are two different parts of the same fruit of the nutmeg tree, *Myristica fragrans*. The fruits are harvested when they split open on ripening. Harvested fruits are handpicked and washed in water to remove dirt and mud adhering to the outer pericarp. Mace which is the outer aril of the nutmeg is separated from the nut and the two spices are dried separately (Krishnamoorthy and Rema, 2001). The nut which is very rich in fat called the nutmeg butter is dried at a low temperature of about 45°C and takes about 5-6 days for drying. Drying is stopped when a rattling sound is heard on shaking the nut (Joy *et al.*, 2000). Mace is dried at a temperature of 55°C and takes about 6-7 h for complete drying (Amaladhas *et al.*, 2002).

Value added products from nutmeg

i) *Nutmeg powder*: Dried nutmeg is ground to fine powder to be used in various end products.

ii) *Nutmeg oil*: Essential oil from nutmeg is steam distilled and the oil percentage varies from 5-15%. The essential oil is highly sensitive to light and temperature and yields a colourless, pale yellow or pale green oil with characteristic odour of nutmeg.

iii) *Nutmeg oleoresin*: Nutmeg oleoresin is obtained by solvent extraction of spices. Oleoresins contain saturated volatile oil, fatty oil and other extractives soluble in the particular solvent. Nutmeg oleoresin is extracted with organic solvent and yields about 10-12% of oleoresin.

iv) *Nutmeg butter*: Fixed oil of nutmeg is known as nutmeg butter with a consistency of butter at ambient temperature. Nutmeg butter contains 25 to 40% fixed oil and can be obtained by pressing the crushed nuts between plates in the presence of steam or hot water.

v) *Mace oleoresin*: When extracted with petroleum ether, mace yields 10 to 13% oleoresin.

vi) *Mace oil*: is obtained by steam distillation of dried aril and yields 4-17% oil. It is a colourless liquid with characteristic odour and flavour. Mace oil is more expensive than nutmeg oil.

Cinnamon

Cinnamon (*Cinnamomum zeylanicum*) is obtained by drying the central part of the bark after the third year of planting. It is harvested from the branches which have attained greenish brown colour indicative of maturity and when the bark peels of easily. The shoots are cut for bark extraction and the rough outer layer is first scraped off with special knife. Scraped portion is polished with brass rod to facilitate easy peeling. A longitudinal slit is made from one end to the other end and the bark is peeled off. Pieces of removed bark are known as quills. Curled quills are placed inside one another to make compound quill and dried under shade to prevent warping. After 4-5days of drying, the quills are rolled on a board to tighten the fillings and then placed in subdued sunlight for further drying.

The Sri Lankan grading system divides cinnamon quills into four main groups based on the diameter. The best known grade of cinnamon is the quills. Quills are long compound rolls of cinnamon bark measuring more than 107cm long. The small pieces of bark (5-20cm long) left after preparing the quills are graded as quilling. Featherings are the very thin inner pieces of bark or twisted bark and the chips are trimmings of quills or the bark of coarser canes that are scraped off, instead of peeling. The bark that is scrapped off without removing the outer bark is known as unscrapped chips and that scrapped after removing the outer bark is scrapped chips.

Value added products from cinnamon

i) *Cinnamon oleoresin*: The dry cinnamon powder on treating with solvents like acetone, hexane, ethyl acetate yields a viscous mass that attribute to the total taste and aroma of cinnamon. Oleoresin content varies from 7-10%. Oleoresin is dispersed on sugar and salt and used for flavouring processed foods.

ii) *Cinnamon bark oil*: It is essentially extracted by steam distillation of cinnamon and the oil percentage varies from 0.5 to 2.5%. Main constituent of this oil is cinnamaldehyde which is about 65% but other compounds like eugenol, eugenyl acetate, ketones, esters and terpenes also impart characteristic odour and flavour to this oil. It is used in flavouring bakery foods, sauces, pickles, confectionary, soft drinks, dental and pharmaceutical preparations and in perfumery.

iii) *Cinnamon leaf oil*: Cinnamon leaf oil is produced by steam distillation of leaves yielding 0.5 to 0.7% oil. Major constituent is the eugenol (70-90%) while the cinnamaldehyde content is less than 5%. The oil is used in perfumery and flavouring and also as a source of eugenol.

Clove

Clove is the small, reddish brown unopened flower bud of the tropical evergreen tree *Syzygium aromaticum*. Trees begin to yield from 7-8 years after planting. Buds are harvested when the base of calyx has turned from green to pink in colour. If allowed to develop beyond this stage, the buds open, petals drop and an inferior quality spice is obtained on drying. Prior to drying, buds are removed from the stem by holding the cluster in one hand and pressing it against the palm of the other with a slight twisting movement. Clove buds and stems are piled separately for drying. Buds may be sorted to remove over-ripe cloves and fallen flowers. Immediately after the buds are separated from the clusters, partial shade drying is followed. In sunny weather, drying is completed in 4-5 days giving a bright coloured dried spice of attractive appearance. During drying, clove loses about two-third of its original fresh green weight. When properly dried, it will turn bright brown and does not bend when pressed. Dried cloves are sorted to remove mother of cloves and khoker cloves, bagged and stored in a dry place.

Value added products from clove

i)   *Clove bud oil*: Clove bud oil contains 14 to 20% essential oil, the principal component of which is the aromatic oil eugenol (85-89%) which is extracted by distillation. Clove bud oil is used for flavouring food and in perfumery.

ii) *Clove stem oil*: Clove stem oil is obtained from dried peduncles and stem of clove buds. The eugenol content ranges from 90-95% and possesses a coarser and woodier odour than the bud oil.

iii) *Clove leaf oil*: Clove leaves on distillation yield 2-3% oil which is a dark brown liquid. On rectification, it turns pale yellow and smells sweeter with a eugenol content of 80-85%.

iv) *Clove oleoresin*: Clove oleoresin is prepared by cold extraction of crushed spices using organic solvent like acetone giving a recovery of 18-22%. Oleoresin is chiefly used in perfumery and used for flavouring when it is dispersed in salt, flour etc.

New technologies/ innovations in value addition of spices

Cryoground spice powder: In the conventional grinding of spices, the mill and the product temperature can rise to as high as 90°C and at high temperatures there is considerable loss of volatile oil. Cryogenic grinding overcomes this problem and helps in retaining more volatile oils besides reducing oxidation, improving fineness and posing minimum distortion in the natural composition of powder. The usual practice during the cryogrindng is to inject liquid nitrogen (-80°C) into the grinding zone. A temperature controller maintains the desired product temperature by suitably adjusting the nitrogen flow rate. The exhausted gas is recirculated for precooling of the spice.

Supercritical fluid extraction (SFC): When a gas is compressed and maintained below its critical temperature and critical pressure, it becomes a supercritical fluid. The solvent free extraction of essential oils and oleoresins by supercritical fluid extraction technology has shown promising results. Though, there are many supercritical fluids, carbon di oxide is the most widely accepted one. The critical temperature and pressure beyond which carbon di oxide behaves as supercritical fluid are 31.3°C and 73.8 bar pressure respectively.

Microencapsulated spice flavour: Microencapsulation is the technique by which the flavouring material is entrapped in a solid matrix and is ready for release as and when required. Encapsulation is achieved mostly by spray drying. In the production of spray dried spices, the essential oils and or oleoresins are dispersed in the edible gum solution, generally gum acacia or gelatin, spray dried and then blended with dry base such as salt or dextrose. As water evaporates from the spray dried particles, the gum forms a protective film around each particle of extractive. The protective capsule prevents the spice extractive from evaporating and from being exposed to oxygen.

Success stories in spice processing sector

Establishment of incubation facility for spice processing at ICAR-Indian Institute of Spices Research and hand holding of spice entrepreneurs

ICAR-Indian Institute of Spices Research has established the spice processing facility at its experimental farm at Peruvannamuzhi, Kozhikode during 2013-14. It was setup with the objectives of encouraging research and entrepreneurship development in spice processing for product and process development. This facility is established to attract entrepreneurs in spice sector by developing integrated processing capabilities, hand holding entrepreneurs, providing training and technical guidance on post harvest operations and quality maintenance of major spices. Processing unit is equipped with state of the art facility for primary as well as secondary processing of spices. The facility has three units, each for cleaning and grading black pepper, curry powder production and white pepper production. The facility has also obtained the "Manufacturing license" from FSSAI (Food Safety and Standards Authority of India) for commercial production of cleaned and graded black pepper, white pepper and spice powders.

Black pepper cleaning and grading unit

Pre cleaning equipment installed in the black pepper cleaning cum grading unit includes a black pepper cleaner cum grader, spiral separator and a metal detector. Fully matured green pepper is harvested when one or two berries in the spike turn orange red. The berries are separated from spike using a thresher and the separated berries are dried in the drying yard for about five days. In this process, there are chances of contamination by dust, dry leaves, sticks and other foreign matters. It is therefore necessary to clean black pepper before it is packaged and used for consumption (Amaladhas *et al.*, 2004). The capacity of the cleaner cum grader, spiral separator and the metal detector is 200 kg/h. Once, the black pepper is cleaned, it is graded according to size and then packaged in clean gunny bags. List of machineries in black pepper cleaning and grading unit installed at ICAR-IISR are listed in Table 2. Fig. 1 shows the internal view of black pepper cleaning and grading unit.

Table 2: Machineries for a black pepper cleaning and grading unit

S. No	Machineries	Capacity	Quantity	Cost,Lakhs
1.	Cleaner cum grader	200 kg/h	1	5.50
2.	Spiral separator	200 kg/h	1	2.50
3.	Metal detector	200 kg/h	1	1.25
4.	Filling machine	200 kg/h	1	4.50
5.	Continuous band sealer	150 packs/h	1	0.30
6.	Weighing machine	100 kg	1	0.50
			Total	14.55

Fig. 1: Black pepper cleaning and grading unit – a. Black pepper destoner with grader b. Spiral separator

White pepper production unit

White pepper is produced from fully matured freshly harvested green pepper or from black pepper. The freshly harvested green pepper spikes are despiked/threshed using a pepper thresher and the berries are graded in a rotary grader. Berries of size 4.0 mm and above are used for white pepper production. The fresh berries are washed in the drum washer and introduced into the fermentation tank where the pepper is fermented, with daily change of water in the tank. After required days of fermentation, the fermented pepper is fed into the pulper-cum-washer for the removal of outer skin. White pepper so obtained is washed and dried for a period of 2-3 days in the solar tunnel drier. Dried white pepper is cleaned, graded and packaged for commercial use. Machineries in white pepper production unit installed at ICAR-IISR (Fig. 2) are listed in Table 3.

Table 3: Machineries for a white pepper production unit

S. No.	Machinery	Capacity	Quantity	Cost, Rs.(in Lakhs)
1.	Drum washer	200 kg/h	1	2.50
2.	Fresh pepper grader	200 kg/h	1	2.00
3.	Fermentation tank	250 kg/batch	2	3.25
4.	Pulper cum washer	200 kg/h	1	2.50
5.	Weighing machine	100 kg	1	0.50
			Total	10.75

Fig. 2: White pepper production unit

Curry powder production unit

Curry powder production unit is equipped with facilities for powdering and packaging spices or spice blends. Spices brought to the unit are first checked for its moisture content and if the moisture content is above 10 per cent, spices are dried in the solar tunnel drier (Fig. 3). Rotary drier can also be used for drying spices. Spices are then crushed in the plate crusher or finely powdered in the micro pulverizer as per the material size of the powder required. Powdered spices are sieved in the vibro sifter and is filled in pouches and sealed. In case of curry powder production *i.e* spice blends; the dried spices are roasted to a definite temperature in the drum roaster before powdering for flavour enhancement and then powdered in the micro pulveriser. Powder is sieved in the sieve shaker and then transferred to the blender for the homogeneous production of spice mix. Curry powder is then weighed automatically and filled in pouches and sealed. Fig. 4 shows the machinery arrangements of the curry powder production unit. List of machineries in curry powder production unit installed at ICAR-IISR is listed in Table 4.

Table 4: Machineries in curry powder production unit

S. No.	Machinery	Capacity	Cost, Rs.(in lakhs)
1.	Solar tunnel drier	100 kg/ batch	5.00
2.	Roaster	10-25kg/ batch	1.50
3.	Micro- pulverizer (Hammer Mill)	100 kg/h	4.50
4.	Sifter	25 kg/ batch	1.50
5.	Ribbon blender	50 kg/batch	1.35
6.	Filling machine	200 kg/h	4.50
7.	Continuous band sealer	150 packs/h	0.30
8.	Weighing machine	100 kg	0.50
		Total	19.15

Fig. 3: Solar tunnel drier for drying spices

Fig. 4: View of curry powder production unit

Cardamom processing unit

This is a new facility being installed at ICAR-IISR, Regional Station at Appangala, Kodagu District of Karnataka. The unit has facilities for washing freshly harvested cardamom capsules to remove dust and soil particles. List of machineries required for a cardamom processing unit are listed in Table 5.

Table 5: Machineries for cardamom processing unit

S. No.	Machinery	Capacity	Quantity	Cost, Rs.(in Lakhs)
1.	Cardamom washer	75 kg/ batch	1	1.50
2.	Cardamom drier	150 kg/ batch	1	2.25
3.	Cardamom grader	150 kg/h	1	1.40
4.	Cardamom polisher	150 kg/h	1	1.85
5.	Continuous band sealer	150 packs/h	1	0.30
6.	Weighing machine	100kg	1	0.50
			Total	7.80

Successful incubatees of spice processing unit of ICAR-IISR

During 2015-16, the facility was licensed to four clients for commercial production of spice/curry powders by entering into a memorandum of understanding with the entrepreneurs. M/s SUBICSHA, the Coconut Producer's Company comprising of 532 women self help groups was the first incubatee to launch their products on 1st January 2016. M/s Abiruchi Food Products, a unit of Kudumbasree, launched their spice products on 18th February 2016. M/s Malu Pure Food mix launched their products on 26 June 2016 and M/s Cookway Foods, Kozhikode is yet to start their commercial production. Apart from this, some entrepreneurs utilize the facilities of white pepper production and black pepper cleaning cum grading units on hired basis by paying the rental charges. Fig. 5-9 shows the utilization of machinery in the spice processing facility by the women entrepreneurs.

The black pepper cleaning cum grading unit is also licensed to two entrepreneurs. M/s Mannil Spices, Kozhikode and M/s. CAMCO, Mangalore has entered into a MOU on 15 October 2016 and 23 February 2017 respectively. The firms are ready to begin their commercial production.

In addition to hand holding the entrepreneurs, the facility is also utilized for conducting trainings for potential entrepreneurs, farmers and students regularly.

Fig. 5: Incubatee utilizing the micro pulveriser

Fig. 6: Incubatee working with the vibratory powder sifter

Fig. 7: Incubatee using the curry powder blender

Fig. 8: Filling of curry powder using automatic filling machine

Fig. 9: Continuous sealing of spice powder pouches

Products from ICAR-IISR spice processing facility

SUBICSHA spice powders

M/S SUBICSHA is involved in the production of chilli, turmeric and coriander powders.

Spice products by M/S Abiruchi food products

M/S Abiruchi food products launched six products processed at the spice processing facility of ICAR- IISR, Peruvannamuzhi.

Spice products by M/S Malu pure food mix

M/S Malu pure food mix, a Kozhikode based new start-up company entered the fast growing curry masala market by launching a slew of products under the brand name 'Malu'. Products marketed by them include chilli, coriander, turmeric and black pepper.

Production from the Incubation Centre

More than 6 tons of spice powder has been processed at the spice processing unit during 2014-15. Fig. 10 shows some of the products processed at the spice processing unit by the entrepreneurs. The incubating firms are extremely satisfied with the support they obtained from the Institute.

Fig. 10: Products marketed by incubatees processed at ICAR-IISR Spice Processing Unit

Safety and quality aspects of spice products

In the international market, quality specifications for trade are laid by the importing as well as the producing countries. The parameters assessed are extraneous matter, light berries, pinheads, bulk density, insects, excreta and microbiological aspects like presence of *Salmonella, E.coli,* aflatoxin etc. American Spice Trade Association (ASTA) or European Spice Association (ESA) or International Organization for Standardization (ISO) specifications are the commonly adopted standards in the international trade. Table 6 gives the ASTA Cleanliness Specifications (Effective April 28, 1999) for important spices.

Packaging of spice products

Spices are hygroscopic in nature and its nature of absorbing moisture during rainy season result in mould attack and insect infestation, as it contains good amount of starch. Mould and insect damage can lead to loss of aroma, caking and hydrolytic rancidity. Efficient packaging and proper storage are essential to ward off these problems. Whole pepper is generally packaged and transported in gunny bags and polyethylene lined double burlap bags. Eco friendly packaging materials such as clean gunny bags or paper bags may be adopted and the use of polythene bags may be minimized. Recyclable/ reusable packaging materials may be used wherever possible.

Spices are to be stored with utmost care as they deteriorate rapidly. The graded produce is bulk packed separately in multi layer paper bags or woven polypropylene bags provided with food grade liners for export or in jute bags. The bags are arranged one over the other on wooden pallets after laying polypropylene sheets. Precautions to be followed during storage are:

- Moisture level in spices is to be in the range of 10-11 per cent before it is stored.

- Store houses should be constructed scientifically and it should be damp-proof, rat proof and bird proof. The room should have controlled ventilation and devised for control of humidity and temperature.

- The room should be properly fumigated before storage.

- The walls should be white washed regularly.

- Proper drainage should be provided.

- Polyethylene lined gunny bags or laminated HDPE are ideal for storing spices.

- A good dehumidifier fitted in the storage rooms can eliminate mould and insect attacks by keeping the atmosphere always dry.

Table 6: ASTA Cleanliness Specifications for Spices

Name of spice	Whole insects, Dead By count/lb	Excreta, Mammalian mg/lb	Excreta other mg/lb	Mould % by wt.	Insect Defiled/ Infested % by wt.	Extraneous Foreign matter % by wt.
Black Pepper	2	1	5.0	1	1	
White Pepper	2	1	1	1	0.5	
Cardamom	4	3	1	1	1	0.5
Turmeric	3	5	5	3	2.5	0.5
Ginger	4	3	3	3	1	
Chillies	4	1	8	3	2.5	0.5
Nutmeg (Broken)	4	5	1	5	0.5	
Nutmeg (whole)	4	0	0	10	0	
Mace	4	3	1	2	1	0.5
Cinnamon	2	1	2	1	1	0.5
Cloves	4	5	8	1	1	1

Source: ASTA (1999)

Conclusion

Traditionally, India has been exporting unprocessed bulk (whole) spices to the world market which earned less price in international market compared to value added products. There is assured market for processed spice products like curry powders, oils and oleoresins. Demand for spices as nutraceuticals is showing an upward trend. Spices and its derivatives offer great scope under food related agriculture industries. Post harvest management of spices has great scope considering present international trade scenario. A huge jump in the export of curry powders and other value added products of spices in the coming decade is expected which will yield more sustainability of agriculture in Kerala.

References

Agarwal, Y. C., A. Hiran & A. S. Galundia. 1987. Ginger Peeling Machine Parameters, *Agricultural Mechanization in Asia, Africa & Latin America* 18(2):59-62.

Amaladhas, H. P., P. Rajesh & S. Subramanian. 2002. Get Better Quality Mace by Blanching, *Spice India* 15(2): 8-10.

Amaladhas, H. P., S. Subramanian., M. Balakrishnan., R. Viswanathan & V. V. Sreenarayanan. 2004. Performance Evaluation of a Cleaner cum- Grader for Black Pepper, *J. Plantation Crops* 32(1): 32-36.

ASTA, 1999. ASTA Cleanliness Specifications for Spices, Seeds and Herbs. *American Spice Trade Association*, NJ. 07632, USA.

Jayashree, E. 2005. Value Addition of Major Spices at Farm Level, *Spice India* 18(8): 33-41.

Joy, C. M., G. P. Pittappillil & K. P. Jose. 2000. Quality Improvement of Nutmeg using Solar Tunnel Dryer, *J. Plantation Crops* 28(2):138-143.

Kachru, R. P. & P. K. Srivastava. 1988. Post Harvest Technology of Ginger. *CARDAMOM* 21(5): 49-57.

Kachru, R. P. & P. K. Srivastava. 1991. Processing of Turmeric, *Spice India* 4(9): 2-6.

Korikanthimath, V. S. 1993. Harvesting and On-Farm Processing of Cardamom, *Proceedings of the National Seminar,* held at RRL, Trivandrum 13-14 May organized by Indian Society for Spices Research and Spices Board, pp. 50-52.

Krishnamoorthy, B. & J. Rema. 2001. Nutmeg and Mace. In: K. V. Peter (ed.) *Handbook of Herbs and Spices,* Woodhead Publishing Limited & CRC Press LLC, pp. 238-248.

Patil, R.T. 1987. Cardamom Processing in South India. *Agricultural Mechanization in Asia, Africa and Latin America* 18(2):55- 58.

12

Entrepreneurship Oriented Processing and Value Addition Technologies of Coconut

Manikantan M R, Shameena Beegum, Pandiselvam R & Hebbar K B

Introduction

Coconut palm (*Cocos nucifera* L.), a perennial horticultural crop, is a symbol of national and international integration involving more than 93 producing countries and more than 140 consuming countries. It is eulogised as 'Kalpavriksha'- the '*tree of heaven*' as each and every part of the palm is useful to mankind in one way or other. There are countless uses of this coconut palm. It is bestowed with multiple benefits like health, wealth and shelter to mankind. It is also denoted as "heavenly tree", "tree of abundance" and "nature's supermarket". India is the largest producer of coconut followed by Indonesia and Philippines. In India, Tamil Nadu ranks first in production followed by Karnataka, Kerala and Andhra Pradesh. These states account for more than 90 per cent of the area and production of coconut in India. However, it is also cultivated with varying success in other states like Assam, Goa, Gujarat, Maharashtra, Nagaland, Orissa, Tripura, West Bengal, Andaman and Nicobar Islands, Lakshadweep and Puducherry.

Health benefits of coconut

People of coconut growing regions of the world have been relied on coconut for health and nourishment since time immemorial. In India, coconut in its many forms is used to treat a variety of health problems and to nourish the body.

Coconut water

Coconut water is the liquid endosperm of the nut. The tender coconut water is consumed as refreshing drink. It is a rich source of B- group vitamins, minerals, sugars etc. It is effective against gastroenteritis, diarrhoea, vomiting and to prevent dehydration of body tissues (Rajagopal and Ramdasan, 1999). Coconut water contains major electrolytes such as potassium, sodium, magnesium, phosphorous and calcium required for the body. In the Indian ayurvedic medicine, it is described as "unctuous, sweet, increasing semen, promoting digestion and clearing the urinary path" (Rethinam and Kumar, 2001). Coconut water is traditionally prescribed for burning pain during urination, gastritis, burning pain of eyes, indigestion and hiccups or even expelling of retained placenta. Presence of L-arginine in coconut water has a cardio protective effect through its production of nitric oxide, which favours vasorelaxation (Anurag *et al.*, 2007). Concerning nutraceutical effects, coconut water reduce histopathological changes in the brain induced by hormonal imbalance in menopausal women (Rundorn *et al.*, 2009). Coconut water contains major phyto hormones such as auxin, various cytokinins, and gibberellins (Fonseca *et al.*, 2009) which can minimize the aging of skin cells, balance pH levels, and keep the connective tissues strong and hydrated. It is an excellent oral rehydration sports beverage - replaces electrolytes lost from exercise, heat stress and illness and aids in exercise performance. Natural isotonic beverage contains the same level of electrolytes found in human blood and has 15 times more potassium (264 mg / 100 ml) than most sports and energy drinks (12.5 mg /100 ml) (Reddy and Lekshmi, 2014).

Coconut oil

Diversified utilities of coconut oil are hair care, skin care, healing, aromatherapy, weight loss, digestion, protection against heart diseases and infections. Coconut oil acts as effective moisturizer on all types of skin. It is effective in preventing dryness and flaking of skin and delays the appearance of wrinkles and sagging of skin which normally accompany aging. It is also effective against psoriasis, dermatitis, eczema and other skin infections (Vala and Kapadiya, 2014).

Lim-Sylianco (1987) studied the anticarcinogenic effects of dietary coconut oil. Antimicrobial activity of the monoglyceride of lauric acid (monolaurin) of coconut oil has been reported since 1966. Coconut oil is useful for slowing down the degenerative process by improving mineral absorption. Coconut oil also helps to supply energy to cells (because it is easily absorbed without the need of enzymes) as well as improve insulin secretion and utilisation of blood glucose (Garfinkel *et al.*, 1992).

Coconut milk

The medium chain saturated fatty acid (MCFA) of coconut milk is converted in the body into a highly beneficial compound called monolaurin, an antiviral and antibacterial agent that destroys a wide variety of disease causing organisms. According to the National Center for Biotechnology Information, lauric acid has many germ fighting, antifungal and antiviral properties that are very effective at ridding the body of viruses, bacteria and countless illnesses (Baldioli *et al.*, 1996).

Coconut inflorescence sap

Coconut sap, normally called as neera, is a natural health drink, which is traditionally collected from coconut spadix. It is the phloem sap, rich in sugars, proteins, minerals, antioxidants, and vitamins. It is also a rich source of phenolics, ascorbic acid and essential elements viz. N, P, K, Mg and micronutrients viz. Zn, Fe and Cu. (Hebbar *et al.*, 2015). It is considered as a nutritious drink for cure of anaemia, tuberculosis, bronchial suffocation and piles. Because of the glycemic index, it is considered good for weight maintenance, preventing over weight and obesity and good for diabetic people (Maier *et al.*, 2003). It also has antihypertensive effect. Bhagya and Soumya (2016) revealed the significant effect of supplementing coconut neera in reducing systolic blood pressure among hypertensive adult women.

Scope of entrepreneurship development in coconut processing sector

Majority of the farmers engaged in coconut cultivation mainly have small and marginal scattered holdings. This hampers the prospects of processing and value addition in coconut. Presently, coconut growers are more exposed to economic risk and uncertainties owing to high degree of price fluctuations. Further, the mindset of traditional coconut grower is attuned to processing for copra and coconut oil. Coconut-based economy can expect a revival only when the profitability of coconut farming is delinked from the price behaviour of coconut oil. This is possible through efficient utilization of land under coconut cultivation and post harvest value addition activities.

India has not made tangible progress in product diversification and by-product utilization of coconut except for the traditional activities such as oil milling and coir processing. As a result, coconut oil continues to be the only major coconut product having influence on farm level price of coconut. Present level of value addition is 8% which needs to be increased to at least 25%. This situation can be transformed only when coconut based both edible and non-edible products gets priority over coconut oil. As compared to tardy growth recorded by the country in processing sector, most of the coconut growing countries are making profit from production and export of diverse coconut products. Philippines, Indonesia and Thailand realizes more than 50% value addition level, export over 40 non-traditional products of which coco chemicals, coconut milk products, coconut water based products, and shell and coir products are important.

Product diversification and value addition play a crucial role in the stabilization of coconut oil driven market and is essential for reorienting and energizing the coconut industry cost effective and globally competitive. Hence, there exists a huge scope for coconut based agri-business in India. Processing and related activities can mitigate the seasonal price variation and generate income and employment opportunities for over two million people in India.

Entrepreneurship development through coconut value addition

Coconut is mainly consumed as fresh nuts, tender coconuts, coconut oil and copra meal. Around 50 per cent of the world production is consumed in the form of fresh nuts and tender nuts. Close to fifty per cent of the nut production is converted into copra and consumed as coconut oil and copra meal. Around 2.52 per cent of the production is consumed as desiccated coconut. In India, annual consumption of tender coconut is about 200 million. Coconut palm also provides a series of by-products such as fiber, charcoal, handicrafts, vinegar, alcohol, sugar, furniture, roofing, fuel, etc. and it has more than 200 diversified local uses. Products and by-products of the crop form vital inputs for many of the industries and support the livelihood of many millions. They contribute a significant amount to the national revenue and country's exports by way of excise and export earnings. They also provide direct and indirect employment to a large number of people in the country. Potential of converting coconut into different emerging value added products such as desiccated coconut powder, virgin coconut oil, coconut chips, coconut milk, preserved tender nut water & coconut inflorescence sap into coconut sugar is realized in view of globalization over the traditional processed products of copra and coconut oil.

Value added food products from coconut can be broadly categorized as, those derived from mature coconut kernel, mature coconut water, tender coconut water, tender coconut kernel and coconut inflorescence sap.

Products derived from mature coconut kernel

Coconut reaches full maturity in about 12 to 13 months time after the opening of inflorescence. Mature raw coconut are used for culinary uses, religious purposes, making copra and manufacturing convenience products such as desiccated coconut, coconut milk, spray dried coconut milk powder, etc. About 40% of the Indian annual coconut production goes to culinary and religious purpose, 35% for copra production, 17% for tender nut purpose, 2% for seed purpose and only 6% is the share of value addition which comprises of desiccated coconut powder (4%), virgin coconut oil and coconut milk/ cream (1% each). Process technology for major value added mature coconut based products is described below.

Desiccated coconut

Shredded and dried white kernel or endosperm is marketed as desiccated coconut. The steps involved in processing of desiccated coconut involves selection, sorting and husking of coconut, shelling, paring, washing, sterilizing, grinding, drying, sieving, packing, and storage (Fig.1). The main uses of desiccated coconut are for the confectionary industry, as a filling for chocolates and candies; bakery industry for biscuits, cake and nut filling products; direct usage to decorate cakes, biscuits and ice cream and preparation of various snacks. During the year 2015-16, India exported 4261 MT desiccated coconut worth Rs. 52.60 crores. In comparison with the export figure of previous year, India achieved an increase to the tune of 60%, which indeed is a remarkable achievement. The flow chart of desiccated coconut powder preparation is described below.

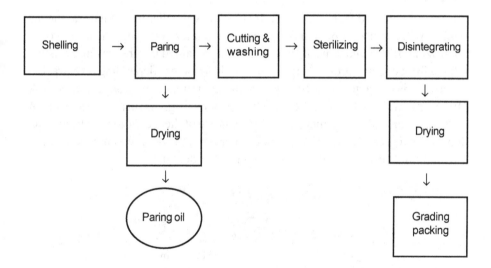

Techno economic details

Machineries required	Coconut dehusker, desheller, testa remover, washing unit, inspection conveyor, blanching unit, pulverizer, fluidized bed dryer, desiccated powder cooler, lump breaker, vibro siever, packaging unit.
Capital investment	Rs.130 lakhs for processing 15,000 nuts per day
Yield	1 ton from 10,000 coconuts

(*Source:* Coconut Development Board)

Fig. 1: Desiccated coconut powder

Virgin coconut oil (VCO)

It is the oil obtained from fresh, mature kernel by mechanical or natural means, with or without use of heat and absence of chemical refining, bleaching or de odourizing. It is called "virgin" because the oil obtained is pure, raw and pristine. It has a fresh coconut aroma ranging from mild to intense depending on extraction process. It is extracted directly from fresh coconut meat or from coconut milk. Different methods involved are hot-processing, natural fermentation, centrifugation and direct micro expelling. Choice of the technology to be adopted depends to a great extent on the scale of operation, the degree of mechanization, the amount of investment available and market demand.

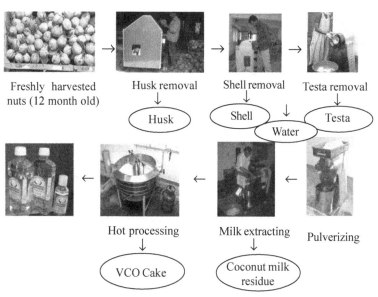

Fig. 2: Flow chart of VCO processing

ICAR-CPCRI has developed processing technologies for production of VCO by hot and fermentation methods. In hot process, coconut milk is cooked in specially designed cooker whereas in fermentation process, coconut milk is allowed to ferment in specially designed fermentation tank for specified period to get VCO. The process protocol is given in Fig.2.

Techno economic details

Machineries required	Coconut dehusker, Coconut desheller, Coconut testa removing machine, Coconut pulverizer, Milk expeller, VCO cooker, Vacuum dryer, Packaging system, Weighing balance, Miscellaneous items such as stainless steel containers, stainless steel containers with trolley attached and other vessels, electrical fittings, electrical water heaters etc
Total investment on machines for processing 500 nuts /day	Rs.15 lakhs (Hot process)Rs.12 lakhs (Fermentation process)
Unit production cost	Rs.420 per litre
Breakeven period	103 days
Net profit percentage	47.17%
Production details per year	VCO – 7500 litres (20% of kernel weight)Milk residue – 7500 kg VCO Cake – 1500 kgTesta – 1000 kg Husk – 60000 kg Shell – 20000 kg Water – 15000 litres

i. VCO based margarine

Technology for preparation of virgin coconut oil based margarine to be operated at small and micro level industries is reported to be patented by Indonesia. The process involves mixing of emulsifiers, stearine, antioxidants, β-carotene, water & salt with VCO, blending at 60°C for 10 minutes, filling, packing and cooling at 16°C (Fig.3). Product can be used as bread spread. It contains high lauric acid and no trans-fats.

Fig. 3: VCO based margarine

ii. VCO based mayonnaise

Mayonnaise is prepared by mixing coconut oil, vinegar or citric acid or emulsifiers. Carbohydrates, spices and flavour enhancers are added to modify the flavour and to avoid crystallization (Fig.4). Final formulation consist of 70% VCO, 6% natural vinegar, 7% fresh yolk and 1% emulsifiers and cooled boiled water.

Fig. 4: VCO based mayonnaise

Mayonnaise production units can be commercially operated at home or micro level to enhance the income of farmer families.

Coconut oil

Coconut oil is one of the major edible products of coconut. This is referred as lauric oil in the world market because of its high lauric acid content. It is very heat-stable, and suitable for cooking at high temperature. It is slow to oxidize and resistant to rancidity for up to two years due to its high saturated fat content. Coconut oil is used in India for culinary/edible, toiletry, soap making and as an illuminant and lubricant.

The fresh kernel is dried to less than 6% moisture (copra) which is then expelled to get coconut oil. Copra is cut into small chips in a copra cutter. Chips are fed into steam jacketed kettles and cooked mildly at a temperature of 70°C for 30 minutes. After proper cooking, cooked material is fed into the expeller continuously and pressed twice. Combined oil from the first and the second pressing is collected in a tank. This oil is filtered by means of a filter press and stored in MS tanks and then bulk packaging is done in tin containers. HDPE containers and polymeric nylon barrier pouches are used for consumer packaging.

Techno economic details

Machineries required	Copra cutter, bucket elevator, steam jacketed kettle, oil expeller, screw conveyor, crude coconut oil storage tanks, filter press, micro filter, filtered oil storage tank, baby boiler, packaging unit.
Capacity	5 tons copra/day
Yield	3 tons oil
Total project cost	Rs.72 lakhs for 3 ton/d capacity
Plant & machinery cost	Rs.25 lakhs
Annual sales turnover	Rs. 315 lakhs
Net profit	Rs. 12 lakhs
Return on investment	28%

(*Source:* Coconut Development Board)

Coconut chips

Coconut chips are ready-to-eat, snowy white crisp and healthy non fried snack prepared from fresh kernel through osmotic dehydration in a forced hot air electrical dryer at 70-80°C for 5-6h to less than 3% moisture content. The kernels undergo paring, blanching, slicing and osmotic dehydration to prepare ready to eat chips (Fig 5&6). It contains 46% carbohydrate, 1.24% protein, 48% healthy fat, 6.13% fibre and 1.36% minerals. Method of drying on the basis of osmosis is used, in which partial dehydration in sliced form is brought about by dipping fresh kernel in sugar solution followed by hot air drying. This is claimed to result in product with better flavour than freeze drying method at comparatively lesser cost. Hence, the resultant coconut chips give health promoting substances and do not pose any health hazard. Nutraceutical and medicated coconut chips can also be made by incorporating juice of beet root, carrot, ginger and pepper.

Fresh kernel of 8-9 months old coconut is to be used for making chips. Here, the index for selection of the nut is that the nut should be matured enough to be sliced. If it is too tender, slicing and testa removal is not possible. Important steps involved in the production of coconut chips are given below.

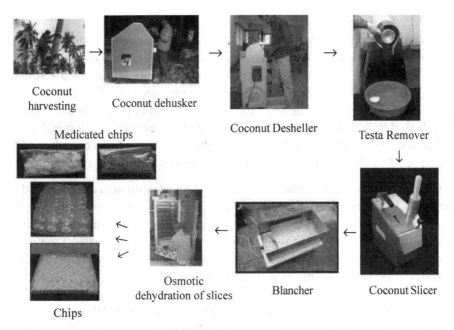

Fig. 5: Process protocol developed for the production of coconut chips

Techno economic details

Machineries required	Coconut dehusker, coconut desheller, coconut testa removing machine, multi commodity coconut slicer, blanching unit, plastic basin, filter, muslin cloth, vessel, gas stove, stirrer, solar dryer, electric dryer, heat sealing machine etc.
Capacity	250 coconuts per day
Total investment on machines	Rs.6 Lakhs
Unit production cost	Rs.8.45 / packet of 25 g
Breakeven period	56 days
Net profit percentage	57.71
Production details per year	Chips – 11250 kg Husk – 30000 kg Shell – 10000 kg Testa – 500 kg Water – 7500 litres

(Manikantan *et al.*, 2015)

Fig. 6: Coconut chips

Coconut milk

Coconut milk is an oil-in-water emulsion, stabilized by naturally occurring proteins (globulins and albumins) and phospholipids (lecithin and cephalin). In comparison with dairy milk, coconut milk is richer in fat, poorer in protein and sugar. Coconut milk can be used in curry preparation, confectionaries etc. It is utilised as a substitute of dairy cream in beverage type milk, and in evaporated and sweet condensed milk and in the preparation of white soft cheese, yoghurt and many other food stuffs (Sanchez and Rasco, 1983). Coconut milk contains 5.8% protein, 38 to 40% fat, 6.2% minerals and 9 to 11% carbohydrates.

First step is breaking the dehusked nuts into halves. Split nuts are deshelled to separate the kernel. Kernel is washed and then blanched by immersing in hot water at 80°C for 10 minutes. It is then pulverized after cutting into small pieces and is subjected to pressing using continuous screw press to extract the milk. Coconut milk thus obtained is filtered by passing through a vibratory screen. Food additives such as emulsifiers and stabilizers are to be added to the milk to obtain a stable consistency and texture. For this purpose, permitted emulsifiers and stabilizers are mixed with hot water separately and mixed thoroughly. This is added to the coconut milk and then subjected to emulsification using a mechanical impeller emulsifier. Emulsified milk assumes a creamy consistency. Coconut cream is then pasteurized at 95°C for 10 minutes in a plate heat exchanger. Pasteurized coconut cream is hot filled in cans using a mechanical volumetric filling machine followed by steam exhausting. Cans are seamed using an automatic can seamer. Seamed cans are sterilized in a rotary retort at 15 psi for 20 minutes. Cans are then cooled in running water (Fig. 7).

Techno economic detail

Machineries required	Hammer mill, elevator, screw press, coconut milk storage tanks, vibrating sieving machine, coconut residue mixer, additive mixing tank, emulsifier, homogenizer, pasteurizer, volumetric filling machine, exhaust box, can seaming machine, horizontal rotary retort, hot air drier, agro waste vertical boiler, sterilization tank, coconut residue storage bins.
Capacity	10,000 mature coconuts/day
Land	1 acre (cost variable)
Building - 6000 sq.ft @ Rs.1000 per sq.ft.	Rs.60 lakhs
Plant & machinery (does not include DG set, weigh bridge, effluent treatment equipment and other items not directly connected with process operation)	Rs.75 lakhs
Electrification	Rs.25 lakhs
Preliminary and pre-operative expenses	Rs.15 lakhs
Margin money for working capital	Rs.40 lakhs
Yield	2,500 kg coconut milk/500 kg coconut cream residue

(Source: Coconut Development Board)

Fig. 7: Packaged coconut milk

Spray dried coconut milk powder

Spray drying method is used for the commercial production of coconut milk powder. The product is packed in laminated foil bags and contains 62 per cent fat, 14 per cent protein and 2 per cent moisture which can be used in place of fresh coconut milk for food preparation/ beverages in household and food industries. Additives such as maltodextrin, casein or skim milk or corn syrup are added to the extracted milk and the mixture is pasteurized and homogenized before spray drying (Gonzalez, 1986). Such additives aid the spray drying process and help to convert a high fat coconut milk into flowable, but cohesive powders through encapsulation of fatty substances (Seow & Leong, 1988). It can be reconstituted into coconut milk by diluting with water. It offers additional advantages like less storage space, enhanced shelf life and reduced packaging cost. Central Food Technological Research Institute, Mysore with the financial assistance of the Board has developed the technology for spray dried coconut milk powder.

Techno economic analysis

Machineries required	Hammer mill, coconut milk extractor/ screw press, coconut milk storage tanks, additive mixing tank, pasteurizer, homogenizer, spray drier, elevator, vibrating sieving machine, volumetric filling machine, horizontal rotary retort, sterilization tank, can seaming machine, agro waste vertical boiler, coconut residue mixer etc.
Capacity	20,000 coconuts per day
Land	60 cents
Building - 6000 sq ft.	Rs.45 lakhs
Plant & machinery	Rs.200 lakhs
Contingencies	Rs.10 lakhs
Preliminary & pre-operative expenses	Rs.15 lakhs
Working capital (margin money)	Rs.25 lakhs
Yield	1 ton of coconut milk powder

(*Source:* Coconut Development Board)

Coconut flavoured milk

Coconut milk is vegan alternative to dairy milk. Coconut milk does not contain lactose and is lower in carbohydrates than dairy milk, which can be consumed by people who are lactose intolerant or just don't enjoy the taste of dairy milk. Milk is extracted from freshly grated coconut of 9-10 months old. Extracted milk is clarified to remove suspended solids which are present in the milk. Coconut milk is then mixed with coconut water and diluted by adding purified drinking water until it is appropriate for flavoured coconut milk production. It is then mixed with 10-12% sugar, 2% stabilizers, emulsifiers and flavours. The flavoured coconut milk is then UHT sterilized at 138-140°C for about 15 seconds, which is then packed in sterilized polypropylene bottles (Fig 8).

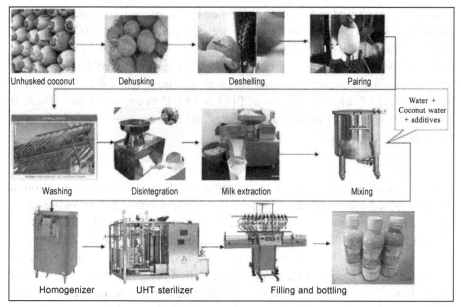

Fig. 8: Flow chart of flavoured coconut milk processing
(*Source:* http://coconutboard.nic.in/process.htm)

Techno economic details

Machineries required	Dehusking, deshelling, testa removing, blanching, pulverizing, milk expelling, filtration, mixing, agitation, homogenization, UHT sterilization.
Capacity	5,000 coconuts/ day
Yield	4000 litres flavoured milk
Cost of plant and machinery	Rs.132 lakhs
Total project cost	Rs.2.23 crores
Pay back period	4.5 years
Internal rate of returns	19%
Breakeven point	49%

(*Source:* Coconut Development Board)

Products derived from mature coconut water

Water obtained from mature coconut is usually disposed after dehusking, while kernel is used for production of coconut oil or coconut milk. Until recently, mature coconut water has been considered as waste, especially in coconut processing plants.

Water when taken out from the nut spoils within a day because of external contamination by microorganisms. Minerals catalyze lipid oxidation and results in free fatty acid (FFA) formation that affects aroma and quality of either fresh or processed coconut water. Even if coconut water is extracted aseptically, its exposure to air initiates oxidation promoted by polyphenol oxidase (PPO) and peroxidase (POD), which are naturally present in coconut water (Duarte *et al.*, 2002). Minerals and electrolytes in coconut water also catalyze lipid oxidation and formation of volatile compounds. Hence, it is recommended that the storage temperature for processed coconut water should not exceed 4°C. Addition of ascorbic acid inhibits the activity of PPO and POD in coconut water. There is a scope for processing mature coconut water into different commercially feasible value added products.

Coconut vinegar

Coconut vinegar is the resultant product of alcoholic and acetic fermentation of sugar enriched coconut water. Coconut water can be converted into vinegar by using vinegar generators. Matured coconut water consisting of 1-3% sugar is concentrated to 15% level by fortifying with sugar after filtration. Pasteurized mixture is then cooled and inoculated with active dry yeast *Sacharomyces cerevisiae* (1.5g/L). After alcoholic fermentation for about 5 to 7 days, clear liquid is siphoned off and inoculated with mother vinegar or starter culture containing *Aceteobactor* bacteria. This acetified vinegar is then aged before bottling. Vinegar generator assembly comprises a feed vat, an acidifier and a receiving vat for collection of vinegar. Vinegar has extensive uses as a preservative in pickle industry and flavouring agent in food processing sector.

Techno economic details

Machineries required	Feed trough, vinegar acetifier, receiving trough, wooden storage drums
Capacity	100 litres coconut water/ day
Yield	100 litres vinegar
Land	25 cents
Total project cost	Rs. 6 lakhs
Building (Area - 750 sq. ft.)	Rs.3.0 lakhs
Plant & machinery	Rs.2.5 lakhs
Preliminary & pre-operative expenses	Rs.0.25 lakhs
Contingencies	Rs.0.20 lakhs
Margin money for working capital	Rs.0.25 lakhs
Annual sales turnover	Rs.4.0 lakhs
Net profit	Rs.0.8 lakhs
Return on investment	20 per cent

(*Source:* Coconut Development Board)

Nata-de-coco

Nata-de-coco is a translucent gelatinous product prepared from matured coconut water by the action of cellulose forming bacteria namely *Acetobacter aceti* sub species *xylinium*. *Acetobacter xylinum* metabolizes glucose in coconut water that act as carbon source and converts it into extracellular cellulose as metabolites. The organism can be cultured either in coconut water or skimmed coconut milk. (Hagenmaier *et al.*, 1974). It is widely used in desserts and confectioneries especially in ice creams and fruit cocktails.

Coconut water is strained and mixed with sugar and glacial acetic acid in stipulated proportions (for every litre of coconut water, 100 g refined sugar and 5 g monobasic ammonium phosphate is added). It is then boiled for ten minutes and cooled. Then, add *Acetobacter xylinium* culture solution (150 ml) along with glacial acetic acid (10 ml) and fill in glass trays or wide mouthed jars covered with a muslin cloth and keep for 2-3 weeks without any disturbance. During this period, a white or cream coloured jelly-like substance forms and floats on top of the culture medium. At this stage, the jelly-like substance or Nata will be about an inch thick. Harvest this surface growth; slice into cubes, approximately 1x3 cm or according to requirement. Then, wash it thoroughly to remove the acid taste smell. Drain the nata and equal quantity of sugar is added, mix thoroughly and keep overnight. Next day, stir the mixture to disperse any undissolved sugar. Add small amount of water. Heat the mixture to the boiling point with occasional stirring. Any flavour material can also be added at this stage. Keep the mixture overnight and repeat the heating process until the nata is fully penetrated with sugar as evident by the clear and crystalline appearance of the sweetened nata and preserve in either tin containers or bottles. Optimum temperature for nata production is in the range of 23-32°C (Fig.9).

Techno economic analysis

Capacity	100 litres mature coconut/ day
Land required	5 cents
Building	Rs. 2 lakhs
Equipment/glassware	Rs.0.5 lakhs
Yield	20 kg Nata-de-coco
Annual sales turnover @ Rs.40 / kg	Rs. 3.75 lakhs
Net profit	Rs. 1 lakh per annum
Return on investment	40%

(*Source:* Coconut Development Board)

Fig. 9: Nata- de- coco

Bottled coconut water

National Institute for Interdisciplinary Science and Technology, Thiruvananthapuram has developed a process for the upgradation and preservation of mature coconut water. Main operation involves collection, upgradation, pasteurization, filtration and bottling. Process essentially consists of upgrading the flavour of mature coconut water to the level of tender coconut water by supplementation with additives including sugar and preserving it by a judicious combination of heat pasteurization and permitted chemicals.

Coconut water squash and ready to serve beverage

Coconut water squash and ready to serve beverage can be prepared from mature coconut water. The process involves filtration, heating, mixing with sugar, lime juice and ginger, cooling, mixing with preservatives (sodium benzoate) and packaging in sterilized bottles. It contains sodium, potassium, vitamin C and carbohydrate with a calorific value of 300 Kcal/100ml. The carbonated and non-carbonated beverages stored in aluminium and poly ethylene laminated packages has a shelf life of six months at room temperature.

Products derived from tender coconut water

Tender coconut water is the liquid endosperm, and is the most nutritious beverage that nature has provided for the people of the tropics to fight the sultry heat. It has a calorific value of 17.4Kcal per 100 g of water. It contains water (95.4%), protein (0.1%), fat (<0.1%), mineral matter (0.4%), carbohydrates (4.0%), calcium (0.02%), phosphorous (<0.01%) and iron (0.5mg/100g) (Fife, 2011).

Minimal processing of tender coconut

Once the tender coconut is detached from the bunch, its natural freshness will get lost within 24 to 36 hours even under refrigerated conditions unless treated scientifically. The bulkiness of tender coconut is due to the husk which accounts for two-third of the volume of tender nut. (Haseena *et al.*, 2010). Technologies for minimal processing of tender coconut have been developed by Kerala Agricultural University (KAU) for retaining the flavour and to prevent discolouration. The process involves dipping (partially) dehusked tender coconut in a solution of 0.50% citric acid and 0.50% potassium metabisulphite for three minutes. The product can be stored up to 24 days in refrigerated condition at 5-7°C. By using this process, tender coconut can be transported to distant places and served chilled like any other soft drink. Optimized uniform size facilitates using of plastic crates and insulated chill boxes for transporting and storage (Fig. 10).

In Thailand, young coconut is trimmed, treated with 1-3% sodium metabisulphite and packaged with opener, straw and spoon. These are commercially produced, marketed and exported. The shelf life of processed young coconut is 45 days at 3-6°C or 3 weeks at 7 –10°C.

Fig. 10: Minimally processed tender nuts

Snow ball tender coconut

Snow ball tender nut is a tender coconut without husk, shell and testa which is ball shaped and white in colour. This white ball will contain tender coconut water, which can be consumed by just inserting a straw through the top white tender coconut kernel (Fig.11). Seven to eight months old nut is ideal for making snow ball tender nut in which there is no decrease in quantity of tender nut water and the kernel is sufficiently soft. Technology for preparing snow ball tender nut (SBTN) has been developed at ICAR-CPCRI, Kasaragod. This is served in an ice cream cup. User can drink the tender nut water by piercing the kernel with a straw. After drinking water, the kernel can be consumed using a fork. Coconut water is not exposed to the atmosphere and is natural and sterile.

The machine consists of a circular blade having 24 teeth of 8 mm width that rotates at a speed of 1440 rpm. The prime mover of the machine is a 0.5 HP single phase electric motor. The prime mover attached with the circular blade is fixed on an angle iron frame with a covering made of mild steel sheet. A stop cutter box of stainless steel with a clearance of 15 mm is used to cover the circular blade. The adjustable stop cutter box helps the user to control the depth of cut and protects the user from possible injury while operating the

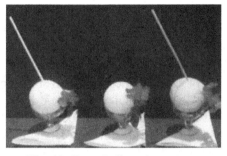

Fig. 11: Snow ball tender coconut

machine. A flexible knife known as scooping tool also has been developed for scooping out the tender nut kernel from the shell. The scooping tool is made of nylon and is flexible at one end. The scooping tool is inserted in between the kernel and shell through the groove and is rotated slowly to detach the entire kernel from the shell.

Packaged tender coconut water

Coconut Development Board (CDB) in collaboration with Defence Food Research Laboratory (DFRL), Mysore has developed a technology for preservation and packing of tender coconut water in aluminium cans/pouches with a shelf life of three months under ambient conditions and six months under refrigerated conditions (Fig.12). A tetra pack technology has also been established in Tamil Nadu. The products are available both in domestic and international markets. Major exporters of the product are Philippines, Indonesia, Malaysia and Thailand (Muralidharan and Jayashree, 2011).

Techno economic details

Machineries required	Mechanical washing system with conveyor, automatic boring and sucking system, ss filter/clarifier, collection tank, treatment tank, pasteurization unit, boiler, filling and sealing machine, shrink wrapping machine, air compressor, coding machine
Capacity	5000 coconuts / day
Total project cost	Rs.131.4 lakhs
Plant & machinery cost	Rs.65 lakhs
Internal rate of return	18%
Breakeven point (sales)	51%

(*Source:* Coconut Development Board)

Coconut water beverages

The processing technologies for coconut water beverages available are given below:

i. RRL technology

Regional Research Laboratory (RRL), Thiruvananthapuram/National Institute for Interdisciplinary Science and Technology, Kerala has developed a process for the upgradation and preservation of tender and mature coconut water. The main operations involve collection, upgradation, pasteurization, filtration and bottling. The process essentially consists of upgrading the flavour of mature coconut water to the level of tender coconut water by supplementation with additives including sugar and preserving it by judicious combination of heat pasteurisation and permitted chemicals. The drink can be carbonated and marketed as beverage.

Fig. 12: Packaged tender coconut water

ii.German technology

Spray evaporation Technique (SET) for making fruit juice concentrate developed by M/s Winter Umwelttechnik, Germany is adopted in this technology. The product retains all the original characteristics of juice such as retention of vitamins and enzymes, aroma, colour, taste etc. This technique was first used by M/s Miracle Food Processors International (P) Ltd. Perinthalmanna, Kerala for concentrating tender coconut water. Coconut water concentrate has a shelf life varying from 6 months to 24 months depending upon the degree of concentration. Ten liters of tender coconut water is required to make about 800g of concentrate. Aerated and bottled ready to drink coconut water beverages also can be made from coconut water concentrate.

Preserved tender coconut water

Preserved tender coconut water involves collection of coconut water, filtration, adjustment of pH, total soluble sugar and taste, pasteurization, filtration and packaging. Ultra filtration system can also be used to clarify tender coconut water. It can be packed in bottles. Bottled drink can be stored for three months at ambient temperature. A system for non thermal preservation of tender coconut water was developed using low ash filter paper and cellulose nitrate membrane which reduced the microbial population and retained the organoleptic properties. Since the tender coconut water is highly susceptible to heating, it is subjected to minimum heating and bio preservatives like Nisin is added, which helped in maintaining the natural pH of 4.9-5.2. Product has a shelf life of three months under ambient storage conditions.

Frozen coconut water

Fresh tender coconut water is collected under hygienic conditions and suspended solids and oil in the sample are removed by means of three-way centrifuge. Salts present in coconut water may be removed if desired, prior to concentration, to produce a very sweet product by centrifugation and passing the centrifuged coconut water through a mixed-bed ion-exchange resin. Ten litres of coconut water will yield about 800 g of concentrate. Concentrate can be frozen or preserved in cans and can be used after dilution to the desired strength. It can be used as base for the production of carbonated and non-carbonated coconut beverages. Concentrated coconut water is also used successfully in the brewery industry.

Tender coconut water jelly

Tender coconut water is a suitable option for the preparation of jelly as its delicate flavour can be well preserved in the form of jelly. Ingredients such as tender coconut water, sucrose and solidifying agent (china grass) are needed to prepare jelly. Standardized quantity and concentration are tender coconut water 1L, sugar-150 g (15% of tender coconut water) and china grass- 10g (1% of tender coconut water).

Tender coconut water is heated in a sauce pan with sucrose and china grass. Care should be taken to continuously mix the content during heating with a stainless steel spoon/ladle to melt the china grass in the tender nut water. Once it is completely melted, remove from the heat, cool it and pour in a wide mouthed vessel/tray and keep inside the refrigerator for about 3 hrs to solidify. After solidification, cut the pieces in cubes or squares and serve along with ice cream/ any other desserts as toppings.

Products derived from tender coconut kernel

Preservation of tender coconut kernel

Tender coconut kernel is a good source of carbohydrate, fiber and other nutrients. Protein content is high in the eight months old fresh coconut meat. Products such as tuty-fruity, candy, preserve and chips can be prepared from the fresh kernels. Tender nut kernel is made into pieces, mixed with cane sugar and subsequently drained and dried are called candied fruits.

Tender coconut is washed and split open to remove the water. Soft kernel is scooped out and cut into cubes. Pricking should be done with stainless steel forks. After pricking, immerse the fruit pieces in dilute lime water (1.5%) or alum (2%) for few minutes before further processing. Wash the pieces 3-4 times with fresh water and blanch for 5 minutes in boiling water to make them soft. This assists in absorption of sugar and prevents enzymatic browning. Spread sugar (50%) on the blanched pieces in alternate layers. Next day, drain the syrup and add enough sugar to raise the concentration of the syrup to about 60°Brix. Citric acid can be added as preservative. Process is repeated every day until the Brix of residual syrup reaches 70-75°. Then, drain the syrup and dry the pieces in hot air and store in glass bottles/polyethylene bags. To prepare crystallized candy the concentration of sugar syrup is continued till Brix value reaches 70-78°. Syrup is drained off and the pieces are rolled in finely ground sugar. Crystallized candy can be stored for 3 months (Lontoc *et al.*, 1973).

Canning of tender kernel

For canning, kernel from 8-10 months old nut is first scooped out, the adhering testa is removed by using a sharp knife and the pared meat is cut into stripes of 0.5 cm thick and 6 cm long after washing. The stripes are put in cans to which is added 50°B syrup with 0.01 per cent sodium metabisulphite. Filled cans are then exhausted at 78°C, sealed and processed at 110°C for about 20 minutes. A jelly like meat formed during the process is scraped out and to every part of the meat, a corresponding amount of refined sugar is added. Mixture is cooked in low heat until the sugar is totally dissolved, hot packed in sterilized bottles and closed tightly.

Tender coconut jam

Tender coconut jam is an intermediate moisture food prepared from the residual pulp left after removal of water from the kernels. It is a high-sugar coconut food product with light to dark brown in colour, thick and spreadable in consistency, with a rich creamy flavour. Coconut jam is prepared by boiling the pulp with sugar, pectin, acid, and other minor ingredients such as preservative, colouring, and flavouring materials, to a reasonably thick consistency. The desired amount of sugar is added to the pulp mixture and heated continuously under low flame. When the total soluble solids reach 60°Brix, pectin (1.25 %) and citric acid (0.5 %) are added to the boiling pulp and the mixture is stirred continuously using a steel ladle. Heating can be stopped when the total soluble solids is 67–68°Brix. The hot mixture is filled into sterilized glass bottles and cooled under ambient conditions. Prepared jam can be stored for a period of 6 months at ambient temperature without compromising the quality. Chauhan *et al.*, (2013) studied the organoleptic properties and shelf stability of mixed fruit jam and the combination of tender coconut pulp and pineapple pulp in the ratio of 75:25 resulted in a jam with good organoleptic and textural characteristics. Jam with increased palatability and sensory acceptability can also be prepared with pineapple pulp and guava pulp with tender coconut pulp.

Coconut pulp ice cream

Tender coconut pulp can be used in ice cream formulation. The product is free from dairy milk, lactose and cholesterol with low fat. Formulation includes the following ingredients: coconut pulp, cocoa powder, sucrose, water, carrageenan gum, guar gum and hydrogenated vegetable fat. Liquid ingredients and the pulp are blended and heated in a tank until the temperature is 45-50°C, when the powdered ingredients are added. Then, the mix is pasteurized at 87°C for two minutes. After 24 hour of ageing, the freezing-whipping step is accomplished using shaved surface heat exchanger. Finally, the product is kept at -5°C to complete the freezing stage and stored at -18°C. Most satisfactory product had 41% coconut pulp, 11% cocoa, 17% sucrose and 31% water. The product contains 65% water, 1.0% fat, 2.4% protein, 0.36% ash and 31.2% carbohydrate (Igutti *et al.*, 2011)

Tender nut pudding

The ingredients for the tender nut pudding followed in ICAR-CPCRI contains coconut milk (100 ml), coconut sugar (15 g), china grass (1%), tender coconut water (200 ml) and tender coconut pulp (50 g). Initially, the china grass is mixed in tender coconut water in a sauce pan and is heated till it completely melts in coconut water. Then, add coconut milk and coconut sugar into it. Heat the contents for 15 min and immediately pour in a pudding dish or a tray. Add tender nut pulp (preferably cut in the form of small cubes) into the pudding mixture and keep inside the refrigerator. Maximum time required for complete setting of tender nut pudding is 1 hr.

Tender coconut water lemonade

Coconut water lemonade is a refreshing drink made of tender coconut water and lemon juice with addition of flavouring ingredients. The product contains tender coconut water (500 ml), tender coconut pieces (100 g/1 cm³), lemon juice (15 ml), ginger juice (2 ml), pepper powder (0.5 g) and coconut sugar (10 g). Tender nut is cut open to collect the water and the pulp is scrapped and cut into uniform sized cubes. Tender nut water is mixed with lemon juice, ginger juice, pepper powder and sugar. Mix well using a hand mixer at a low speed for 2 minutes. Then, add tender nut cubes into the lemonade and serve under chilled condition.

Products derived from coconut inflorescence sap

Coconut sap popularly known as neera is highly prone to fermentation, and collection of unfermented sap is a challenging task. This has been resolved with the development of CPCRI developed 'Coco sap chiller'. The sap collected by coco-sap chilller at low temperature is observed to be entirely different from the neera collected by traditional method with or without preservatives; hence, it was christened as "Kalparasa". Sap collected using the coco-sap chiller is golden brown in colour, delicious and free from contaminants like insects, ants and pollen as well as dust particles.

Coco-sap chiller is a portable device characterized by a hollow PVC pipe of which one end is expanded into a box shape to house a sap collection container bound by ice cubes and the other end is wide enough to insert and remove a collection container of 2 to 3 litres capacity (Fig.13). Each side wall of the pipe from outside is covered with an insulating jacket excluding the portion of spadix holder which retains the internal cool temperature for a longer period. This coco-sap chiller is lighter in weight, water proof, easy to connect to the spadix, requires less ice and retains low temperature for longer period as compared to commercially available ice boxes.

Kalparasa collected by coco-sap chiller under low temperature meets the Codex Alimentarius (International Food Standards WHO/FAO) definition of juice as "unfermented but fermentable juice, intended for direct consumption, obtained by the mechanical process from extractable fluid contents of cells or tissues, preserved exclusively by physical means". Thus, it is amenable to be sold as fresh juice under local market with the adherence to quality standards prescribed by CPCRI. It does not require lot of machineries but requires cold chain or refrigerated system

Fig. 13: Coco sap chiller

CPCRI has developed simple quality standards to check the quality of sap. Fresh sap has a pH above 7 to 7.5. Depending on the pH, sap can be used for different purposes. pH >7 is ideal for health drink, Ph >6.5 is good for preparation of sugar, pH >6.0 is used for jaggery and pH >5.5 is used for concentrate. pH below 5.5 can be used for the preparation of vinegar. Other quality parameters easily judged are brix around 14; colour golden brown; and taste sweet and delicious.

Distinct differences are noticed between the sap collected by traditional method and CPCRI technique (Table 1 and Fig.14).

The collected sap can be stored for any length of time under sub-zero temperature. Deep freezers are used for the purpose. Sap gets frozen and just before use it is thawed to get the original liquid form. However, under refrigerators the quality gets deteriorated within few hours.

Table 1: Quality attributes of sap collected by CPCRI technique and traditional technique

Attribute	CPCRI technique	Traditional technique
Soluble solids (°Brix)	15.5 to 18	13 to 14
pH	7 to 8	6 or low
Colour	Golden brown or honey	Oyster white
Defects, decay, insects, pollen, dust	Absent	Present
Flavour	Sweet and delicious	Harsh odour
Pathogens, chemicals and extraneous matter	Absent	Present
Microbial load	Low	High

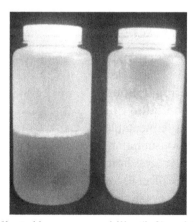

Fig. 14: Coconut sap collected by coco-sap chiller (left) and traditional method (right)

Techno economic details

Machineries/ devices required	Tapping gear (knives, tapping stick, scissor, mallet etc), o-sap chillers, neera collection ice box, ice carrying box, pH meter, measurement jug, neera storage container, neera transport box, freezer, neera dispenser etc.
Capacity	1000 liter of sap per day
Capital investment	Rs. 35,10,200
Operational cost per month	Rs. 75,875
Total cost of production	Rs. 1,06,91,785
Total sap production (l)	Rs. 3,65,000
Selling cost	Rs. Rs 50/ L
Unit cost of production	Rs. Rs. 29/L
Breakeven period	Rs. 176
Net profit %	Rs. 41.41

Coconut sugar

The hygienic, zero alcoholic sap collected by CPCRI method is easy to process in a natural way without the use of chemicals into various value added products which fetches premium price both in domestic and international markets. Very good quality coconut sugar, jaggery, nectar or syrup can be produced in double jacketed cookers with temperature regulation and stirring facility.

Coconut sugar (Fig.15) is the best natural sweetener with several health benefits and thus has a high market potential. It contains essential amino acids, minerals, electrolytes, dietary fibers and phenolics . Moreover, its glycemic index (GI) is low in the range of 35 to 54 GI/ serving.

Fig. 15: Coconut sugar

Techno economic details

Labour cost	Rs. 5,04,000
Total fixed cost	Rs. 20,19,975
Total variable cost	Rs. 1,01,72,500
Total cost of production	Rs. 1,21,92,475
Total sugar production (kg)	Rs. 54,750
Selling cost	Rs. 275/kg
Unit cost of production	Rs. 223
Breakeven period	Rs. 150
Net profit %	20

Kalpa bar

It is a coconut sugar based chocolate purely made from plant based ingredients without milk. It is a joint venture between ICAR-CPCRI and CAMPCO (Central Arecanut and Cocoa Marketing and Processing Cooperative Ltd.) (Fig. 16). It contains cocoa powder, coconut sugar, natural vanilla extract and GMO free sunflower lecithin. It is low in glycemic index. It is a delicious dark chocolate for a healthy life and can be stored under room temperature.

Fig. 16: Kalpa bar dark chocolate from coconut sugar

Kalpa drinking chocolate

It is an instantised blend of coconut sugar crafted from fine cocoa powder formulated to produce the delicious drinking chocolate (Fig.17). It is to titillate the taste buds of drinking chocolate lovers who want a healthier life style. The product is produced by a unique technology of instantisation and agglomeration technique that makes the product soluble instantly in hot or cold milk releasing the chocolate aroma.

Fig. 17: Kalpa drinking chocolate

Methodology for the preparation of fresh coconut inflorescence sap (Kalparasa) based milk sweets have been standardized at West Bengal (Fig 18). It is a way of transporting neera to long distance in the form of sweets. These sweets impart minerals, vitamins, valuable fiber which will not be available in normal cane sugar based milk sweets and their glycemic index is low.

Thus, various value added products can be prepared from Kalparasa which have huge demand in both domestic and international markets. As these products are nutritious, they fetch premium price which will in turn empower farmers/ growers who are dependent on coconut for their livelihood.

Fig. 18: Sweets prepared from Kalparasa

Diabetic friendly cookies

Diabetic friendly cookies are made with whole wheat, desiccated coconut or grated coconut and neera jaggery. Different types of cookies are possible by varying the main contents like oats, multigrain, arrow root, corn, whole wheat and spices. Cookies made with neera jaggery have a low glycemic index (GI 35). Ingredients to prepare wheat based cookies are wheat flour (5 kg), butter (3.25kg), powdered jaggery(5 kg), grated coconut/desiccated coconut powder (1.87 kg), baking powder (0.200kg), vanilla essence-(0.100kg) and salt.

Techno economic details

Machineries required	Oven and mixing unit
Cost of ingredients/cookie	Rs. 3.55/-
Selling price of cookies	Rs. 85/ packet of 7 cookies
Land	10 cents
Building (2000 sq. feet @ Rs. 1000/sq. feet)	20 Lakhs
Other civil works (internal roads, compound wall, water tanks+ neera jaggery making unit)	Rs. 2 Lakhs
Machinery and equipment	Rs. 13.49 Lakhs
Electrification	Rs. 0.50 Lakhs
Preliminary & pre-op expenses	Rs. 1.11
Working capital margin	Rs. 0.90
Net profit after tax on sales	Rs. 17.20 %
Payback period	3 yr 10 months
Selling price	Rs. 85 per pouch
Breakeven point (sales)	60.96%

(*Source:* Coconut Development Board)

New technologies/ innovations in value addition of coconut

Kalpa krunch

Kalpa krunch is a coconut milk residue enriched ready to eat extruded snack. It is prepared from 60% rice flour, 25% corn flour and 15% coconut milk residue (CMR) flour (Fig 19). It is coated with natural and healthy flavours. The flavours are formulated from ten different types of spices and vegetables including coriander, garlic, turmeric, clove, cinnamon, chilli, mint, cardamom, tomato and celery. Kalpa krunch is rich in dietary fiber, protein, fat and carbohydrate with antioxidant activity. The steps involved in extrusion process are mixing, extrusion (140°C extrusion temperature and 220 rpm screw speed), drying (130°C for 20 min), flavour coating and packaging. The torque should be maintained around 12-14 for uniform and high expansion ratio.

Mix all the raw materials in a laboratory mixer (Basic Technology Pvt. Ltd., India) for 15 minutes. Determine the initial moisture content of blend using infrared moisture analyzer. Spray calculated amount of water onto the flour blend before extrusion so as to achieve the required moisture content of 14% and blend again for 10 min.

Fig. 19: Kalpa krunch

Extrude the prepared homogenous blends in a co-rotating twin screw extruder (Basic Technology Private Ltd, Kolkata, India). An extruder die with a diameter of 3 mm can be used for this purpose. The screw speed and barrel temperature of the last zone will be 220 rpm and 140°C. Collect the extrudates after 5 min of steady state processing and dry in a coating machine (M/s Pharma Fab Industries, Mumbai, India) at 130°C for 20 min. The dried extrudates can be coated with different flavours. The oil should be sprayed before coating flavours.

Techno economic details

Cost of machinery	Rs. 44,00,000
Working capital	RS.47,80,000
Selling cost	Rs. 5/packet
Unit cost of production	Rs.3
Breakeven period	Rs. 131.9
Net profit %	21

VCO cake based muffins

Muffin batter formulations can be made by progressively replacing the refined wheat flour with VCO cake. The optimized formulation consist of refined wheat flour (26 g/100g) which can be replaced with 40% VCO cake flour, sugar (26%), egg (21%), full fat milk (13%), shortening (12%), sodium bicarbonate (1.1%) and salt (0.1%) (Fig. 20).

Muffin incorporated with 40% VCO Cake

Muffin made of refined wheat flour (Control)

Fig. 20: VCO cake based muffins

Conclusion

Coconut has the greatest importance in the national economy as a potential source of employment and income generation among the plantation crops. The demand for coconut is high because of its usage and the adaptability of coconut palm to grow under various climatic and soil conditions. With the use of coconut oil in the production of soap and margarine in Europe in the 19th century, it was converted into a commercial crop. In the beginning of 20th century copra was the king among the oil seeds. In East Indies it was known as green gold. However, the period after the Second World War saw the substitution of vegetable oils and oleo chemicals for coconut oil in international trade. Price of coconut oil fluctuated heavily due to frequent short supply situations. A campaign against coconut oil alleging that it causes cardiovascular diseases aggravated the situation. The newly industrialized countries in the East such as Taiwan, South Korea are fast emerging as key importers of coconut products. One of the main reasons for the fall in price of coconut and its products is dependency of price of coconut oil which again depends on the cost of other vegetable oils. Thus, product diversification of coconut and development of value added products become very important in the coconut industry. Effective market promotion activities are to be organized by way of organizing exhibitions, workshops and trade fairs in order to create consumer awareness and boost the demand of coconut products to keep the wheel of the coconut industry moving fast for doubling the income of coconut farmers for their sustainable livelihood.

References

Anurag, P., V. G. Sandhya & T. Rajamohan. 2007. Cardioprotective Effect of Tender Coconut Water, *Indian Coconut J.* 37: 22–25.

Baldioli, M., M. Servili., G. Perretti & G. F. Montedoro. 1996. Antioxidant Activity of Tocopherols and Phenolic Compounds of Virgin Olive Oil. *J. American Oil Chemists' Society* 73(11): 1589-1593.

Bhagya, D. & G. Soumya. 2016. Effects of Coconut Neera (*Cocos nucifera* L.) on Blood Pressure among Hypertensive Adult Women. *International J. Applied & Pure Science & Agriculture* 2(9): 1-6.

Chauhan, O. P., B. S. Archana., A. Singh., P. S. Raju & A. S. Bawa. 2013. Utilization of Tender Coconut Pulp for Jam Making and its Quality Evaluation during Storage. *Food & Bioprocess Technol.* 6(6): 1444-1449.

Duarte, A. C. P., M. A. Z. Coelho & S. G. F. Leite. 2002. Identification of Peroxidase and Tyrosinase in Green Coconut Water. *Cienciay Tecnología Alimentaria* 3(5): pp.266- 270.

Fife. 2011. Coconut Water: A Natural Rehydration Beverage. *Cocoinfo international* 18(2): . 9-11.

Fonseca, A. M., F. J. Q. Monte., M da Conceic & F. de Oliveiraaao.2009. Coconut Water (*Cocos nucifera* L.) – A New Biocatalyst System for Organic Synthesis. *J. Molecular Catalysis B: Enzymatic* 57: 78-82.

Garfinkel, M., S. Lee., E. C. Opara & O. E. Akwari. 1992. Insulinotropic Potency of Lauric Acid: A Metabolic Rationale for Medium Chain Fatty Acids (MCF) in TPN Formulation. *J. Surgical Research* 52(4): 328-333.

Gonzalez, O. N. 1986. State of the Art: Coconut Utilization for Food. In: *Proc Philippine Coconut Research and Development Foundation (PCRDF) Planning and Workshop*, Los Banos, Philippines, pp. 3–4.

Hagenmaier, R., K. F. Mattil & C. M. Cater. 1974. Dehydrated Skim Milk as a Food Product: Composition and Functionality. *J. Food Science* 39(1): 196-199.

Haseena, M., K. V. K. Bai & S. Padmanabhan. 2010. Post-Harvest Quality and Shelf-Life of Tender Coconut. *J. Food Science & Technol.* 47(6): 686-689.

Hebbar, K. B., M. Arivalagan., M. R. Manikantan., A. C. Mathew., C. Thamban., G. V. Thomas & P. Chowdappa. 2015. Coconut Inflorescence Sap and its Value Addition as Sugar – Collection Techniques, Yield, Properties and Market Perspective. *Current Science* 109(8): 1411-1417.

Igutti, A. M., A. C. I. Pereira., L. Fabiano., A. F. S. Regina & E. P. Ribeiro. 2011. Substitution of Ingredients by Green Coconut (*Cocos nucifera* L.) Pulp in Ice cream Formulation. *Procedia-Food Science* 1: 1610-1617.

Lim-Sylianco, C. Y. 1987. Anticarcinogenic Effect of Coconut Oil. *The Philippine J. Coconut Studies* 12: 89-102.

Lontoc, A. V., O. N. Gonzalez & L. Dimaunahan. 1973. Development of Storage Characteristics of Canned Coconut Syrup. *Philippines J. Nutrition* 26(3-4): 243-252.

Maier, J. J., B. Gallwitz., S. Salmen., O. Goetze., J. J. Holst., W. E. Schmidt & M. A. Nauck. 2003. Normalization of Glucose Concentrations and Deceleration of Gastric Emptying after Solid Meals during Intravenous Glucagon-like Peptide 1 in Patients with Type 2 Diabetes. *The J. Clinical Endocrinology & Metabolism* 88(6):2719-2725.

Manikantan, M. R., T. Arumuganathan., M. Arivalagan., A. C. Mathew & K. B. Hebbar. 2015. Coconut Chips: A Healthy Non-Fried Snack Food. *Indian Coconut J.* 58(2): 34-36.

Muralidharan, K. & A. Jayashree. 2011. Value Addition, Product Diversification and By-Product Utilization in Coconut. *Indian Coconut J.* 54(7): 4-10.

Rajagopal ,V. & A. Ramadasan. 1999. *Advances in Plant Physiology and Biochemistry of Coconut Palm*. APCC, Jakarta.

Reddy, E. P. & M. Lekshmi. 2014. Coconut Water -Properties, Uses, Nutritional Benefits in Health and Wealth and in Health and Disease: A Review. *J. Current Trends in Clinical Medicine & Laboratory Biochemistry* 2(2): 6-18.

Rethinam, P. & T. B. N. Kumar. 2001. Tender Coconut –An Overview, *Indian Coconut J.* 32: 2-22.

Rundorn, W., J. R. Connor., F. Saleh., N. Radenahmad., B. Withyachumnarnkul & K. Sawangjaroen. 2009. Young Coconut Juice Significantly Reduces Histopathological Changes in the Brain that is Induced by Hormonal Imbalance:A Possible Implication to Postmenopausal Women. *Histology & Histopathology* 24(6): 667-674.

Sanchez, P. C. & P. M. Rasco. 1983. Utilization of Coconut in White Soft Cheese Production. *Research at Los Banos* 2(2): 13-16.

Seow, C.C. & K. M. Leong. 1988. Development and Physical Characterization of Artificial Milk (Santan) Powder. *Food Science & Technol. in Industrial Development* 1: (edited by P. Maneepun ., Varangoon & B .Phithakpool). pp. 54-60. Bangkok: Institute of Food Research and Product Development, Kasetsart University.

Vala, G. S. & P. K. Kapadiya. 2014. Medicinal Benefits of Coconut Oil. *International J. Life Sciences Research* 2(4): 124-126.

13

Scope of Entrepreneurship Development in Non-edible Value Added Products of Coconut

Manikantan M R, Pandiselvam R, Shameena Beegum & Mathew A C

Introduction

Coconut is a versatile crop with several uses to mankind. Apart from the main product, the by-products obtained from the coconut crop have many alternative uses, thus, adding to the total value of the crop (Popenoe, 1969). A tremendous scope exists for use of coconut in a variety of non-food products. The development of cottage industries to produce such products is recommended to increase income of coconut growers. For historical reasons, cultivation of coconut and value addition of non-food products from coconut have taken deep roots in the state of Kerala. The rapid expansion of coconut cultivation in non-traditional areas increased the production of coconut and the industry has also developed gradually in the states of Tamil Nadu, Karnataka, Andhra Pradesh and Orissa. Among the non-food products of coconut, charcoal, activated carbon, and shell powder and coir or coconut fibre assume commercial importance. Other parts of the palm especially coconut wood and leaves are recently gaining attention.

Scope of entrepreneurship development in coconut by-product sector

Large number of people from the economically weaker sections of the society depends on coconut based non-food industry. Production of value added products is less with respect to the availability of raw material and market requirements. At the current level of production of activated carbon, charcoal and coir, the industry utilizes about 40% of the annual yield of coconut by-products in the country. There is possibility to increase the utilization to at least 60%. Therefore, there exists vast potential for stepping up of production of non-edible value added coconut products in India. The increased utilization of coconut husk, coconut shell and wood in the coconut growing states of India provides scope for development of fibre processing and charcoal/activated carbon processing sector and thereby augmenting rural employment. For example, the coir products like mats, rugs, carpets, cordages, ropes, fishing nets, etc are having both domestic and export demand (Gopal and Gupta, 2001). By proper utilization of coconut husk, the coconut farmers could augment the farm level income and employment. It has been shown that value addition of coir fibre, enhance income of fibre manufacturers by minimum of 20% and consequently increase the income of coir workers in fibre extraction units by minimum of 10% (Anonymous, 1960).

Value added products of coconut

Charcoal

In the developed world, charcoal is an almost indispensable industrial commodity, especially in metallurgy and as an adsorbent. With the development of the chemical industry and the increasing legislation concerned with control of the environment, the application of charcoal for purification of industrial waste has increased markedly. In the barbecue fuel market, charcoal has little competition and in almost all other applications charcoal could be substituted by coal, coke, petroleum coke or lignite. Charcoal produced from coconut products are listed below.

Coconut trunk charcoal

Coconut trunk and other saw mill residues are readily usable for charcoal making and for the production of energy (Romulo and Arancon, 2009). Coconut wood is similar to other woods in its characteristics as fuel, although the range of densities within the stem leads to variation in the energy potential (Anonymous, 1985). Charcoal and charcoal briquettes have higher heating value. They are easily handled and produce less smoke compared to wood.

For fuel purposes, coconut trunk charcoal (Fig.1) must be converted into briquettes to increase its strength and density as well as to improve its shipping properties. The briquettes have good crushing strength and burning properties. Sorghum grain is an effective binder for charcoal briquettes of coconut trunk.

Fig.1: Coconut wood charcoal
(*Source:* http://www.thegreenhead.com/2013/04/afire-koko-all-natural-coconut-charcoal.php, Accessed on 22-03-2017)

Coconut shell charcoal

Most important produce derived from shell is charcoal. Coconut shell charcoal is recognized as one of the best fuels for cooking because of its pleasant smell. Yield of shell charcoal is about 30% of the weight of the shells used, and it is generally reckoned that about 17,000-24,000 whole shells makes one metric ton of charcoal (Anonymous, 1969). In general, shell charcoal is made by burning coconut shells in a limited supply of oxygen. Coconut shell used should be clean, fully dried and mature in order to get high quality charcoal. Among the different methods of producing coconut shell charcoal, the pit method and drum method are most widespread.

In pit method, the shells are often burnt in the pits (Fig.2a). Some dry shells, clean and free from adhering fibres of the husk, are placed at the bottom and set on fire. When the shells emit flame, they are slowly piled together, and more and more shells are added until the whole pit is filled. Pit is then covered with a zinc or iron plate, and the hole is made airtight by packing earth around the edges. It is preferable to use fire-resistant bricks for the lining, but, locally made bricks will stand up for a considerable time. Mud mortar is found to be more satisfactory than cement. Circular pits which are narrow at the top compared to the bottom or bottle-shaped pits are preferable as the firing is more easily controlled.

The drum kiln (Fig.2b) is one of the most widespread methods of coconut shell carbonization. It has 3 sets which consist of 6 holes, middle and upper layers and a lid. The drum also includes a chimney which is placed on the lid of the drum (Fig. 2c). Optimum carbonization of shell in a limited supply of oxygen provides good quality charcoal.

Raw shells must be put into the drum, leaving a 4 inch space in the centre of the drum, which plays an important role during carbonization allowing the flow of smoke. To start carbonization, a fire should be started in the middle of the circle using a piece of a coconut shell. Then, all the free space in the drum should be filled with raw materials. When the flame flares up, the chimney and the lid should be attached. The middle and the upper sets must be closed. When some of shells shrink to the bottom, more coconut shells should be added up to the brim of the drum. The process of carbonization begins at the bottom of the drum and goes up. When the carbonation is finished, a glow in the 6 holes of a set of the particular zone can be seen. When the holes of the bottom zone indicate this glow it means that the bottom set is closed, while the middle set is opened. When the carbonization in the middle zone is completed, its holes are closed and the bottom holes are opened. Closing of the top set of the holes after the full carbonization in the top region stops the airflow into the drum. The charcoal is collected and packed after about 8 hours when the drum is cooled. In general, 30 thousand coconut shells are needed to produce 1 ton of charcoal using drum method. Composition of coconut shell charcoal and coconut husk charcoal is given in Table 1 and economics of charcoal production is given in Table 2.

Table 1: Composition of coconut shell and coconut husk charcoal

Particular	Coconut shell charcoal	Coconut husk charcoal
Calorific value (MJ/kg)	27.0 – 31.8	26.0 to 27.0
Fixed carbon (%)	80.6 -88.5	76.6 – 80.0
Volatile matter (%)	11.6 -14.80	8.0 -10.0
Ash content (%)	3.0 – 4.7	7.0 -12.0
Moisture content (%)	2.0 – 3.5	3.0 – 5.5

(a) Pit method   (b) Drum method

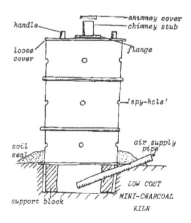

(c) Structure of drum

Fig. 2: Coconut shell charcoal production methods
(*Source:* http://ukrfuel.com/news-how-to-make-coconut-shell-charcoal-39.html, Accessed on 22-03-2017).

(a) Carbonization   (b) Milling and sieving

(c) Packing in sacks
Fig. 3: Charcoal process flow chart
(*Source:* http://charcoalshell.blogspot.in, Accessed on 22-03-2017).

The processing of coconut shell charcoal briquettes includes the carbonization of the coconut shells and crushing which allows manufacturing different shape and sized charcoal briquettes (Fig.3). For this purpose, charcoal drum or stove and the charcoal powder making machine are used. Drum or the stove should maintain the temperature of 500 - 900°F.

Table 2: Economics of charcoal production one ton / day (30,000 coconut shells)

Items	Amount (Rs)
Land (cost variable)	35 cents
Building 1000 sq. ft.	2.5 lakhs
Plant machinery	10.5 lakhs
Preliminary & pre-operative expenses	2.0 lakhs
Contingencies	0.5 lakhs
Margin money for working capital	2.0 lakhs

(*Source:* http://coconutboard.nic.in/charcoal.htm, Accessed on 22-03-107)

Prior to the industrial revolution, charcoal was occasionally used as a cooking fuel. Historically, charcoal was used in great quantities for melting iron in bloomeries and later blast furnaces and finery forges. Charcoal can be used for the production of various syngas compositions. The syngas is typically used as fuel, including automotive propulsion or as a chemical feedstock. Charcoal may be used as a source of carbon in chemical reactions. It is mainly used for the production of carbon disulphide through the reaction of sulphur vapours with hot charcoal.

Charcoal may be activated to increase its effectiveness as a filter. Activated charcoal readily adsorbs a wide range of organic compounds dissolved or suspended in gases and liquids. In certain industrial processes, such as the purification of sucrose from cane sugar, impurities cause an undesirable colour, which can be removed with activated charcoal. It is also used to absorb odours and toxins in air. Charcoal filters are also used in some types of gas masks. Charcoal is also used in drawing and making rough sketches in painting.

Coconut Shell Charcoal

Coconut Shell

Granulated Activated Carbon

Powdered Activated Carbon

Fig. 4: Activated carbon from shell
(*Source:* http://sakthi-coir.com, Accessed on 24-03-2017)

Activated carbon

Activated carbon is a carbonaceous, highly porous adsorptive medium that has a complex structure composed primarily of carbon atoms. Coconut shells are mainly used to manufacture activated carbon (Fig.4). Activated carbon plays a very important role in solvent recovery processes, water and effluent treatment and in treatment of flue gas before discharge into the atmosphere. The intrinsic pore network in the lattice structure of activated carbon allows the removal of impurities from gaseous and liquid media through adsorption. This is the key to the performance of activated carbon.

Chemical activation or high temperature steam activation mechanisms are used in the production of activated carbon. In the activation process, shell charcoal is fed continuously into a retort. Normal activation process involves the use of steam at selected temperature for the selective oxidation of material, resulting in production of carbon with pores of molecular dimension. Shell carbon, having a cellulose base produces material with a finer pore structure than obtained from coals. Approximately, three tons of shell charcoal is needed to produce one ton of activated carbon. Retorts designed to produce activated carbon usually operate in one of the three ways-vertically, horizontally, or by means of a series of hearths. Vertical retort utilises steam and activation is controlled by the rate at which the material is withdrawn from the discharge hopper. Activation can be carried out with a variety of gases, including oxides of carbon, chlorine, and mixtures of steam and air. After withdrawal from the retorts, the material is cooled and passed through a series of granulators and screens, thereby attaining carbon of a known quality, available in variety of grade sizes to suit many applications.

For certain specific purpose, different process is used to prepare the activated carbon. This process consists of the treatment of crushed coconut shell with surface active chemicals followed by drying and subjecting the material to carbonization. The carbonized material is activated with steam followed by air to facilitate oxidation.

The activated material is subjected to steam quenching to reduce the bed temperature and is then discharged in a receptacle. The material is subsequently subjected to acid treatment to adjust the pH value. Acid treated activated material is then washed with water, dried and stored. Granular activated carbon produced from shell charcoal is an important industrial material, and the prospects for the intermediate charcoal appear to be good as long as quality is maintained. In general, activated carbon is used where the compound to be absorbed has a small molecular diameter or, if it is a gas, when a boiling point is below 100°C. The use of this type of carbon is also specially indicated where the concentration of the absorbate is very low. Shell based activated carbon (Fig.5) is considered superior to those obtained from other sources because it is generally dense, very hard, and highly retentive. They have a very fine pore structure, and their rate of absorption is generally faster than coal carbon. Economics of activated carbon production is given in Table 3. Process flow chart for production of activated carbon is given in Fig.6.

Fig.5: Coconut shell based activated carbon
(*Source:* http://sjzkzcable.en. Coconut-Shell-Activated-Carbon for-Water-Treatment.html, Accessed on 24-03-2017)

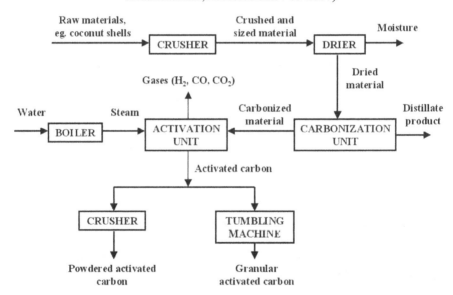

Fig. 6: Process flow chart for production of activated carbon from coconut shell
(*Source:* http://www.corecarbons.com/process.html, Accessed on 22-03-2017)

Table 3: Economics of activated carbon production

Items	Amount (Rs.)
Land	Two acres (cost variable)
Building -12000 sq.ft. @ Rs.600 per sq.ft.	72 lakhs
Plant & machinery	275 lakhs
Preliminary & pre-operative expenses	15 lakhs
Electrification	20 lakhs
Working capital (margin money)	50 lakhs

(*Source:* http://coconutboard.nic.in/activatd.htm, Accessed on 22-03-107)

Applications of activated carbon

- Activated carbon is playing an important role in the decolourization of sugar solution before it is crystallized to make granulated sugar in pure white colour.

- It is very useful in the manufacture of wine, alcohol, beer, rum, whisky, vodka etc. for purification of the ingredients as well as the final products. They are also used for decolourization, prevention of turbidity during ageing and for the removal of congeners that affect the taste or odour of the alcoholic beverages.

- The activated carbon is used in the fruit juice industry in the decolourization caused by complex compounds like polyphenols, melanoidins etc. They are also used to remove undesirable taste causing substances, colour changing chemicals that get added to the fruit juices during their manufacture.

- Activated carbon products are used to remove molasses compounds from citric acid, lactic acid and other forms of food substances made using bio-chemical processes. By using the activated carbon products it is possible to increase conversion of bio-chemical process rate during fermentation. Purification of gluconates and lactates is also achieved to the highest possible standards with the use of activated carbon products.

- Activated carbon is used in the manufacture of many of the starch based sweeteners like glucose, maltose, fructose, dextrose etc. They are mainly used for decolourization, polishing of syrups to comply with the highest standards and the requirements of the soft drink industry.

- Activated carbon is used for purification of natural glycerin. It is extensively used for the removal of organic impurities like odour causing substances and coloured matter to ensure the purity of glycerin that is necessary for other processes.

- Activated carbon products are finding their great use in the conversion of edible lactose into pharmaceutical grade lactose by effective removal of riboflavin from lactose.

Shell flour

The coconut shells are available from all coconut producing states in India. Kerala, Tamil Nadu, Karnataka and Andhra Pradesh contribute more than 90% of the production in the country (NIIR Board of Consultants and Engineers, 2012). A second important product derived from shell is shell flour. It is prepared by grinding clean coconut shells to a fine powder, the particle size depending on the end use. ISI specifications for coconut shell powder are given in Table 4.

Table 4: ISI specifications for coconut shell powder

Parameters	Specification
Appearance	Clear light brown free flowing powder
Moisture	Upto 10 per cent
Apparent density	0.6 to 0.7 g/cc
Ash content	Upto 1.5
Sieve analysis	Retained on 200 mesh sieves not to exceed 0.1%

Process

Coconut shells free from contamination are broken into small pieces (5 cm pieces) with the help of Jas Pounding Machine and fed into a pulverizer. The powder from the pulverizer is fed into a cyclone, where they are separated into coarse and fine particles and the ultra fine particles are collected in bag filters. From the cyclone, the coarser particles pass to the second hammer mill, and the ground products are subjected to the same air separation as the particles from the first grinding. The fine particles from the cyclone are fed into a vibrator-sieving unit and graded into the required mesh size for various end uses. The rejects from the sieving machine is recycled in the pulverizer for size reduction. Important requirements for consistent good quality coconut shell powder are proper selection of shell of proper stage of maturity and efficient machinery. Waste granules are used as fuel in solvent factory, sugar industries and boiler units. About 10 tons of coconut shell gives 7 – 8 tons of shell powder. The price is determined based on sieve size (grade) and it ranges from Rs. 4.50/kg for 80 mm grade to Rs. 8/kg for 300 mm grade shell powder. Process flow chart for the production of coconut shell powder is given in Fig.7. Details pertaining to the economics of production are furnished in Table 5.

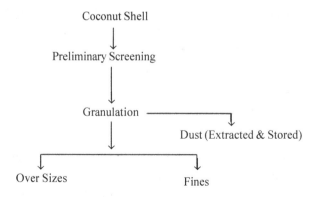

Fig. 7: Process flow chart for production of coconut shell powder
(*Source:* http://cpreec.org/77.htm, Accessed on 23-03-2017)

Table 5: Economics of coconut shell powder

One ton / day capacity (12000 shells yield one ton of shell powder)	
Land required (cost variable)	40 cents
Building (2000 sq. ft. building area)	Rs.10 lakhs
Plant and machinery	Rs.14.5 lakhs
Preliminary & pre-operative expenses	Rs.3.0 lakhs
Contingency	Rs.1.5 lakh
Working capital (margin money)	Rs.5.0 lakhs

(*Source:* http://coconutboard.nic.in/shelpwdr.htm, Accessed on 23-03-2017)

Uses of coconut shell powder

The shell powder is obtained as by-product from coconut oil industries and individual households (Fig.8). The powder has various uses as a filler in synthetic resin glues, filler and extender in phenolic moulding powders, mosquito repellent coils, mastic adhesives, resin casting, bituminous products etc., (Anonymous, 2015). Coconut shell powder finds its application in manufacturing mosquito coils as a burning medium (Fig. 9).

Coconut shell powder is used as thermo set moulding powder such as phenol formaldehyde moulding powder or bakelite and synthetic resin glues. Coconut shell powder of a particle size of 90-100 mesh is suitable as filler in the thermo set moulding powder and powder of 200-300 mesh size is used for synthetic resin glues. The demand in this sector is always on the increase (Resmi, 2015).

It is used mainly as filler, replacing wood flour either partially or wholly in the manufacture of phenolic moulding powders by the thermoplastic sector. The inclusion of shell flour results in an improvement in the surface finish of the mouldings,

Fig. 8: Coconut shell powder

Fig. 9: Use of shell powder

and because of its higher resinous content and lower absorption properties, it can be used in higher concentrations than wood flour. Shell flour is also used as filler in phenolic glues for plywood and laminated sheet manufacture, filler for mosquito incense coils and filler in specialized surface finishes, resin castings, etc. As a mild abrasive, it is used as a soft blast to clean piston engines. It has been incorporated into hand cleaners and used as a diluent for potent insecticides (Grimwood, 1975).

Coconut husk

Coconut husk is one of the important by- products of coconut tree and coconut-based activities. Husk is the outer fibre (35%) of the nut, followed by the hard protective shell (12%). Dehusking of coconut is done at various stages in the marketing network. Except the husk obtained during dehusking at household level, the entire husk, including the unorganized marketing sector, reaches the coir industry, where it fetches a market value (Anonymous, 1970). In the coir industries, fibre is extracted from coconut husk. The thickness of the husk of an ordinary nut varies from 2.5-3.0 cm in the case of thin-husked nuts and 4.0-5.0 cm for thick husked ones.

Husk is a useful source of potash and valuable mulch for the conservation of moisture. These are often burnt to produce ash which is used to fertilize the trees. Burying the husk in the soil is more beneficial than burning. A layer of husk is placed in a ring, convex side upwards from about 0.3 m up to a distance of 1.8- 2.1 m from the base of the palm. This method is beneficial during period of drought. Husk can also be used in planting holes during coconut seedling transplantation.

Coir industry in India is one of the important rural industries. There are two methods of processing the coconut husk: namely, the manual and the mechanical processes (Fig. 10). The manual process is simple and no investment is needed for equipment.

Utilization of coconut husk

Coconut husk is generally removed from whole coconut at the farm site in close proximity to the trees from which they are harvested. After dehusking, the husks are piled and left to rot in the fields or normally burnt as waste. A greater portion is used as fuel in farm site copra making. To a certain extent, husks are also utilized in handicrafts, floor polishers and other minor applications.

Fig. 10: Extraction process of coir fibre
(*Source:* http://swapsushias.blogspot.in/2014/09/focus-coir-board-of-india.html#.
WNN-oNR96t8, Accessed on 23-03-2017)

i. Coconut husk particle boards

Husk of matured coconut is the unique raw material to prepare particle boards (Fig.11). Usually, wood particle boards use 8-10% adhesives on weight basis, while coconut husk boards require only 0.25 % adhesives.

Fig. 11: Particle board from coconut husk
(Source: https://www.wur.nl/en/show/Ecocoboard-a-new-material-made-from-coconut-husk.htm, Accessed on 23-03-2017)

About 325 million coconut trees will yield 12 billion nuts from which about 4 million tons of particle board can be produced. Particle board is manufactured industrially by mixing coconut husk with resin and forming the mix into a sheet. Though, several types of resins are commonly used, formaldehyde based resin is the best in terms of cost and ease of use. Urea melamine resin or phenol formaldehyde resin is used to offer water resistance. Once, the resin has been mixed with the particles, the liquid mixture is made into a sheet. The sheets formed are hot-compressed under pressure in between 2 and 3 mega pascals and temperature between 140°C and 220°C. This process sets and hardens the glue. Boards are then cooled, trimmed and sanded. They can then be sold as raw board or by improving the surface through the addition of a wood veneer or laminate surface. Important machineries required to manufacture particle board are illustrated in Fig.12.

Machinery required

1. Hydraulic hot press
2. Power generator set
3. Boiler
4. Aluminium or steel plates
5. Blender (mixer)
6. Mat formation machine
7. DD Saw machine
8. Dryer
9. Screening machine

(a) Hydraulic hot press   (b) Saw machine

(c) Boiler

Fig. 12: Machineries required for manufacturing particle board

ii. Coconut fiber-cement board (CFB)

The coconut fiber-cement board (CFB) is relatively a new product (Fig 13 a) that makes use of coconut waste and can be combined with coconut wood (Anonymous, 2015). It is manufactured from fibrous materials like coconut coir, fronds, spathes, coconut top logs, or even shredded wood from small diameter fast-growing trees growing along the borders of coconut plantations. Manufacturing CFBs can be a good investment for the suppliers of construction materials, building contractors and private agencies involved in building low-cost houses (Fig 13b). The cement-fiber mixture is formed into mats and pressed to the desired thickness.

Fig.13a: Coconut fiber-cement board

Fig. 13b: House made of CFB and coconut **wood**

iii. Production of coir in industries

Coconut husk is the basic raw material of coir industry. At present, only 35% of total husk available is utilized by the industry while there is scope for utilizing at least 50% of the husk produced in the country. Husk of one coconut gives 90 g of coir fibre and 180 g of coir pith (Krishnamurthy *et al.,* 2009). There are two distinct varieties of coir namely white fibre and brown fibre (Fig.14). White fibre is extracted from retted coconut husk. Kerala produces mostly white fibre, which is used for making traditional coir products like mats, matting, rugs and carpets, which have an export market. Brown fibre is extracted from unretted husk. It is mainly used for the manufacture of curled coir. Curled coir is used in the rubberized coir mattresses, sofa cushion, bolsters, pillows, carpet underlay etc. The bristle fibre is a thick and long variety and is used for brush making. The mattress fibre which is a shorter staple fibre finds use in the upholstery, mattresses etc., for stuffing purpose. White fibre is extracted from green husks by mechanical defibring process.

Fig. 14: White and brown coir fibre

Process of producing coir fibre

Method of producing coir fibre from coconut husk is divided into three categories: traditional manual method, semi mechanical method and modern mechanical method. Traditional method with little mechanization is used to prepare yarn fibre, while bristle and mattress fibres are usually prepared by semi mechanical or mechanical methods.

Traditional and semi mechanical methods

i. Production of yarn fibre

Southern India is the chief source of coir yarn fibre, although small quantities are also prepared in southern parts of Asia. It is produced by traditional labour-intensive methods, including the natural bacteriological process of retting, where by the husks are soaked in water until the pith decomposes. Coconut is harvested every forty five days or so and then husked immediately in usual manner. A pit measuring 2.5 X 1.25 m will contain 1000 husks. Husks are then retted for eight to ten months in coastal lagoons or back water which are quite and undistributed but have ebb and flow ensuring a constant change of water. When the husks are sufficiently soft, they are washed and squeezed in water to remove the mud and bad smell. After removing the tough exocarp, remaining fibre is placed on hard wood and beaten traditionally with a strong round rod of tamarind heartwood until all the pith is removed. Dry fibres are then beaten and put through a winnowing machine, consisting of a number of knives with saw like teeth fixed to a shaft which is rotated by hand in a drum which will remove any small particles or pith.

The non-retting process involves direct decortication or mechanical extraction. Fresh husk before drying gives white fibre and the dried and retted husk gives brown fibre. Yield from retted husk is more than that from unretted husk. Average yield of white fibre from 1000 full husks in India is estimated as 81 kg with an average of about 50 kg of brittle fibre and 100 kg of mattress fibre.

Fibre is now spun into yarn either by hand or by using a wheel. Spinning on a wheel gives better quality yarn with a hard twist that is suitable for the manufacture of matting and other such materials. Two wheels are required for spinning, one is stationary and carries two spindles driven by the centre wheel and the other is mounted on three castors and has one spindle (Grimwood, 1975).

Grading of coir yarn

Two main classes of coir yarn are hard-twist or machine-twist and soft-twist (Fig. 15) or hand-twist; the grade names come from the locality where it is produced.

Soft- twist yarns are classified as beach and vycome both of which are subdivided into different numbers which vary in prices. Beach yarn is used for making mats and finer variety. Vycome yarn is used mainly for matting manufacture (Grimwood, 1975). Better quality fibres can be obtained if the nuts are harvested before they are fully ripe and the husk processed without delay. Retting can be carried out in almost any area where stagnant water is available. Some of the more modern mills have their own well-built concrete retting tanks but these are very expensive. The tanks are generally built in series by each unit and measures roughly 8×2.7×1.8 m. The water may be changed by pumping from one tank to another or by adding from a reserve supply. Retting time is reduced in some factories by soaking the husk for 30 to 40 h in hot water during which time they are weighted down with boards and stirred frequently. Periodically, the foul water is run off and replaced by clean water at the same temperature. Before soaking, the husk is crushed between fluted rollers to facilitate the penetration of water through the exocarp. At the large mill, the total capacity of the tank is said to be 2, 00,000 crushed husks or 1, 40,000 normal husk. If the husk is crushed previously, retting in tank is completed within three to seven days, otherwise they require seven to ten days. Retting in pits where the husks are not crushed and the water is not changed will take three to six weeks depending on the position of the husk of the pits. Those which are completely submerged require a shorter time than those on the surface.

After retting, next stage is extraction of fibre from the husk. Extraction of fibre requires the breaking down and removal of both the connective tissue and pith between the fibre and the outer exocarp. The process is called milling, and is carried out with specially constructed machine called drums. They are usually arranged in pairs and one is called the breaker drum and the other is the cleaner drum. Husk segments are first treated at the breaker drum which consists of a wooden wheel of about 0.9 m in diameter with 0.3 m wide and 1 m long into which iron nails have been bolted 3.8-5 cm apart. Main part of the wheel is enclosed in a wooden guard or casting with an opening of about 30 cm wide and is protected by a pair of iron bars. Nails are replaced every two months, because of wear and tear, since worn nails tend to split and damage the bristle fibre. Lower part of the casing of the wheel takes the form of a chute, through which the extracted fibre is delivered to the ground below. As the wheel

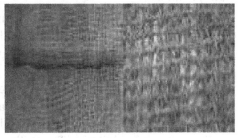

Fig. 15: Soft-twist yarn

revolves, first end of the husk and then the other are pressed between the bars. Nails tears away the short mattress fibre, which passes down the chute leaving the longer bristle fibre pith in the hands of the operator. It is generally accepted that one pair of drum can handle 2000 husks in a working day of eight hours and produce 100 kg of bristle and 200 kg of matress fibre. Therefore, a small mill with three or four pairs of drums should be able to produce 300-400 kg of bristle and 600-800 kg of mattress fibre per day.

Fibre grades

The fibres are used for spinning into yarn to manufacture mats and matting's, ropes, twines etc. Bristle fibre is long and stiff and is used for brushes and brooms. Coir is graded according to the colour and length of the fibre as also its refraction content. Four grades (Table 6) are recognized in India based on the specifications of Bureau of Indian Standards. First grade is mainly utilized for making superior quality fibre mats. Second grade constitutes fibre of white lustrous colour and the third slightly reddish or greyish coir containing pith. Fibre in fourth grade is mainly dark in colour and contains more pith and is used to make cheap yarn known as Beach yarn.

Table 6: Grading of coir

Sl No.	Grade	Maximum impurities	Length of fibre, proportion of medium and short fibres
1	I	2.0	70% by weight is long and remaining medium and short
2	II	3.0	50% by weight is long and remaining medium and short
3	III	5.0	30% by weight is long and remaining medium and short
4	IV	7.0	20% by weight is long and remaining medium and short

Coir yarn is woven into mats, matting's, carpets and rugs. Mats and matting's are woven on wooden handlooms; wheel spun yarn is used for warp and hand spun yarn for weft. Power-looms are seldom employed in India; on an account of coarseness of fibre, the yarn produced is coarse, and shuttles take only short length of yarn and need frequent replacements.

Coir products

Even though coconut palm is grown abundantly in all countries in the tropical belt, India and Sri Lanka stands first and second in the utilization of coconut husk for the manufacture and marketing of coir and coir products (Fig.16). India accounts for 71% of the world fibre production while Sri Lanka's share is 23%. Almost entire production of white fibre is from Kerala. Tamil Nadu, Karnataka, Andhra Pradesh and Orissa are the major brown fibre producing states. It is estimated that about 27% of the coconut husks produced in Kerala is utilized by the coir industry. The ability of coir yarn and ropes to withstand the prolonged action of sea water makes them especially suitable for use on boats

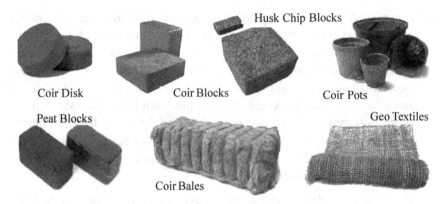

Fig. 16: Uses of coconut coir
(*Source:* http://www.usesofcoconut.com/benefits-and-uses-of-coconut-coir/,
Accessed on 23-03-2017).

and ships. Coir fibre has also been used successfully in the manufacture of shockproof packing materials, hard board suitable for tabletops, doors, and pencil and battery containers. Hard boards obtained possess an attractive glaze, high tensile strength and high density and are suitable for railway coaches and cuttings.

i. *Artificial animal hair:* Bristle fibre and decorticated fibre are boiled for one hour in caustic soda solution. Then, the fibre is immersed in a dye bath composed of direct black dye, soda ash and salt for two hours and then cooled for 12h. Then, the fibre is dried, polished with emulsion of paraffin,

Fig. 17: Coir product- brush

washed with soap and soda ash and twisted. By these processes simulated animal hair (animal – hair like fibre) is produced. This is used as filling material for upholstery.

ii. *Curling:* Mattress fibre, bristle fibre and decorticated fibre are twisted into ropes to produce curled fibre. Curling imparts special resilience to the fibre and the curl is permanent. Curled fibre is impregnated with rubber latex to produce rubberized coir, which is used for making car seats, filter pads, carpet underlay, cushion etc.

iii. *Flagging:* Ends of trimmed bristle fibre are immersed in a chemical solution and then split lengthwise by rotating pins or similar devices. This gives a soft feathery feel to the ends and improves sweeping efficiency of the brooms and brushes (Fig.17).

iv. *Spinning*: Mat variety of coir fibre is spun into yarn. In India, coir spinning is organized on a cottage industry basis and three methods of spinning are currently followed viz., (i) hand spinning, (ii) wheel spinning and (iii) mechanized spinning by using treadle operated machines. There are no power machines for spinning coir in India. Hand spun yarn is soft, and has even twist and thickness. Wheel spun yarn has higher strength and more uniformity in size and twist. Yarn produced by treadle-operated machine is less hairy, more regular in twist and has continuous length than wheel-spun yarn. Hand – twist yarn is used for matting and ropes. Soft-twist yarn is used for matting.

v. *Rope making*: Ropes and cordages are made out of coir yarn. Plain, hawser-laid and cable laid ropes are made in India.

vi. *Mats*: According to quality of yarn and method of weaving, mats are classified into three classes namely (i) coir mats (ii) fibre mats in which unspun coir is used for piles and (iii) speciality mats.

vii. *Mattings*: Patterns requiring up to 8 treadles are usually woven on ordinary looms. For designs requiring more than 8 treadles, Jacquard machines are employed.

viii. *Rugs*: These are mattings in rug sizes in attractive stenciled patterns.

ix. *Mourzouks*: Method of weaving mourzouks is different from that employed for matting. In this, special cross – weaving looms are used. Surface and patterns are formed by weft and not by warp. This method of weaving enables production of intricate geometrical and floral designs. Aloe and jute yarn may also be used as warp yarn.

x. *Carpets*: These are woven on matting looms. Warp strands are varied in thickness and number to produce thick and heavy fabric with a ribbed finish. Required design is secured by inserting coloured weft yarn.

xi. *Poly coir*: The Central Coir Research Institute of the Coir Board in collaboration with the Regional Research Laboratory, Thiruvananthapuram has developed "poly coir" which is made out of brown fibre. The coir felt, a non-woven material from coir fibre, is cut into appropriate width and coated with desired quantity of phenol formaldehyde resin by weight to form rolls of prepeg. The composite products from prepeg sheets are prepared by hot press moulding. The prepeg sheet is cut into the required size and stacked one over the other. The number of layers used is decided on the requirement of thickness of the component and the pressure applied for moulding which varies depending on the density and surface finish of the product. Trimming and polishing of the edges and wastage can be minimized by taking care of the size of prepeg sheet used for moulding. Major advantages of poly coir over plywood are (i) termite proof, (ii) water resistant, (iii) fire resistant, (iv) mouldable to desired shapes and (v) very good aesthetic appeal.

xii. *Coir matting decorative boards*: Coir matting, cut into required piece is treated with phenol formaldehyde resin and hot pressed to make the boards. The number of matting pieces can be suitably increased so as to make high-density boards.

Dyeing and printing of coir products

Colour and design are important for marketing coir products. For dyeing to bright tone, prior bleaching is necessary. Coir fibre may be bleached by SO_2 fumes by burning sulphur. Coir yarn may improve its colour and give it some amount of brightness. For dyeing of coir, different classes of dyes are used, viz, natural colouring matters like logwood, acid dyes, acid – mordant dyes, basic substantive dyes, sulphur dyes etc. Dyed yarn is exported from India to Australia for manufacture of matting.

Simple geometrical patterns and floral designs are printed on coir products by employing stencils and screens. Dye paste, resin thickness and chemicals to fix colours are applied through stencils. Dried prints are steamed for fixing. Fine designs of intricate configurations cannot be applied because of stiffness of coir fibre and roughness of coir products.

Geo-textiles

Coir geo-fabrics are woven coir nettings or mesh matting which is inexpensive, ready-to-use and effective items for a variety of applications including control of soil erosion and landslides, slope stabilization, seepage of water through canals and in other civil engineering applications like road embankments etc. In these applications, coir is used, because, it is natural, hard fibre with high tensile strength, durability and moisture resistance. Coir matting (mesh mat) are firmly laid on the slopes of canals, railway embankments, and road embankments and sown upon with grass seeds or slips are planted. With passage of time, the grass takes root and furnishes a permanent coverage, thereby stabilizing the soil. Coir matting also degrades and merges with the solid adding to nutrient content of the soil. Coir matting serves to hold the seed and soil intact, thereby preventing erosion during heavy rainstorms. It also serves to dampen the kinetic of flowing water and keep both soil and seeds in place.

Coir Pith

Coir pith constitutes as much as 70% of the husk and is now a waste product of coir industry. Accumulation of this waste in industrial yards causes environmental pollution and fire hazard. It is assessed that in India, 7.5 million tons of coir pith is produced annually. To obviate this problem, green technologies called coir pith composting units are promoted through formation of Self Help Groups. Every month, around 105 tons of coir pith is being converted to compost. (Resmi, 2015).

Uses of coir pith

i. In moisture conservation

The continuous application of coir dust will lead to a reduction in bulk density; improve the water holding capacity and organic carbon status of soil resulting in early flowering of palms. Coir pith has the ability to absorb and retain 10 times its weight of water (Tejano, 1984). Use of coir dust for a long period, will considerably improve the water holding capacity of the soil and the soil becomes more porous which allows better root penetration.

ii. Densification of composed coir pith

Tamil Nadu Agricultural University has developed a pelletizer for making pellets from coir pith compost. Compost is extruded into pellets of 6 to 8 mm diameter and 10 to 12 mm length. The unit is operated by a 5 hp electronic motor and had a capacity of 100 kg/h.

iii. As an energy source

a. Briquetting of coir pith

A continuous extruder type briquetting machine, consisting of screw shaft, barrel housing, extruder die pipe and gearbox has been developed at Tamil Nadu Agricultural University. The unit has a capacity of 125 kg per hour. Cow dung and molasses at various proportions to the coir pith were added (0, 10, 15, 20 and 25%) as binder. Briquettes produced had a calorific value in between 3000 to 3200 kcal/kg and cow dung mixed at 15% of coir pith resulted in better stability. This can be utilized as an alternative source of fuel.

b. Biogas

Pith can be added as a substitute in biogas generation, thus saving other energy producing materials.

c. Power gas

Coir pith can be used as a starting material for the production of gas by controlled combustion, which has been tried in industrial engines on a small scale

Miscellaneous products from coir pith

Many commercial products such as card boards, insulators, expansion joint filters etc. can be prepared using coir pith (Handreck, 1993). Polymers and composites can be prepared from pith by co-polymerising the lignin present in pith with either formaldehyde or phenols. Pith can also be used along with rubber to make composite flooring, ceiling floors and other similar products.

The pith in combination with cement has been found to be an excellent thermal insulating material. It is much lighter, easier to apply and gives much better thermal insulation for equal cost with lime concrete. Compared to modern methods, the cost is found to be only half to two-third. In the National Institute of Technology, Calicut, Kerala, coconut pith was successfully utilised in the production of a variety of light weight high strength bricks by the partial replacement of clay. In India, cashew nut shell liquid filled pith composite has been used as joint filler between concrete slabs in roofs, roads, and runway with a view to accommodate thermal movements. The pith joint fillers are resistant to alternate heating and wetting and also freezing and thawing. They are also resistant to termite and fungi and superior in qualities to those of bituminised fibre boards.

Coconut leaf

Coconut leaf is another product of importance for domestic use. Plaited and unplaited leaves are used for thatching houses, fencing and for making baskets. Leaves soaked in saline water before painting withstands climatic influences better than the unsoaked ones. Lifespan is only 1-2 years. Leaves being lignocellulosic in nature are susceptible to attack by sunlight, fungi, insects, rain and air. Thick walled sclerenchyma cells which impart mechanical strength to the tissues are relatively scarce in coconut leaf (Pillai et al., 1981). Mathew (2004) could extract a useful fibre from the leaves. Here, the fresh leaves are boiled in water and separated into upper and lower halves. Each half is made into strips of convenient width and again boiled in 5 to 8 percent sodium carbonate solution for one to two hours. After thorough washing, they are immersed in a bleaching solution for one to three days with periodic stirring. Then, they are washed and dried in shade. These strips, which form smooth, semi transparent, water-proof threads are excellent for making hats, bonnets, mats, bags and slippers.

Midribs of the leaves are used for stiff brooms, bird cages and lobster and fish traps. The petioles, bunch stalks, spathes and stipules etc. are used as fuel. Roots have medicinal properties and hence the decoction of the roots is used as mouth wash. Roasted roots can be used as a dentifrice.

Coconut leaf weaving

Coconut leaf weaving is a traditional method to make some useful items for house hold things (Fig.18). Coconut leaf can be utilized for making hats, head band, fans, mats, bracelets and baskets as well as decorative items such as roses, grasshoppers, whips and fish (Rivera and Reuney, 2010).

Fig. 18: Coconut leaf weaving

Conclusion

Coconut is one of the important fruit trees in the world, providing food for millions of people, especially in the tropical and subtropical regions. Apart from the main products, the by-products obtained from this crop have many alternative uses, thus adding to the total value of the crop. Products obtained add revenue to the farmers, processors and entrepreneurs who are involved in processing and marketing of the produce. Products obtained from the by-products are utilized in different sectors like food, feed and shelter, and to manufacture decorative materials. Thus, every part of the palm is beneficial to mankind and hence it is aptly described as "Kalpa Vriksha".

References

Anonymous. 1960. *Coir and its Extraction Properties*. Council of Scientific & Industrial Research, New Delhi.

Anonymous. 1969. *Note on the Market for the Coconut Shell Charcoal and Flour*. International Trade Centre, New York, US.

Anonymous. 1970. *The Preparation of Coir or Coconut Fibre by Traditional Methods*. Tropical Products Institute, London.

Anonymous. 1985. *Coconut Wood Processing and Uses*. FAO, UN, Rome.

Anonymous. 2015. *Innovative Technologies in Coconut Processing Sector*. Coconut Development Board, Ministry of Agriculture, Govt. of India.

Gopal, M. & A. Gupta. 2001. Coir Waste for a Scientific Cause. *Indian Coconut J.* 31(12): 13-15.

Grimwood, B. E. 1975. *Coconut Palm Products: Their Processing in Developing Countries*. FAO, Rome, P. 142.

Handreck, K. A. 1993. Properties of Coir Dust and its Use in the Formulation of Soilless Potting Media. *Communications in Soil Science and Plant Analysis* 24: 349-363.

Krishnamurthy, K. C., R. Maheswari., Udayarani & V. Gowtham. 2009. Design and Fabrication of Coir Pith Prequetting Machine. *World Applied Sciences J.* 7(4): pp. 552-558.

Mathew, M. T. 2004. Coconut Industry in India: An Overview. *Indian Coconut J.* 35(7): 3-17.

NIIR Board of Consultants and Engineers. 2012. *The Complete Book on Coconut and Coconut Products (Cultivation & Processing)*. Asia Pacific Business Press Inc., New Delhi, P. 472.

Pillai, C. K. S., P. K. Rahatgi & K. Gopakumar. 1981. Coconut as a Resource for Materials and Energy in Future. *J. Scientific Industrial Research* 40: pp. 154-165.

Popenoe, J. 1969. *Coconut and Cashew, North American Tree Nuts*. Humphrey Press, Geneva, P.115.

Resmi, D. S. 2015. *Major Coconut Products and Assistance under Technology Mission on Coconut*. Coconut Development Board, Kochi, India.

Rivera, M. N. & K. Reuney. 2010. Using Coconut Weaving in Guam Classrooms to Improve the Language Arts and Mathematics Skills of Local Students. *29th Annual International Pibba Conference*. Majuro, June 22-25.

Romulo, N. & J. Arancon. 2009. The Situation and Prospects for the Utilization of Coconut Wood in Asia and the Pacific. *Working Paper No. APFSOS II/WP/2009/15, Asia-Pacific Forestry Sector Outlook Study II*, FAO, UN, Regional Office for Asia and the Pacific, Bangkok, Thailand.

Tejano, E. A. 1984. State of the Art of Coconut Coir Dust & Husk Utilization (general overview). Paper presented during the *National Workshop on 'Waste Utilization, Coconut Husk'* held on November 12, 1984 at the Philippine Coconut Authority, Diliman, Quezon City, Philippines.

14

Entrepreneurial Opportunities in Tuber Crops Processing

Sajeev M S, Padmaja G, Sheriff J T & Jyothi A N

Introduction

In an era of liberalization of world trade where there is a continuous inflow and outflow of goods into and from a country, the need for developing competitiveness to sustain in global markets is extremely important. Technological progress and scientific advancements go hand in hand for the economic growth of a country and the Global Competitiveness index (GCI) which was revised in 2006 includes several factors such as institutions, infrastructure, health, primary and higher education, business sophistication, innovation, training etc. of nations to categorize them on the basis of competitiveness (Baig, 2006). Successful entrepreneurship development is a primary driver for the sustainable economic progress of any nation. The term 'entrepreneurship' has been defined by Cole (1968) as "'Entrepreneurship is the purposeful activities of an individual or a group of associated individuals undertaken to initiate, maintain or organize a profit oriented business unit for the production or distribution of economic goods and services". An entrepreneur is an enthusiastic individual committed to bring change in the society through the successful implementation of innovative ideas. However, bringing together a package of resources by an entrepreneur for effective transformation into a profitable venture may require inputs from several institutions such as research and development sectors, financing organizations, and training and education providers. The role of R & D institutions in providing innovative technologies having commercial value and the requisite capacity building for prospective entrepreneurs is of prime significance. In this context, ICAR-Central Tuber Crops Research Institute under the aegis of the Indian Council of Agricultural Research, New Delhi is a premier research organization dedicated to the research and development of tropical tuber crops and has developed a plethora of technologies for the processing and value addition of tuber crops.

Tuber crops, though branded as poor man's crops have remarkable unrealized potential for processing into novel and high value products for food, feed and industrial uses. Cassava, sweet potato, yams, aroids, minor tuber crops like coleus, tannia, yam bean, arrowroot etc. are the important sources of starch after cereals and are used as staple or supplementary food. They are the third most important food crops after cereals and grain legumes and have high dry matter production and capacity to withstand the vagaries of climate change. Having high adaptability to wide range of climatic conditions of the tropics and sub tropics and requiring minimum crop husbandry measures, tuber crops can be very well fitted into the prevailing cropping system. Among the tropical tuber crops, cassava and sweet potato are the most important and other tubers are grown as vegetable crops in homestead or semi commercial scale.

Cassava is cultivated in 20.73 million hectares, spread over the continents of South America, Africa and Asia, producing 276.72 million tons of tubers with a productivity of 13.35 tons per hectare and that in India it is 35.65 ton per hectare from an area of 0.23 million hectares amounting to 8.14 million tons of total production (FAOSTAT, 2017). Sweet potato is the second most important root crop with a world production of 106.60 million tons from an area of 8.35 million hectare (FAOSTAT, 2017) and china is the leading producer accounting for almost 80% of the global production. India ranks tenth position in sweet potato production in 2014 having an area, production, and productivity of 0.11 million hectares, 1.09 million tons and 10.28 t/ha, respectively (FAOSTAT, 2017).

The perishable nature of tropical tuber crops and the difficulties in long distance transport, storage and marketing constitutes major problems for their post harvest utilization. In order to overcome this problem, *in situ* value addition near the farm site is recommended. The produce will also ensure promotion of cottage and small-scale industries besides ensuring food security by incorporating tuber flour/starch to a certain extent in various food preparations. Central Tuber Crops Research Institute, a pioneer in the R&D activities of tropical tuber crops evolved number of value addition technologies suitable for home, farm and industrial front. Technologies are available for making fried chips with good colour and texture, rava similar to wheat semolina, quick cooking dehydrated tubers etc. Tuber flour is a major ingredient in composite formulations to produce several fried products, extruded products, pasta etc. Starch, besides being used as an industrial raw material in paper and textile industries, can be used for making sago, wafers, alcohol, adhesives, glucose, biodegradable plastics, modified starches, super absorbent gels etc. Agro-industrial transformation of these crops by linking improved production and processing technologies, marketing techniques and institutional innovation in processing technologies ensure food security, entrepreneurship development, rural employment and adequate remuneration to the producers. Various innovative and low cost value addition technologies are available in tuber crops suitable for the micro and macro level food ventures to elevate the status of these crops from subsistence level to a commercial commodity. This will in turn help the grass root level to generate additional income and employment; helping them to make the rural population self reliant by providing remunerative prices to their bio-produce and value added products.

Health benefits of tuber crops

Most of the tuber crops have one or more functionally active principles, which can help combat various diseases. Tuber crops such as sweet potato, yams and chinese potato offer immense scope as health protectants, therapeuticals and biocolours. Tuber crops contain an array of bioactives, nutraceuticals and phytochemicals which are actively being researched globally for their healthcare potential. Anthocyanins, carotenoids and chlorophylls have dual effects as natural colourants as well as therapeutic antioxidant agents. Dietary fiber has many reported effects like reducing serum cholesterol, preventing colon cancer, maintaining good intestinal health as well as prophylactic action on cardiovascular diseases, diabetes and obesity. Among the tuber crops, cassava, yams, elephant foot yam and sweet potato are known to be rich sources of dietary fiber. Sweet potato is a wonder tuber crop having many nutraceutical principles like carotenes, anthocyanins, antioxidant flavonoids etc. in tubers and the eye protectant xanthophyll, lutein in its leaves. Taro is rich in mucilage, which has reported cholesterol and triglyceride lowering activities. Chinese potato is a rich source of flavonoids having potent antioxidant activity and hence beneficial as a free radical scavenger for the body.

Sweet potato

Sweet potato is increasingly recognized as a health food, due to several of its nutraceutical components such as polyphenols, anthocyanins and dietary fiber, which are important for human health. The non-starch polysaccharides comprising cellulose (2% dry matter), hemicellulose (2-4%) and pectin (2.5-5.1%) contribute towards the 'dietary fibre' fraction of sweet potato roots. Sweet potato is reported to have a low glycaemic index of <55, which makes it an ideal food for diabetics. The roots are considered as a highly functional, low calorie food, with antidiabetic effects (Kusano and Abe, 2000). Caiapo, an extract of white skinned sweet potato was reported to improve glucose tolerance, by reducing insulin resistance, without affecting body weight or insulin secretion in human volunteers.

The most studied nutraceuticals in sweet potatoes are carotenoids and anthocyanins. Purple-fleshed sweet potatoes are a rich source of anthocyanins, which have medicinal value as antioxidant and cancer preventing agent. Besides, in Japan, the coloured roots are used for extracting the pigment, which is further used in various food products. Anthocyanin rich purple-fleshed sweet potatoes were reported to have multiple physiological functions ranging from antioxidants (radical scavenging), anti-mutagenic, hepatoprotective, hypoglycaemic and hypotensive effects (Suda *et al.*, 2003). Besides, phenolic acids especially caffeoylquinic acid and vitamins such as ascorbic acid (vitamin C) and α-tocopherol (vitamin E) also contribute to the antioxidant activity of purple fleshed sweet potato. Consumption of orange-fleshed sweet potatoes, the cheapest source of β-carotene (provitamin A), becomes significant in combating night blindness in children. ICAR-CTCRI has a collection of deep orange-fleshed accessions like S43 (11928 I.U of b - carotene/100 g), S47 (9889 I.U) and S58 (9195 I.U), besides about 60 yellow to orange-fleshed sweet potato accessions.

Sweet potato leaves and tops are also nutritionally rich materials that could enhance the nutritional security of malnourished people especially those living in poverty-stricken areas. The nutritive value of sweet potato leaves is mainly attributed to the presence of high levels of antioxidants such as phenolic compounds (Islam *et al.*, 2003; Yoshimoto *et al.*, 2003). Lutein being the major pigment of human retina and is not synthesised *de novo*, dietary consumption is essential to prevent or delay the degenerative process (Ribaya-Mercado and Blumberg, 2004). As compared to common leafy vegetables such as broccoli, spinach and lettuce, sweet potato leaves are reported to contain much higher levels of lutein (0.38- 0.58 mg/g fwb) capable of preventing macular degeneration in aged people (Menelaou *et al.*, 2006).

Carotene rich sweet potato

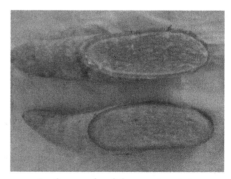

Anthocyanin rich sweet potato

Yams

Yams are sources of essential micronutrients and phytochemical compounds. *Dioscorea alata* has been reported to contain several antioxidants such as phenolics, anthocyanins, flavonoids, vitamins etc. Several physiological effects including lowering of lipid and sugar levels in blood, antioxidant activity, anti-mutagenic activity and anti-allergic activity have been reported in yam extracts due to the presence of phenolic phytochemicals. Purple fleshed cultivars contain anthocyanins, which are antioxidants and wide variation exists in the colour of the tubers, contributed mainly by the bioactive pigments. Cheng *et al.*, (2007) isolated five estrogenic compounds from *D. alata* tubers and there are several studies on the antihypertensive, bone protective, immune-stimulatory and antidiastogenic effects of *D. alata* tubers (Chen *et al.*, 2008). The purple yam *(D.alata)* extensively utilized in Philippines was reported to contain several anthocyanins such as cyanidin-3-gentiobioside, alatanins 1 and 2 and alatanins A-C (Shoyama *et al.*, 1990; Yoshida *et al.*, 1991). Moriya *et al.* (2015) isolated and characterised four new acylated anthocyanins from purple yam (*D. alata*) having high antioxidant activity.

Elephant foot yam

The various medicinal properties of elephant foot yam have been extensively studied, which include gastroprotective, analgesic, antibacterial, anti-inflammatory, antioxidant and hepatoprotective activities (Nataraj *et al.*, 2009 Dey *et al.*, 2010). The presence of antioxidant vitamins such as vitamin C and vitamin E could further enhance the health value of elephant foot yam tubers (Basu *et al.*, 2014). Konjac (*Amorphophallus konjac syn. A. rivieri*) is native to warm subtropical to tropical eastern Asia (from Japan and China south to Indonesia). It is an important commercial crop grown in China, Japan, Indonesia, and elsewhere in subtropical Asia. Though, it has long been used as a food in China and Japan, it is becoming increasingly popular as a health food in many countries (Misra, 2013). Konjac corms have unique feature of containing high levels of glucomannan (over 45%), which is the best known natural edible and water soluble fibre capable of preventing constipation. Regular consumption of glucomannan is reported to reduce total cholesterol, prevent constipation, reduce blood glucose, regulate blood sugar levels and lipid metabolism, control obesity and immune functions of human body and function as a mycotoxin adsorbent.

Scope of entrepreneurship development in tuber processing sector

Because of the predominant importance of cassava and sweet potato over the other tuber crops, research efforts have been largely concentrated on these two crops. Despite the ability to combat biotic and abiotic stresses, adaptability to marginal lands and high photosynthetic efficiency, cassava and sweet potato production is declining in India, over the last few years. Other tuber crops like yams and aroids are only gaining importance. In order to maintain the rhythm in the supply of food materials and to keep pace with the geometrically increasing population, secondary or tertiary staple food crops like tuber crops have to be retained within the cropping system of marginal farmers. Better post harvest management and diversification for the production of value added products are ways to sustain tuber crops within the cropping system. Though, the traditional modes of consumption will continue to be significant from the food security dimension, there exist immense potential to add value to sweet potato and cassava and introduce a whole new range of food products that can cater to the needs of rapidly urbanizing societies.

Creating awareness about the immense scope of tuber crops for value addition and product diversification is absolutely essential to properly exploit the developmental elasticity of these crops. A shift from the rural markets to semi-urban and urban markets is also visualised globally, which also can enhance the market demand for value added, ready-to-eat food products. Transition from subsistence agriculture to market-oriented agriculture has its base in agri-business units set up in the rural and producing areas. Besides providing opportunities for income and employment generation, they also ensure value addition to horticultural produce.

Value added food products from tuber crops

Cassava

Cassava rava/semolina: Cassava is one of the important tuber crops valued for its high starch content (20-35%). Cassava rava or semolina is a pre-gelatinized granular product similar to wheat semolina and finds use as a breakfast recipe product. For the preparation of cassava rava, the tubers are peeled and sliced into round chips. It is then partially cooked by boiling in water for 5 min, decanting the steep water, sun-drying the parboiled pieces and powdering coarsely in a hammer mill. This is then sieved through fine sieve to separate out the finest fraction which can be converted to porridge powder by flavoring with cardamom and fried powdered cashew nuts. The residue is sieved through larger mesh size sieve to obtain rava. The uneven large pieces are again powdered to recover rava.

Fried cassava chips: Fried cassava chips presently available in the market are often too hard to bite and bear no comparison with potato chips. This leads to poor acceptability of the product and lower price. Research at CTCRI has shown that excellent quality fried chips can be made from cassava tubers, by soaking the chips in acetic acid-brine solution for 1 h, parboiling for 5 min, surface drying and deep frying in oil. This facilitates the removal of excess starch and sugars from the cassava slices, with the result that light yellow crispy chips can be obtained, having soft mouth feel and good texture.

Cassava based extruded products: The demand for extruded snack products is expanding at a phenomenal rate in developed and developing countries. Extrusion cooking is a high temperature short time cooking process designed for processing of starchy as well as proteinaceous materials. Being a concentrated source of starch, cassava can be extruded to obtain a variety of nutritionally enriched, ready to eat/cook products. Cassava tubers after washing, peeling and slicing into chips are dried and powdered in a hammer mill. The dry flour after conditioning to 12-15% moisture content is extruded by maintaining appropriate temperatures at different sections of the barrel and die of the food

extruder. Cassava being rich in carbohydrates and lacking in protein content, addition of low cost protein sources like wheat, finger millet, soy flour etc. could give more nutritional and market value products.

Fried snack foods from cassava: Fried food products from composite flour based on cassava have high nutritional and textural qualities as well as longer shelf life and could easily capture the urban markets. Technology for making fried snack food products viz., hot fries, hot sticks, sweet fries, sweet dimons, salty fries, salty delight, murukku, crisps, nutrichips with and without egg were perfected at ICAR- CTCRI. These include:

i)   *Cassava pakkavada*: This is a hot snack food having good texture and

taste made out of cassava flour. The other ingredients include *maida*, bengal gram flour, salt, chilli powder, asafoetida, baking soda and oil. The ingredients are thoroughly mixed and made into dough with hot water (50°C), proofed for 1h and extruded through hand extruder having flat rectangular holes, into hot oil.

ii) *Cassava sweet fries*: This is a sweet snack food made out of cassava flour, *maida*, baking soda and oil. The ingredients are mixed well and made into dough with hot water (50°C). The dough after proofing for 1h is hand extruded through die having round holes, into hot oil. The fried product is then coated with sugar by dipping for a few minutes in sugar syrup having thick consistency.

iii) *Cassava nutrichips*: This is a high protein snack food made out of cassava flour by mixing with other ingredients like *maida*, groundnut paste, egg, salt, sugar, *sesame*, coconut milk, baking soda and oil. After mixing the ingredients, hot water is added and mixed to form smooth dough. The dough after proofing is made into small balls which are then spread into sheets of 0.2 cm thickness. This is then cut into dimon shape using a sharp knife and deep fried in oil.

iv) *Cassava crisps*: This is a soft and good textured crispy snack food made from cassava flour, *maida*, rice flour, bengal gram flour, salt, baking soda, turmeric powder and oil. The dough made with hot water is proofed for 1h and then extruded through the small pore size die having round holes. The deep fried material is mixed with fried nuts, curry leaves etc. before packing.

v) *Cassava starch, sago and wafers*: The process of extraction of starch consists of peeling, rasping, screening, settling and drying. Peeled roots of cassava are disintegrated into pulp by a rasper which releases the starch granules from the fibrous matrix. The resulting slurry is pumped onto a series of vibratory screens  and the fibrous waste (thippi) is retained on them and the starch milk passing through the sieves are channeled into sedimentation tanks. After at least 8 hours of settling, the supernatant liquor is run off and the starch cake settled at the bottom is scooped up for sun drying on a cement floor.

Sago (Saboodana) is manufactured from the partially dehydrated (35-40% moisture) starch cake. The lumps are broken in a spike mill and then globulated on a gyratory shaker. The globules are graded according to size and then partially gelatinized by roasting on shallow metal pans. Finally, the sago pearls are dried in the sun on cement floor. The agglomerates are separated by means of a spike beater and polished before bagging.

Wafers

Wafers are made by arranging the wet granules in suitable die and steaming. The steamed granule take the shape of the die and after drying, it can be separated out from the dies and packed.

Sweet potato products

Sweet potato based composite flours have been used in many countries for making small baked goods like cakes, cookies, biscuits, doughnuts etc. Sweet potatoes are consumed at home level, mainly after cooking, baking or converting into fried chips. The roots are often converted to canned or pureed form, to enhance the shelf life. Sweet

potato based baby foods are preferred in many countries as the first solid food for infants. Canning of sweet potato is widely practised in the United States, to enhance the storage life and ensure round- the year availability of the product.

Sweet potato roots are transformed into more stable edible products like fried chips, crisps, french fries etc. Sweet potato puree, is a primary processed product from the roots, which is used directly as a baby food or used for mixing various food items like patties, flakes, reconstituted chips etc. High quality puree can be made from white, cream or orange- fleshed sweet potatoes and also from tubers of any size or shape.

Sweet potato roots can be termed as a '3-in-1' product, as it integrates the qualities of cereals (high starch), fruits (high content of vitamins, pectins etc.) and vegetables (high content of vitamins, minerals etc.). The beneficial effects of these ingredients have been appropriately put to use by converting the roots into a number of intermediary food products like jam, jelly, soft drinks, pickles, sauce, candies etc

Quick cooking dehydrated cassava and elephant foot yam tubers

Quick cooking dehydrated cassava and *Amorphophallus* tubers were developed to reduce bulk during export and to increase the shelf life of the product with a cooking time of only 2-5min.

Value added industrial products

Modified starches

Starch is the most important value added product from cassava and is a versatile and inexpensive raw material used for a variety of industrial applications viz., paper, textile, adhesive, sweetener, pharmaceutical, and food industries. Native starches because of their inability to withstand extreme processing temperature, diverse pH, high shear rate and freeze-thaw variation are undesirable for many applications. Lack of stable viscosity is a disadvantageous property of cassava starch, restricting its use in food, textile, pharmaceutical and paper industries. In order to improve on desirable functional properties, native starches are often modified by physical, chemical and enzymatic techniques. Modified starches having altered viscosity, better stability in viscosity, higher gel strength, improved film forming capacity, clarity, lower retrogradation tendency and higher tack find wide applications as binders, fillers, emulsion stabilizers, consistency modifiers and adhesives.

Chemically modified starches: The major chemical modifications of starch include oxidation, esterification with reagents such as succinic anhydride, alkenyl succinic anhydrides, citric acid and sodium orthophosphate, cross-linking and etherification.

i)   *Oxidised starches:* Oxidised starches are synthesised by the reaction of starch with oxidising agents such as sodium hypochlorite in alkaline medium. Oxidised starches are useful in the paper industry due to their low viscosity, better film strength and clarity.

ii) *Starch esters:* Succinylation by reacting starch with succinic anhydride is a commercially used chemical modification method. Formation of starch succinates increases the hydrophilicity of starches and high densities of starch-side chain carboxylic groups provide useful properties like metal chelation. Chemical modification of cassava starch with alkenyl succinic anhydrides produces highly versatile starch products with amphiphilic side chains. Starch succinate is an important food ingredient. It has been used as a binder and thickener in soups, snacks, canned, and refrigerated foods; and tablet disintegrant for pharmaceutical applications. Other applications include surface sizing agents and coatings and binders in paper and textile industries.

Starch citrates, synthesized by reaction of starch with citric acid, are mainly used to increase the dietary fiber content in food in the form of RS IV type resistant starch. Phosphorylation is an important chemical modification used to improve the pasting and water holding characteristics of starch. Phosphorylated starches exhibit excellent water binding capacity, solubility and paste clarity. Starch phosphates are used in food as emulsion stabilizers for vegetable oil in water and as thickening agents with good freeze-thaw stability. These can also be used as flocculants for water clarification. Methods have been developed for the preparation of carboxymethyl starch using monochloro acetic acid and finds use in pharmaceutical industry as tablet disintegrants for immediate release tablet formulations.

iii) *Cationic starch***: Cationic starch has been prepared from cassava starch by treating starch with 3-chloro 2-hydroxypropyl trimethyl ammonium bromide. Paper, textile, cosmetic and water treatment industries are the major fields which consume cationic starches. Cationic starches are used in large scale by the paper industry as wet-end additives to improve retention and drainage rate of the pulp and strength of the finished sheets, surface size, and coating binders, wrap sizing agents in textile manufacture, binders in laundry detergents, and flocculants for suspension of inorganic or organic matter containing a negative charge.

iv) *Cross-linked starches*: Cross-linked starches constitute a major class of modified starches. Cross-linking reinforces the hydrogen bonds already present in the granules with new covalent bonds. Cross-linked starches are used in food, pharmaceutical and textile industries. At low levels of cross-linking, these can be used as food ingredient in salad dressings, frozen foods, canned foods and puddings due to high paste consistency and stability. Cross-linked starches have been applied in soups, gravies, sauces, baby foods, fruit filling, pudding, and deep fried foods. Highly cross-linked starch is thermally stable and can find application as surgical dusting powder and in sizing of textiles.

v) *Hydroxypropylated starch:* The reaction of cassava starch with propylene oxide has been standardized to produce hydroxypropylated starch. Due to high freeze-thaw stability, hydroxypropyl starches can be used in frozen food industry and in convenience foods. One of the largest areas of application is as thickener in a number of food and food-related products. These are used in canned sauces and gravies; as thickener in a number of food and food-related products, and as edible film coating on various foods.

Physically modified starches: Physically modified starches exhibit improved stability of viscous solutions, higher pasting temperature, lower swelling power and lower *in vitro* digestibility in comparison to their native counterparts. Since, no chemicals are involved in these modifications, these starches are especially useful in food applications. Major physical treatments for modifying starch are hydrothermal modifications, i.e., heat-moisture treatment and annealing. In heat moisture treatment, starch is heated to a temperature, which is generally higher than its gelatinization temperature, under a semi-dry condition, whereas annealing involves treatment of starch slurries with excess water at temperatures below the gelatinization temperature of starch. These find application as thickener/ viscosifier in various food products such as pie filling, puddings, ice creams, soups etc. These are used in the preparation of resistant starch and also in paper sizing.

Enzyme-modified starches: Enzymatic modification of starch is used to produce derivatives with varying viscosity, gel strength, thermo reversibility and sweetness. In enzymatic modification technique, gelatinized starch is subjected to degradation by enzymes resulting in various products (Alexander, 1992, Kennedy *et al.*, 1995). Selective enzymatic hydrolysis of starch produces a range of products like glucose, maltose, oligosaccharides and polysaccharides with varying chain length and dextrose equivalent (DE) (Taggart, 2004). Major enzymatically modified products are linear dextrins of varying DE, high fructose syrups, glucose syrups, dextrose, maltodextrins and cyclodextrins

(Blanchard and Katz., 2006). The commonly used enzymes are α- and β- amylases, isoamylase and pullulanase. The a-amylase selectively attack the β-(1, 4)-linkages of starch and produce maltodextrins and low DE dextrins. However, α-amylase hydrolyze every other 1, 4-linkages to give lower molecular fragments and higher DE syrups like maltose. Isoamylases and pullulanase are debranching enzymes and attack at specific sites such as 1, 6-linkages in starch and produce high DE syrups. Cyclodextrins are produced by enzymatic hydrolysis of starch using cyclodextrin glycosyltransferases.

New technologies/innovations in value addition of tuber crops

Functional pasta from sweet potato and cassava

Pasta, as a food rich in complex carbohydrates with low glycaemic index is gaining wide acceptance in the recent years. Pasta products, largely consumed all over the world are traditionally manufactured from durum wheat semolina, known to be the best raw material suitable for pasta production due to its unique colour, flavour and cooking qualities. Nevertheless, wheat semolina proteins are deficient in lysine and threonine leading to low biological value for the product and hence several fortified pasta products have been attempted. Whilst the global prevalence of diabetes is projected to increase from 4% in 1995 to 5.4% by 2025, approximately 170% increase in diabetic population has been predicted in developing countries, with India topping the list, followed by China. FAO-WHO Expert Consultation recommends the increased consumption of low glycaemic foods rich in resistant starch, non-starch polysaccharides and oligosaccharides. Cassava is known to be a high starch, high glycaemic food, while sweet potato is a low glycaemic health food. Considering the projected rise in diabetic population in India to 80 million by 2030 and in an attempt to diversify the use of these root crops, studies were made at ICAR-CTCRI to develop an array of pasta products having high functional value coupled with low starch digestibility. Since, pasta is getting wide popularity among the young Indians and in the metros as a convenient food, transformation to new health and wellness food is essential to add value to cassava and sweet potato and sustain their cultivation in India.

Protein-enriched pasta: As protein content and gluten strength are critical factors deciding the cooking quality of pasta and since cassava and sweet potato lack gluten, fortification with protein sources like whey protein concentrate (WPC), defatted soy flour (DSF) and fish powder (FP) have been used to give an appropriate texture for the products, besides enhancing their protein content. The study showed that slowly digestible functional pasta with good quality could be developed from cassava or sweet potato, which also has high protein content (Jyothi *et al.*, 2011).

Dietary fiber enriched pasta: Rapid increase in lifestyle diseases have led to an increasing awareness among the consumers about the health benefits of dietary fiber, and several reports indicate that diets rich in fiber could reduce the risk from coronary heart disease, cancer, obesity, and diabetes. Physiological effects of dietary fiber depend on the type of fiber, its chemical and physical composition, solubility, etc., and have reported effects like prevention of colon cancer, constipation, etc. Dietary fibre enriched pasta were prepared from cassava/sweet potato - *maida* blends using various fibre sources like oat meal, wheat bran etc. (Jyothi *et al.*, 2012; Renjusha *et al.*, 2015a).

Edible gums such as guar gum (GG), xanthan gum (XG) and locust bean gum (LBG) were also used as fiber sources to enhance the dietary fiber content in pasta, which also had 10-13% protein due to whey protein fortification. It was found that gum fortification significantly enhanced the Swelling Index for the pastas, while cooking loss was considerably reduced. *In vitro* starch digestibility was slow and progressive over a period of 2h for the gum fortified pastas, with a high retention of resistant starch after digestion (Jyothi *et al.*, 2014)

Resistant starch, is gaining lot of importance recently as a food additive due to its number of physiological effects such as prevention of colonic cancer, hypoglycaemic action, reduction in gall stone formation, hypocholesterolaemic effect, control of obesity etc. NUTRIOSE is a partially hydrolyzed starch from wheat or corn which is reported to be rich in dietary fiber and NUTRIOSE fortified foods are reported to reduce hunger and promote satiety in humans. Being a product with high digestive tolerance, low glycaemic index (< 25) and good stability to high temperature process, NUTRIOSE was used to develop cassava pasta with low glycaemic index. The study showed that NUTRIOSE is a highly preferred additive for making cassava pasta with low glycaemic index and addition at 10% level was optimum.

Functional pasta from cassava with carrot and beet root as additives: Carotenoids are natural pigments occurring in plants and are abundantly present in many fruits and vegetables. Beta-carotene is the most important carotenoid due to its provitamin A and antioxidant activity. Antioxidants have ability to prevent chronic diseases such as cancer, age related problems and cardiovascular diseases. Betanin is the most common pigment that occurs in high concentrations in beet root. According to the regulation on food additives, betanin is permitted as a natural red food colourant (E162). Functional pasta was developed using beet root and carrot as additives.

Betanin enriched pasta

Low glycaemic spaghetti from tuber crops: High incidence of metabolic diseases among people consuming foods rich in carbohydrates has led to increased research efforts in the development of low glycaemic foods. Effect of fortification of sweet potato flour with banana and legume starches as well as sweet potato starch itself in producing low glycaemic spaghetti from sweet potato was investigated. Also, three gum sources such as guar gum, xanthan gum and locust bean gum were used to develop slow digestible sweet potato spaghetti (Renjusha *et al.*, 2015b).

Legume flours such as black gram (urad dhal), chickpea (besan) and green gram (mung bean) which are known to be rich sources of non-starch polysaccharides (having fibre like action) have been used to manufacture medium digestible spaghetti from sweet potatoes. Low glycaemic spaghetti was also developed successfully using the resistant starch source, NUTRIOSE.

Starch noodles from sweet potato: Starch noodles, presently known by different names such as glass noodles, cellophane noodles, vermicelli, *bihon* noodles etc. have been a favourite food in China since 1400 years. They are now produced from purified starches of various plant sources and have become popular in several Asian countries as well.

Starch noodles differ in their quality and texture from Asian or Italian pasta/spaghetti, as the latter are made from flour or semolina, although many fortified pasta and noodle products have been attempted by subsequent researchers with a view to improve the quality. Traditionally, starch noodles made from mung bean starch is considered as the best owing to the transparent appearance, fine threads, high tensile strength and low cooking loss.

Although, starch noodles have been developed form sweet potato starch and are very popular in China, Korea and the Philippines, the product has a bland taste and low nutritive value as starch is the only ingredient. Hence, protein rich starch noodles were developed using whey protein concentrate as additive. Sweet potato starch noodles fortified with banana, and green gram having slow starch digestibility were also developed.

Gluten-free pasta from cassava and sweet potato: Coeliac disease is an immune-mediated disease affecting approximately 15 million people round the globe. Coeliac disease patients have intolerance to gluten, the proteins in wheat, rye and barley. Consumption of gluten leads to inflammation of the small intestine ultimately damaging the villi and affecting the absorption of iron, vitamins, minerals etc. A strict gluten-free diet as a lifelong diet strategy is the only treatment for coeliac disease.

As gluten containing substances have to be totally avoided in the development of gluten-free spaghetti, the networking ability of gluten has to be imparted to the dough using other substances that mimic gluten and this could be achieved using whey protein concentrate which has gluten mimicking property. Technologies were developed for making gluten-free pasta from sweet potato/cassava flour-rice flour blends along with additives such as WPC and guar gum.

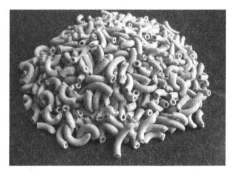

Protein and fibre enriched functional foods

Keeping an eye on the health conscious consumers, mini- papads or wafers were developed from cassava flour by adding fibre sources like wheat bran, oat meal, rice bran and cassava fibrous residue. Fibre sources are added to gelatinized cassava slurry and mixed thoroughly. The spicy condiments are also added and spread on plastic sheets which are then dried in the sun for 36h. Papads are peeled off from the sheets and packed. Deep fried products have

soft and crisp texture. Mini- papads with high protein content (7-15%) could be made from cassava flour by adding protein sources like cheese, defatted soy flour, prawn powder and whey protein concentrate along with other spicy condiments. Papads are allowed to dry for 36h, after which they are separated from the sheets and packed. This could be deep fried in oil before use.

Sweet potato based functional snacks

A number of novel food products with functional value have been developed worldwide from sweet potatoes. Sweet potato tubers with their low glycaemic index have additional value as a food for diabetics. There are a range of primary food products that could be made from sweet potato like chips, flakes, frozen products, french fries, puree etc., while it is also the raw material for a host of secondary products like noodles, sugar syrups, alcohol, pasta etc. Fried chips retaining their colour and functional value were developed from purple (anthocyanin-rich) and orange fleshed sweet potatoes (carotene-rich) through vacuum frying technology.

Sweeteners from starch

Starch is an important raw material for the production of sweeteners and functional oligosaccharides. Liquid glucose and high fructose syrup are made from cassava starch by liquefying the starch and saccharifying it using enzymes or acids. After saccharification, decolourisation is done using charcoal to produce glucose syrup. Fructose syrup is obtained from glucose syrup using glucose isomerase at 60-62°C for 2h. This is then decolourised and concentrated. HFS is a highly valued product for the confectionery industries, due to its specific properties like non-crystallising nature, extra sweetness etc. Maltose syrup is another product which is commercially prepared from starch, which finds application in food and confectionery industries. Essential composition includes maltose, glucose, maltotrioses, and higher oligosaccharides. Process consists in hydrolysing the starch using -amylases from native sources like rice seedling extracts. CTCRI has also standardized techniques for the production of maltodextrins with different dextrose equivalents (DE), which can be used as fat replacers in low calorie foods. The process involves the treatment of starch with heat-stable bacterial amylase and purification of the product.

Bioethanol from starch

Fresh cassava tubers, dry chips/flour or starch can be used for the production of ethanol. Process consists of three steps viz., liquefaction, saccharification and fermentation. During liquefaction, the cooked starch slurry is hydrolysed to maltose and low molecular weight dextrins using either acids or α-amylase. In the saccharification step, hydrolysis to glucose is achieved using acids or glucoamylases. During fermentation, glucose formed is fermented to ethyl alcohol using yeast, *Saccharomyces cerevisiae*. Optimum concentration of sugars for ethanol fermentation is 12-18% and optimum pH and temperature are 4.0 to 4.5 and 28-32°C respectively. After 48-72h of fermentation, alcohol is recovered through distillation.

This process is highly energy-intensive and involves use of costly enzymes and hence the overall energy balance is negative. However, presently, improved enzymes are available and a global shift has already occurred. Research under a Department of Biotechnology (DBT) funded project led to the development of new technology for ethanol production from cassava. This new process has a saving of time and energy over the earlier process patented in 1983, as the process could be done at room temperature (30°C) and completed in 48h. (Shanavas *et al.*, 2011). Yield of alcohol in the new process is 680 L/t of starch against 450 L/t in the old process.

Starch based adhesives

Liquid adhesives and gum pastes can be made from cassava starch using simple low cost technologies. Starch is cooked with water, cooled and preservatives like formaldehyde or copper sulphate are added. Shelf life of the gums can be improved by adding borax, urea, glycerol, carboxymethyl cellulose etc. These chemicals help in improving and stabilising the paste viscosity. Another potential area is the development of corrugating adhesives from starch.

Biodegradable plastics

One of the innovations in the potential use of cassava starch is in the production of biodegradable plastics. In the context of increasing risk of pollution from plastics, use of biodegradable plastics can help in a big way to provide a healthy and pollution less environment for the future generations. Research has shown that cassava starch incorporated into polypropylene granules can yield products having good strength and enhanced biodegradability. Starch incorporated plastic films (up to 25-40%) possess adequate mechanical strength and flexibility and can be processed just like normal plastics, i.e., heat-sealed, printed, coloured etc. Granules and finished products can be stored almost like synthetic plastics and biodegradable under soil burial conditions. From the outdoor weathering and soil-burial tests, biodegradation time has been reported to vary from 6 months to 5 years depending on its composition and soil conditions.

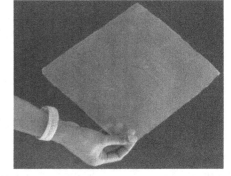

Starch-graft-copolymers and superabsorbent polymers

Among the various modifications of starch, graft copolymerization with vinyl monomers is a fascinating field for research with unlimited possibilities for improving starch properties.

A number of graft-copolymers have been prepared by grafting different monomers onto starch which are gaining increasing importance in the manufacture of natural biodegradable polymer based plastics, ion-exchange resins and cosmetics.

Grafted starches possess very high water absorption capacity and thermal stability. Superabsorbent polymers (SAP) can absorb a large amount of water ranging from hundred to thousand times of that of the polymer. These have been extensively used as absorbents in personal care products and incontinence products. Starch can be modified as superabsorbent polymers through radical graft polymerization with vinyl monomers and proper cross-linking. Due to the biocompatibility, biodegradability and abundant nature, natural biopolymer based superabsorbent can act as potential replacers for petroleum based products.

Superabsorbent gel polymer

Starch based edible films and composite films

Cassava and sweet potato starch based edible films are developed by incorporating various hydrocolloids and glycerol in starch. These films can be utilized as edible coatings for various food products. Development of starch based nanocomposite films for specialized applications such as edible coatings for food, packaging films and slow release matrices for tablets is currently underway.

Scope of by-product utilization/effluent treatment

Food processing industry is the most rapidly growing industry in the world and hence is the one that generates enormous amounts of waste products, making their efficient disposal a major concern for various countries. Although, the major mode of utilization of agro-industrial by-products is as feed or fertilizer, as it eliminates the economically limiting factors like the cost associated with drying, storage and transportation, techniques have been developed to appropriately convert the by-products to value added functional additives, nutraceuticals, microbial proteins, biogas, bioethanol etc. through biotechnological and other means. Channeling of by-products of food processing industries through environmentally sound methods will also help to ensure a clean unpolluted environment. Setting up of waste treatment plants along with the main industrial unit is a mandatory requirement in various countries.

The commonest forms of waste generated from these root crop based industries include peels, liquid effluent, solid bagasse and distillery stillage. Besides, secondary waste generated from these crops includes foliage, stems, vines etc. Utilization of the primary waste (directly from the raw materials) for the production of several commodity chemicals, biofuels, energy and animal feed is only dealt with in this treatise.

Value addition of cassava starch factory residue (CSFR)

Bioethanol: Cassava starch factory residue (CSFR) is a solid by-product discharged from cassava starch industries, causing major pollution problems in India due to lack of appropriate disposal strategies. This is usually discharged in to the premises of the factories causing environmental pollution. Cassava starch factory residue is a cellulo-starch material with a composition of starch (56-60%), cellulose (15-18%), hemicellulose (4-5%), lignin (2-3%), protein (1.5-2.0%), reducing sugars (0.40-0.50%) and pentosans (2%). Despite the high starch content in CSFR, it has been reported as a poor substrate for ethanol production, due to the difficulty in the release of starch granules from the fibrous matrix and the viscous nature of the slurry at high solid concentration. Studies were taken up to tap its potential as a biofuel feedstock using improved cellulolytic enzymes.

The potential of CSFR as a feedstock for ethanol production was investigated at ICAR- CTCRI. Very high ethanol yield of approximately 390 L/ton of dry CSFR have been obtained in the new technology, making CSFR an alternative substrate for ethanol production (Divya *et al.*, 2012).

Microbial exopolysaccharides : Microbial exopolysaccharides (EPS) are carbohydrate polymers, which are secreted by microbes into the growth medium and are not bound to the cell walls of the organisms making their extraction easier. These have typical gel characteristics which make them useful in food and pharmaceutical industries. Technologies were perfected at ICAR- CTCRI to produce pullulan from CSFR. Pullulan is an exopolysaccharide produced by the fungus, *Aureobasidium pullulans* and is a α-D-glucan having maltotriose and maltotetrose units linked by $1 \rightarrow 4$ bonds and are coupled through α-$1 \rightarrow 6$ bonds (Singhal and Kulkarni 1999). Due to the unique structure, biodegradable nature and characteristic physical properties, it has a wide range of industrial applications like packaging films, adhesives, molded articles, coatings, fibers etc.

Production of pullalan from cassava starch factory residue (CSFR) using *A. pullulans* MTCC 1991 was studied by Ray and Moorthy (2007). Solid support was prepared using 20 g CSFR in 40 ml basal medium containing sodium glutamate (0.35%) and salts like K_2HPO_4, KH_2PO_4, $MgSO_4$, NaCl and ferrous sulfate $7H_2O$ at 0.05% level. Fermentation was carried out on thermally pre treated CSFR at pH 6.5 and 28 ± 2 C for 7 days. The authors observed that pullulan production was negligible without the supplements and EPS yield was the highest from CSFR supplemented with the salts and glutamate and as high as 27.5 g/Kg of pullulan could be extracted, corresponding to a yield of 18.6 g/L of biomass. It was reported that the higher yield of pullulan from CSFR than wheat bran or rice bran was due to the wavy and undulated surface of CSFR, giving a larger surface area for the EPS to be deposited. Ease of operation and low investment make CSFR an attractive substrate for production of pullulan.

Protein enrichment for animal feed: Cassava starch factory residue (CSFR) is constrained by its high water and fiber content, with low protein. Fermentation through microbial inoculation is a low cost technique to enhance the utility of this waste and also to reduce the environment threats. Protein enrichment of cassava starch factory residue using *Trichoderma pseudokoningii* Rifai was studied by Padmaja & Balagopalan (1992). A 1:1 mixture of cassava flour and CSFR was enriched with 0.15% ammonium sulfate and exposed to steam in a vegetable steamer for 2h and cooled to room temperature. One week old inoculum of *T. pseudokoningii* grown on cassava flour-CSFR mix (1:10 w/w) was added to the pre heated and cooled mix. Solid substrate fermentation was carried out by spreading on trays for 6 days, with occasional sprinkling of water to replenish the lost moisture.

Value addition of waste water from cassava processing plants

One of the major processing wastes from cassava starch factories is the liquid waste, which causes enormous environmental pollution due to its indiscriminate disposal to nearby lakes, rivers or fields. Processing of one ton of cassava roots generates around 4000 litres of waste water. Primary effluents from the starch factories contain unextracted starch, cellulose, nitrogenous compounds and cyanoglucosides (Balagopalan *et al.*, 1994). Secondary effluents contain much less amount of total suspended solids, sugars, nitrogen, ash etc. Waste water coming out of the starch settling tanks contains unextracted starch, cellulose, carbohydrates, nitrogenous compound and cyanoglucosides. Effluents are usually discharged to surrounding rivers, ponds, lakes, drainage channels or fields. Effluents have a high COD, BOD and cyanide contents and cause serious environmental problems and damage to crop growth. Physical treatment with charcoal (0.03-0.05μm) was found to reduce efficiently the levels of cyanogens in effluents.

Microbial approach involves treatment with *Bacillus* sp., *Saccharomyces* sp. and *Aspergillus* sp. and the bacterium was found to be better than both yeast and fungus in reducing cyanide levels in the effluents. Biomethanation of effluents is an efficient method for biogas production and 130 litres biogas/kg dry matter with an average methane content of 59% was reported (Manilal *et al.*, 1990). Addition of cow dung as source of inoculum bacteria resulted in enhanced generation of biogas. In addition to environmental benefits, this method generates fuel for home and industrial utilization. Recycling of the treated waste water for irrigation and aquaculture is also possible.

Techno-incubation centre: The way forward for entrepreneurship development in tuber crops

The techno-incubation centre at ICAR-CTCRI is financially supported by Small Farmers' Agribusiness Consortium, Department of Agriculture, Govt. of Kerala. Main objectives of the centre are: to organise awareness training programmes on value addition of tuber crops to stake holders viz., farmers, entrepreneurs, officers of agrl/horti. Departments, to provide hands-on training on preparation of value added products, to provide incubator facility to prospective entrepreneurs for the production of value added products, to provide technical assistance to innovative entrepreneurs for product development and to act as a production/ processing unit of tuber crops based products for its widespread popularisation. Overwhelming response from the farmers and entrepreneurs from all the districts of the state towards attending training at the incubation centre shows the interest created on the value addition of tuber crops. The centre was regularly used by the entrepreneurs for the production of various value added products from cassava to test the market viability of the products by selling in the local markets. After realizing the market potential, young entrepreneurs started establishing their own units. Thus, the techno incubation centre acts as a stepping stone for the new entrepreneurs to flourish with tuber crops.

Machineries for small scale processing

Chipping machines

The simplest and the most common mode of processing cassava is conversion of tubers into chips. Dried cassava chips are made into flour and are used for value added products. Under the conventional practice, cassava tubers are sliced with the help of hand-knives with or without peeling the outer skin and rind. Conventional method of chipping cassava tubers by hand-knives produces about 10 to 40 kgh⁻¹ for chip thickness in the range of 2.7 to 12.5 mm. The outturn of the hand-operated cassava chipping machine is from 40 to 120 kgh⁻¹ for chip thickness in the range of 2.3 to 6.9 mm.

The basic parts of chipping machines are two concentric mild steel drums; the annular space is divided into compartments for feeding the tubers. A rotating disc at the bottom of the drum carries the knives assembly. Thickness of chips can be changed by introducing spacing washers between the disc and the blade. Tubers are fed into the compartments from the top and the chips are collected at the bottom.

The capacity of the pedal operated machine ranged from 80 to 770 kg/h for chip thickness of 0.9 to 6.9 mm. The output of the motorized machine ranges from 290- 1090 kgh⁻¹ for chip thicknesses of 2.5 to 9.9 mm. Cost comes to about Rs.20, 000/-, Rs.30, 000/- and Rs.60, 000/- for hand, pedal and motor operated chipping machines respectively.

Mobile starch extraction plant

For starch preparation, tubers are washed by hand and peeled with hand knives. These are then crushed or rasped by adding enough water to get a pulp or mash and sieved with excess water to remove the fibrous residues. The starch milk passing through the sieve is channeled for settling in settling tanks. When starch granules settle down, the supernatant water is decanted and the moist starch is crumbled and sun dried.

Major components of the machine are a hopper to feed the tubers, crushing disc or cylinder with nail-punched protrusions rotating inside the crushing chamber to crush the tubers, sieving tray to remove the fibrous and other cellulosic materials, stainless steel or plastic tanks to collect the sieved starch suspension, tuber storage chamber, handle and wheels for easy transportation from place to place and a frame to support these components. Addition of water during the processing is controlled through a water pipe with holes fixed inside the hopper along its length and during sieving by a shower attachment connected to the water line. An electric motor (¾ hp) or a generator (kerosene–petrol) attached to the frame can be used as the energy source to operate the machine. Extraction efficiency ranges from 75-85% with overall crushing capacity in the range of 400-500 kgh⁻¹. Cost of the machine comes to about Rs.1,00, 000/-.

Snack food manufacturing

Machineries required for the fried chips/snack food manufacturing are flour mill (Rs. 50,000/-), dry blender (Rs. 80,000/-) for mixing the raw materials, dough mixer (Rs. 50,000/-) for wet mixing, screw type or hydraulic type mixture making machine (Rs.85,000/-), industrial burners (Rs. 25,000/-), packing/sealing machine (hand sealer: Rs. 2,500/-, pedal sealer: Rs.15,000/-, band sealer: Rs. 30,000/)- and form fill packaging machine (Rs.4,00,000/-.)

The chip cutters used for making chips are made specially for tuber crops, cost of which varies from Rs. 40, 000/- to Rs. 85, 000/- depending on the mechanism of cutting and capacity.

Pasta machine and food extruder

For pasta making, dry mixer (Rs. 80, 000/-), pasta making machine with different types of dies (Rs.8, 00,000/-), dryer/oven (Rs. 3, 00,000/-) and packing machine are essential. The ready to eat extruded products are manufactured with the help of food extruders (Rs.18, 00,000 to 20, 00,000/-), the cost of extruder varies depending on the capacity. The extrudates could be coated with desired flavours using seasoning machines with manual spraying mechanism (Rs.1, 00,000/-) or with automatic flavour spraying device (Rs. 4, 00,000/-).

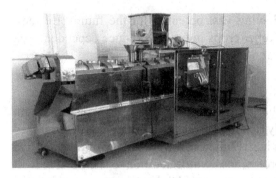

Extruder machine   Pasta making machine

Future scope of research

Even though, earlier laboratory studies have led to the development of technology for bioethanol production from cassava starch, the cost of ethanol was approximately Rs.90/litre. As per the National Policy on Biofuels, India wants to promote only non-food crops for bioethanol and in this context the non-edible varieties of cassava having very high cyanide levels could be promoted in non-traditional States also as a biofuel crop. Further, there is also the need to explore the scope of developing herbal products with medicinal effects for various tuber crops, bio-insecticides, natural food colourants, etc. Cheap and easy availability coupled with its biodegradable nature make cassava starch an ideal raw material for making biodegradable composite materials. Other alternative areas with good developmental elasticity include thermoplastic starches, starch composite foams for making disposable plates, cups and containers and also for cushioning material for the protection of fragile products during transportation and handling. Biodegradable polymers have enormous applications in agriculture for controlled release of fertilizers. Cassava starch based hydrogels with slow water releasing properties could find extensive use in future for water conservation and fertigation in tropical and subtropical regions. Despite being a low glycaemic food rich in complex carbohydrates, dietary fiber, beta-carotene and several vitamins and minerals, sweet potatoes have been seldom used for regular consumption in native form or in processed forms in India. A whole range of health foods having high nutraceutical content like carotene and anthocyanin as well as weaning foods for children could be developed from orange and purple-fleshed sweet potato. There is immense scope for developing functional foods having high prophylactic applications from purple fleshed yams or sweet potato rich in antioxidant anthocyanins or elephant foot yam rich in dietary fibre. Currently, these issues are addressed at ICAR-CTCRI, there is a need to validate the findings using clinical studies. Sweet potato leaves are reported to be one of the richest sources of lutein capable of preventing macular degeneration of the eye and cataract development. Value addition of agricultural and processing residues of tuber crops is another area having vast potential for the future. Besides, mechanization in cultivation of tuber crops in order to reduce dependence on labour is also the need of the hour.

Conclusion

The significance of root and tuber crops for adding value to transform them into nutritionally and functionally rich food products is increasingly recognized globally in recent years and this has resulted in a whole range of diverse food products suiting regional food preferences. Value addition to shelf-stable food products is especially important for cassava which has very low post harvest life and undergoes rapid deterioration after harvest. The enormous possibilities for enhancing the nutritive value of tuber crop products through fortification are brought out in this chapter. Starch being a major polysaccharide in tuber crops, there is vast scope for industrial utilization for making modified and functional starches, starch-based innovative products like biodegradable films, bioethanol, graft polymers etc. and these have been elaborated. A number of processing machineries have been developed to reduce the tedium associated with processing and also to make it more efficient and cost-effective. Expansion of non-edible, high cyanide cassava varieties to non-traditional areas is a major challenge for the promotion of cassava as a biofuel crop in India. Small scale entrepreneurship development is the key to the success and promotion of tuber crop cultivation in India and this necessitates adequate capacity building and creation of awareness on the vast scope of value addition in tuber crops. The Techno-Incubation Centre started at ICAR- CTCRI takes a lead in capacity building for the promotion on agri-business in tuber crops.

References

Alexander, R. J. 1992. Maltodextrins: Production, Properties, and Applications. In: Starch Hydrolysis Products. (eds.) Schnek F. W. and R. E. Hebeda, *Starch Hydrolysis Products. Worldwide Technology, Production and Applications*, VCH Publishers, New York, pp.233-75.

Baig, A. 2006. Entrepreneurship Development for Competitive Small and Medium Enterprises. *Report of the APO Survey on Entrepreneur Development for Competitive SMEs* (05-RP-GE-SUV-41-B).

Balagopalan, C., R. C. Ray., J. T. Sheriff & L. Rajalekshmy. 1994. Biotechnology for the Value Addition of Wastewater and Residues from Cassava Processing Industries. In: *Proceedings of the Second International Scientific Meeting of the Cassava Biotechnology Network*, Bogor, Indonesia, 22-26 August 1994.

Basu, S., M. Das., A. Sen., U.R. Choudhury & G. Datta. 2014. Analysis of Complete Nutritional Profile of *Amorphophallus campanulatus* Tuber Cultivated in Howrah District of West Bengal, India. *Asian J. Pharm. Clin. Res.* 7: 25-29.

Blanchard, P.H. & F.R. Katz. 2006. Starch Hydrolysates. In: Stephen A.M., G.O. Phillips., P. A.Williams (eds.) *Food Polysaccharides and their Applications*, 2nd edn. Taylor & Francis, Boca Raton, FL, pp. 119–145.

Chen, H., L. Hong., J. Lee & C Huang. 2008. The Bone-Protective Effect of a Taiwanese yam (*Dioscorea alata* L. cv. Tainung No. 2) in Ovariectomised Female BALB/C mice. *J. Sci. Food Agric.* 89: 517-522.

Cheng, W., Y. Kuo & C. Huand. 2007. Isolation and Identification of Novel Estrogenic Compounds in Yam Tuber *(Dioscorea alata Cv. Tainung No. 2). J. Agric. Food Chem.* 55: 7350-7358.

Cole, A.H. 1968. Meso-Economics: A Contribution from Entrepreneurial History. *Explorations in Entrepreneurial History* 6(1): 78-86.

Dey, Y. N., S. De & A. K. Ghosh. 2010. Anti inflammatory Activity of Methanolic Extracts of *Amorphophallus paeoniifolius* and its Possible Mechanism. *Inter. J. Pharm. Biosci.* 1: 1–8.

Divya, M.P., G. Padmaja., M.S. Sajeev & J. T. Sheriff. 2012. Bioconversion of Cellulo-Starch Waste from Cassava Starch Industries for Ethanol Production: Pretreatment Techniques and Improved Enzyme Systems. *Indus. Biotechnol.* 8: 300-308.

FAOSTAT. 2017. http://www.fao.org/faostat/en/#data/QC Accessed on 20 March, 2017.

Islam, M. S., M. Yoshimoto., K. Ishiguro & O. Yamakawa. 2003. Bioactive and Functional Properties of *Ipomoea batatas* L. Leaves. *Acta Hortic.* 628: 693–699.

Jyothi, G.K., M. Renjusha., G. Padmaja., M. S. Sajeev & S. N. Moorthy. 2011. Nutritional and Functional Characteristics of Protein-Fortified Pasta from Sweet Potato. *Food Nutr. Sci.* 2: 944-955.

Jyothi, G. K., M. Renjusha., G. Padmaja., M.S. Sajeev & S.N. Moorthy. 2012. Evaluation of Nutritional and Physico Mechanical Properties of Dietary Fibre Enriched Sweet Potato Pasta. *Eur. Food Res. Technol.* 234: 467–476.

Jyothi, G.K., M. Renjusha., G. Padmaja & M.S. Sajeev. 2014. Comparative Studies on Quality and Starch Digestibility of Hydrocolloid Fortified Sweet Potato Pasta Dried at Low and High Temperatures. *J. Root Crops* 40(2): 40-48.

Kennedy, J. F., C. J. Knill & D.W. Taylor. 1995. Maltodextrins. In: *Handbook of Starch Hydrolysis Products and their Derivatives*. Blackie Academic & Professional, Glasgow, pp.65-82.

Kusano, S. & H. Abe. 2000. Anti-Diabetic Activity of White Skinned Sweet Potato (*Ipomoea batatas* L.) in obese zucker fatty rats. *Biol. Pharm. Bull.* 23(1): 23- 26.

Manilal, V.B., C.S. Narayanan & C. Balagopalan. 1990. Anaerobic Digestion of Cassava Starch Factory Effluent. *World J. Microbiol. Biotechnol.* 6: pp.149-54.

Menelaou, E., A. Kachatryan., J.N. Losso., M. Cavalier & D. L .Bonte. 2006. Lutein Content in Sweet Potato Leaves. *Hort. Sci.* 41: pp. 1269-1271.

Misra, R. S. 2013. Konjac Needs Domestication. *Ind. Hortic.* 58(3): 22-24.

Moriya, C., T. Hosoya., S. Agawa., Y .Sugiyama., I. Kozone., K. Shinya., N. Terahara & S. Kumazawa. 2015. New Acylated Anthocyanins from Purple Yam and their Antioxidant Activity. *Biosci. Biotechnol. Biochem.* http://dx.doi.org/10.1080/09168451.2015.1027652.

Nataraj, H. N., R. L. Murthy & S.R. Setty. 2009. *In vitro* Quantification of Flavonoids and Phenolic Content of Suran. *Int. J. Chem. Tech. Res.* 1: 1063-1067.

Padmaja, G. & C. Balagopalan. 1992. Chemical Pre-Treatment of Cassava Starch Factory Waste for Optimal Lignocelluloses Hydrolysis by *Trichoderma pseudokoningii* Rifai. In *New Trends in Biotechnology*, (eds.) N.S. Subba Rao, C. Balagopalan, & S.V. Ramakrishna, Oxford & IBH Publishing Co. PVt. Ltd. New Delhi, pp. 418-434.

Ray, R.C. & S. N. Moorthy. 2007. Exopolysachharide (Pullulan) Production from Cassava Starch Residue by *Aureobasidium pullulans* strain MTCC 1991. *J. Sci. Indus. Res.* 66: 252-255.

Renjusha, M., G. Padmaja & M.S. Sajeev. 2015$_a$. Cooking Behavior and Starch Digestibility of NUTRIOSE_ (Resistant Starch) Enriched Noodles from Sweet Potato Flour and Starch. *Food Chem.* 182: 217–223.

Renjusha, M., G. Padmaja & M.S. Sajeev. 2015$_b$. Ultrastructural and Starch Digestibility Characteristics of Sweet Potato Spaghetti: Effects of Edible Gums and Fibers. *Intern. J. Food Prop.* 18(6): 1231-1247.

Ribaya-Mercado, J.D. & J.B. Blumberg. 2004. Lutein and Xeaxanthin and their Potential Roles in Disease Prevention. *J. Amer. Coll. Nutr.* 23: 567-587.

Shanavas, S., G. Padmaja., S.N. Moorthy., M.S. Sajeev & J.T. Sheriff. 2011. Process Optimization for Bioethanol Production from Cassava Starch using Novel Eco-friendly Enzymes. *Biomass Bioener.* 35: 901-909.

Shoyama, Y., I. Nishioka., W. Herath., S. Uemoto., K. Fujieda & H. Okubo. 1990. Two Acylated Anthocyanins from *Dioscorea alata. Phytochem.* 29: 2999–3001.

Singhal, R.S. & P.R .Kulkarni. 1999. Production of Food Additives by Fermentation. In *Biotechnology: Food Fermentation*, (eds.) V.K. Joshi & A. Pandey, Asiatech Publishers Inc., New Delhi, pp. 1144-1200.

Suda, I., T. Oki., M. Masuda., M. Kobayashi., Y. Nishiba & S. Furuta. 2003. Physiological Functionality of Purple-Fleshed Sweet Potatoes Containing Anthocyanins and their Utilization in Foods. *Japan Agric. Res. Quart.* 37(3): 167–173.

Taggart, P. 2004. Starch as an Ingredient: Manufacture and Implication. In: Elliason A.C. (ed.) *Starch in Food*, England: CRC Press. pp. 363-392.

Yoshida, K., T. Kondo., K. Kameda., S. Kawakishi., A. J. M. Lubag., E. M. T. Mendoza & T. Goto. 1991. Structures of Alatanin A, B and C Isolated from Edible Purple Yam *Dioscorea alata. Tetrahedron Lett.* 32:5575–5578.

Yoshimoto, M., S. Okuno., M. S. Islam., R. Kurata & O. Yamakawa. 2003. Polyphenol Content and Antimutagenicity of Sweet Potato Leaves in Relation to Commercial Vegetables. *Acta Hortic.* 677–685.

15

Hand Holding Entrepreneurs in Honey Processing and Value Addition

Mani Chellappan

Introduction

India is one of the major producers and exporters of honey. Diversity of bee flora and varied agro-climatic conditions in the country offers enormous potential for the development and growth of apiculture. Honey is a supersaturated sugary material produced mainly by honey bees from the sweet substances of plants (floral or extra-floral nectaries) or insect. Honey bee forager (field bee/worker honey bee) collects the nectar and converts the same into honey by a process of regurgitation and evaporation and stores in the honey comb. The sweetness of the honey is largely by monosaccharide *viz.*, fructose and glucose. Honey is either monofloral or polyfloral as it is collected by the bees. The physical properties of the harvested honey depend upon the source from it is derived, water content, temperature, proportion of the simple sugars present and the method of processing. At low temperature (below 17°C), honey gets crystallized. Honey is also hygroscopic, capable of absorbing water from the atmosphere. India produces approx. 65,000 mt of honey per year (Kejariwal, 2015).

History of beekeeping

Honey bees have a very long history. Sources show fossilized remains of honey bees of 150 million years. Usefulness of honey bees has been recognized by the ancient human both as source of food and their role on the environment. Collecting honey from wild bee colonies is one of the most ancient human activities and is still practised in parts of Africa, Asia, Australia and South America. Some of the earliest evidence of gathering honey from wild colonies is from rock paintings in Spain dating back to 15,000 years. At some point, humans attempted to domesticate the honey bees in artificial hives made from hollow logs, wooden boxes, pottery vessels and woven straw 'skeps'. While the Egyptians knew how to keep bees, they really prized wild honey and, just like those Spaniards 10,000 years before, they would risk their own safety to steal from wild colonies. The real credit for the modern beehive construction and beekeeping, however, goes to Lorenzo Lorraine Langstroth, the "Father of American Beekeeping" of 19th century.

In India, honey and beekeeping have a long history. Ancient Indian inhabiting rock shelters and forests used honey as the main sweet food. India has some of the oldest records of beekeeping in the form of paintings by prehistoric man in the rock shelters. Recent past has witnessed a revival of the honey industry in the rich forest regions along the sub-Himalayan mountain ranges and the Western Ghats, where it has been practised in its simplest form. There are several types of indigenous and traditional hives in India including logs, clay pots, wall niches, baskets and boxes of different sizes and shapes. In modern beekeeping, the combs are built on wooden frames that are moveable. This facilitates inspection and management of bee colonies.

Types of honey bees in India

Giant rock bee (Apis dorsata)
Largest bees found all over India in sub mountainous regions upto an altitude of 2700 m which cannot be domesticated. They build single comb nest with an area up to 1-2 m or more. They are good honey gatherers with an average yield of 50-80 kg/colony/year.

Little honey bee (Apis florae)
This species of honey bee is the smallest of the true honey bees found in plains of India up to the altitude of 500 m and amenable for partial domestication. They build single vertical comb from which a maximum of 1000 g honey per colony could be collected in a year.

Indian honey bee (Apis cerana indica)
Asian honey bee or the native bee as it is otherwise called is the widely domesticated species in India. The bees construct multiple parallel combs and yield an average honey of 8-10 kg per colony per year.

European honey bee (Apis mellifera)
This species is the most widely distributed and commercially reared honey bee species in the world. The bees are larger and have similar habits as that of Indian bees, which build parallel combs. The average honey production per colony is 25-40 kg/year.

Dammer bee/stingless bee (Tetragonula iridipennis)

The bees are not truly stingless but the sting is vestigial. Nests are made in hollowed up area in the ground, trees, bamboo, rocks or walls. Honey yield is rather poor (200-500 g/colony/year).

Honey bee hive

A beehive is any container provided for honey bees to nest in. The idea is to encourage bees to build their nest in a way that is easy for the beekeeper to manage and extract the honey from the honeycomb. Movable frame hives are the most developed way of keeping bees. These are used to maximize the honey crop each season with the least disruption to the bee colony. High populations of bees can be kept in this type of hive and honey stores build up during the flowering season of variety of plants. A typical bee hive has made up of wood and has the following parts.

Bottom board (or) Floor board

It forms the floor of the hive made up of a single piece of wood or two pieces of wood joined together. Wooden beading are fixed on to the lateral sides and back side. There is a removable entrance rod in the front side with two entrance slits to alter the size of the hive entrance based on need. The board is extended by 10 cm in front of the hive body which provides a landing platform for bees. Size of alighting board is 40x28 cm (BIS hive).

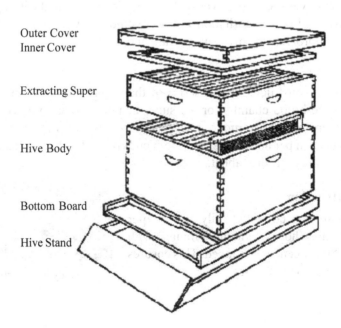

Outer Cover
Inner Cover

Extracting Super

Hive Body

Bottom Board

Hive Stand

Brood chamber

It is a four sided rectangular wooden box of cross section without a top and bottom. It is kept on the floor board. A rabbet is cut in the front and back walls of the brood chamber. Brood frames rest on the rabbet walls. Notches on the outer surface of the side walls are useful for lifting. Four sides of the chamber are joined by special joints. In brood frames, bees develop comb to rear brood. Size of brood frame is (outer dimensions) 29x29x17 cm. There will be 8 frames. Length and height of frame is 20.5x14.0 cm (BIS hive).

Super chamber

It is kept over the brood chamber and its construction is similar to that of brood chamber. Super frames are hung inside. Length and width of this chamber is similar to that of brood chamber. Height may also be similar if it is full depth super or the height will be only half, if it is in a shallow super. Surplus honey is stored in super chamber. The height of the chamber is 9.5 cm. Inner height of the frame is 6.0 cm (BIS hive).

Hive cover

It insulates the interior of the hive. In Newton's hive it has sloping planks on either side. On the inner ceiling plank there is a square ventilation hole fitted with wire gauze. Two holes present in the front and rear also help in air circulation. Hive cover consists of a crown board or inner cover and an outer cover. Inner cover is provided with a central ventilation hole covered with wire gauze. Outer cover is covered over with a metallic sheet to make it impervious to rain water. Circular ventilation holes covered by wire gauze help in air circulation. It protects the hive against rain and sun.

Frames

Frames are so constructed that a series of them may be placed in a vertical position in the brood chamber or the super chamber so as to leave space in between them for bees to move. Each frame consists of a top bar, two side bars and a bottom bar nailed together. Both ends of the top-bar protrude so that the frame can rest on the rabbet.

Honey extraction

Honey should be collected only from super chamber frames (honey containing frames) with more than 80 per cent cells capped. This ensures harvesting the ripened honey.

Honey in Superframe

Stainless steel manual or automatic centrifugal extractors can be used to extract honey from the uncapped combs. High-quality honey can be distinguished by its fragrance, taste, and consistency. Colour of the honey varies with botanical origin, age and storage conditions, but transparency or clarity depends on the amount of suspended particles such as pollen. Darker honeys are more often for industrial use, while lighter honeys are marketed for direct consumption. The steps invoved in the process of honey extraction from super frame is depicted in Fig. 1.

Honey in extractor

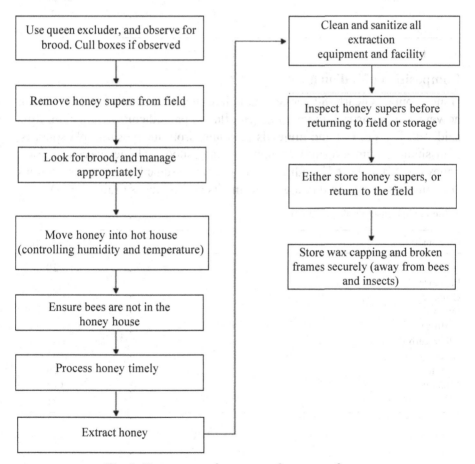

Fig. 1: Honey extraction process from super frames

In many countries with a large honey market, consumer preferences are determined by the colour of honey and thus, next to general quality determinations, colour is the single most important factor determining import and wholesale prices which is measured in Pfund scale (mm). USDA honey colour standards are given in Table 1.

Table 1: USDA honey colour standards in Pfund scale (mm)

S. No.	Parameters	Pfund scale (mm)
1.	Water white	0 to 8
2.	Extra white	> 8 to 17
3.	White	> 17 to 34
4.	Extra light amber	> 34 to 50
5.	Light amber	> 50 to 85
6.	Amber	> 85 to 114
7.	Dark amber	> 114

(Krell, 1996)

Composition of Indian honey

Honey contains nearly 80 per cent carbohydrate, no fat and a meager protein; however, honey is rich in vitamins (riboflavin, pantothenic acid, niacin, folic acid, Vit. B_6, Vit. C) and minerals (calcium, iron, magnesium, phosphorus, potassium, sodium and zinc). Though, one tablespoon of honey has 64 calories, honey is still a preferred healthier choice over table sugar because of its vitamins and minerals that can aid in digestion and its anti-oxidants (Table 2 & 3)

Table 2: Composition of Indian honey

Composition	Quantity (%)
Water	17 - 20
Fructose	38.29
Glucose	31.28
Sucrose	1.31
Maltose	7.21
Other carbohydrates	1.54
Acid	0.57
Protein	0.26
Minerals	0.17
Enzymes, vitamins etc	2.21

Table 3: Nutrient composition of honey (per 100g)

Nutrient	Average amount
Energy (Kcal)	304
Vitamin A (I.U.)	304
Thiamine (mg)	-
Riboflavin (mg)	0.004 - 0.006
Niacin (mg)	0.002- 0.06
Pyridoxine (mg)	0.11.- 0.36
Pantothenic acid (mg)	0.008 - 0.32
Folic acid (mg)	0.02 - 0.11
Vitamin B_{12} (mg)	-
Vitamin C (mg)	-
Vitamin D (mg)	2.2 - 2.4
Tocopherol (I.U)	-
Biotin (I.U)	-
Calcium (mg)	4 – 30
Chlorine (mg)	0.01 - 0.1
Copper (mg)	1. - 3.4
Iodine (mg)	2 – 60
Iron (mg)	0.6 – 40
Magnesium (mg)	1000
Phosphorous (mg)	0.15
Potassium (mg)	400
Sodium (mg)	-
Zinc (mg)	4 – 30

Honey standard in India

As per AGMARK Standard of honey (Schedule II- Grade designation and quality of honey; rules 3 and 4) honey shall be,

- well ripened, natural product produced by honey bees
- of sweet flavour, pleasant odour and taste and characteristic aroma
- of uniform colour through out and may vary from light to dark brown
- free from visible mould, inorganic or organic matters such as insects, insect debris, brood or grains of sand, dirt, pieces of beeswax, the fragments of bees and other insects and free from any other extraneous matter
- free of any added food additives such as colour, vitamins, minerals and saccharin
- free of toxic substances arising from the microorganisms or plants which may constitute a hazard to health
- free of any objectionable flavour, aroma or taint absorbed from foreign matter during its processing and storage
- free from suspended particles

Honey standards have been prescribed under Prevention of Food Adulteration (PFA) Rules, 1955. The Department of Agriculture and Cooperation has laid down standards of honey under the Grading and Marking Rules (AGMARK) (Table 3) which lays down the grades, designation of honey as special grade (<20% moisture), grade A (20-22% moisture) and standard grade (22-25% moisture) to indicate the quality of honey for the purpose of certification. It also specifies the method of packing, marking and labelling of honey.

Table 3: AGMARK standards of honey

Particular	Specification
Specific gravity at 27°C	not less than 1.35
Moisture	not more than 25 per cent by mass
Total reducing sugar	not less than 65 per cent by mass
Sucrose	not more than 5.0 per cent by mass
Fructose : glucose ratio	not less than 0.95
Ash	not more than 0.5 per cent by mass
Acidity (as formic acid)	not more than 0.2 per cent by mass
Fiehe's test	Negative
Hydroxyl methyl furfural (HMF)	not more than 80mg/kg of honey

Well ripened (less than 20% moisture), freshly collected, high-quality honey at room temperature should flow without breaking into separate drops. The honey, when poured, should form small, temporary layers that disappear fairly quickly, indicating high viscosity; otherwise, it indicates excessive water content in it which leads fermentation. Unless the honey is consumed raw, it needs processing for prolonged storage.

Honey adulteration

Honey can be adulterated with sugars, syrups, or compounds to change its flavour or viscosity, make it cheaper to produce, or increase the fructose content to reduce crystallization. Most common adulterant is corn syrup, which, when mixed with honey, is often very difficult to distinguish from unadulterated honey. Only isotope ratio mass spectrometry can detect the adulterant.

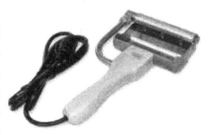

Electrical uncapping plane

Honey processing

Honey processing involves the removal of wax and any other foreign materials from honey. In India, there is a growing awareness on the nutritional, medicinal and industrial uses of honey. Consequently, demand for pure and good quality honey is increasing. Apiary honey produced in coastal areas or tropical humid regions contains more moisture which may lead to fermentation in storage. Such honey has to be processed scientifically with modern processing plants. Recent data shows that more and more entrepreneurs are entering into the honey processing business.

First step of honey processing is the uncapping or removal of thin layer of wax that seals the cells of honey combs using sharp honey knife or electrical uncapping plane (UCP costs $ 145. Uncapping plane is a tool for small or large operations and has an adjustable heated copper cutting blade to remove thin layer of wax from the honey comb).

Large numbers of honey frames are more rapidly processed with partially or completely automated uncapping machines which cut or chop the wax caps with blades, chains or wires. Honey frame processing proceeds, after uncapping, to centrifugal extraction. Honey extractors can be either electrically or manually driven machines which operate on the principle of centrifugal force. Extractors vary in size ranging from small two- frame units to big ones holding up to 85 frames. Manual extractors are equipped with either a hand crank or a bicycle chain while the electrical ones are motor driven. More commonly, 24 to 72-frame radial extractors are used for commercial enterprises. Modified centrifugal extractors can also be used (Krell, 1996).

After extraction of honey, simple straining is done for freshly harvested honey. Liquid honey is allowed to settle overnight and the scum (cream) from the surface of honey is removed by using a spoon before the honey is packed. If honey is intended to be stored it should be either batch or bulk processed.

Batch processing

It is also called water bath method. This method is suitable for semi-processed honey which has been stored for some time and possibly crystallized. Here, honey is first heated in a water bath (indirect heating), up to about 45°C – 50°C. Honey is heated to facilitate both straining and fast handling and to destroy yeast that may be present and may cause fermentation particularly if the moisture content is more than 17 per cent. Indirect heating method involves the use of two containers; the smaller one containing honey is placed inside a bigger one containing water and a piece of wood placed at the bottom so that the smaller one does not touch the bottom of the bigger container. Honey that is being warmed must be stirred to distribute the heat evenly. A glass thermometer is used to measure the temperature of honey till it reaches 65°C. A straining cloth is then folded twice (forms four layers) and firmly tied onto a clean, dry suitable container and honey is poured over. Once, all the warm honey has passed through the cloth, cover the bucket with a lid, and allow it to settle for a minimum of three days to allow the scum to collect at the top of the strained honey.

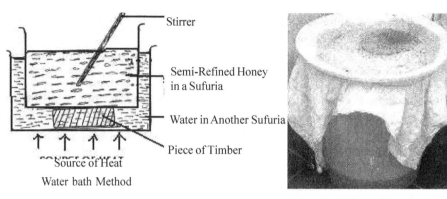

Water bath Method

Straining of honey

Bulk processing

In this method, honey is made to flow through a series of sieves of various sizes. The sieves are arranged in a concentric form, the finest mesh being on the outside and coarser on the inside. The semi-refined honey is heated to 45-50°C in a sump tank and then flows by gravity through the sieves usually referred to as strainers; into a settling tank and is left there for at least 3 days. The scum collects on top of the strained honey, it is then removed and honey is packed. This method is used for large quantity of honey.

In order to bring uniformity of honey collected from different sources, honey is usually blended during processing so that the final product becomes homogenous and has the same physical and chemical properties.

Honey processing plant

Removal of moisture content

Moisture content is the most important quality parameter, since it affects storage life and processing characteristics. Even though, moisture can be removed after extraction, only completely ripe honey should be harvested, i.e. combs with more than 80 per cent of the honey cells capped/sealed. Post- harvest reduction of moisture content is achieved by using boilers in which the temperature of honey is elevated so as to evaporate excess moisture.

Removal of scum

The honey is then poured out into steel cans and then left undisturbed. The scum gets accumulated on the top surface and it can be removed manually.

Packaging

The bottle should be leak proof and airtight so as to safely contain the product, but also present the product in an attractive form, enticing the consumer to buy it.

Storage

Storage containers for honey should be made of stainless steel coated with beeswax. Nothing should be allowed to impart any odour to the honey. Storage conditions should be provided so as to prevent fermentation through either low temperature storage or by preventing further absorption of moisture.

Honey crystallization

Crystallization is an important problem in honey marketing. At low temperature, most honeys crystallize. This is due to the fact that honey is an oversaturated sugar solution, *i.e.* it contains more sugar than can remain in solution. The crystallization results from the formation of monohydrate glucose crystals, which vary in number, shape, dimension and quality with the honey composition and storage conditions. The lower the water and the higher the glucose content of honey, the faster the crystallization. Temperature is an important criterion in this process and at around 14°C fast crystallization occurs, but also the presence of solid particles (*e.g.* pollen grains) and slow stirring result in quicker crystallization. Usually, slow crystallization produces bigger and more irregular crystals.

Method of packing (AGMARK)

- The Honey shall be packed in new clean glass containers, china-ware lacquered cans, acid resistant lacquered tin container, cartons, pet jars of food grade quality or any other containers or packing material of food grade quality as may be approved by the Agricultural Marketing Adviser from time to time.

- All packing materials shall be securely closed and sealed in a manner approved by the Agricultural Marketing Adviser.

- Honey shall be packed in pack sizes as per instructions of the Agricultural Marketing Adviser issued from time to time.

- The containers shall not be composed wholly or partly of any poisonous or deleterious substances which renders the contents injurious to health.

- The containers shall also be free from insect infestation, fungus contamination or any obnoxious and undesirable smell. The screwed caps shall be of non-corrosive and non-reactive material to honey.

Health benefits of honey

Honey is a functional food and has different biological properties such as antibacterial (bacteriostatic properties), anti-inflammatory, wound and sunburn healing, antioxidant, anti-diabetic and anti-microbial activities.

Honey has been consumed for thousands of years for its supposed health benefits. The *vedic* civilization considered honey as one of nature's most remarkable gifts to mankind. Traditionally, according to the texts of Ayurveda, honey is a boon to those with weak digestion. Also it has been emphasized that the use of honey is highly beneficial in the treatment of irritating cough. Honey is regarded by Ayurvedic experts, as valuable in keeping the teeth and gums healthy (Eteraf-Oskouei and Najafi, 2013). Honey is used for both internal and external applications for variety of ailments and predominantly as a vehicle for faster absorption of various herbal drugs. It is mainly used for the treatment of eye diseases, cough, thirst, phlegm, hiccups, blood in vomit, leprosy, diabetes, obesity, worm infestation, vomiting, asthma, diarrhoea and healing wounds. It is also used as a natural preservative and sweetener in many ayurvedic preparations and for delivering some medicines thereby improving the efficacy or to mitigate the side effects of the other medicines it is mixed with. All natural honey contains flavonoids (*viz.,* apigenin, pinocembrin, kaempferol, quercetin, galangin, chrysin and hesperetin), phenolic acids (such as ellagic, caffeic, p-coumaric and ferulic acids), ascorbic acid, tocopherols, catalase, superoxide dismutase, reduced glutathione, maillard reaction products and peptides. Most of those compound works together to provide a synergistic antioxidant effect (Rakha *et al* 2008). Enzyme glucose oxidase produces hydrogen peroxide (which provides antimicrobial properties) along with gluconic acid from glucose which helps in calcium absorption. Modern science is finding that many of the historical claims that honey can be used in medicine may indeed be true. Honey has been reported to have an inhibitory effect on 60 species of bacteria (Olaitan *et al.* 2007). It also serves as a general tonic for children, the young and the elderly, the convalescent and hard working people.

Treating bacterial, fungal and viral infections

Natural honey kills bacteria three times more effectively. Some studies have revealed that a certain type of honey is useful in the fight against multi-drug resistant bacterial infections

Treatment of wounds and burns

There have been some cases in which it is reported positive effects of using honey in treating wounds due to the presence of trace amounts of hydrogen peroxide and methylglyoxal. The enzyme glucose oxidase of honey provides glucose to leucocytes, which is essential for respiratory burst to produce hydrogen peroxide leading to the antibacterial activity. Honey in its pure and unprocessed form is also used in pharmaceutical preparations applied directly on open wounds, sores, bed sores, ulcers, varicose ulcers and burns. It helps against infections, promotes tissue regeneration, and reduces scarring also

Gastro intestinal benefits of honey and treatment of gastroenteritis in children

Honey has been used as a possible remedy for the gasteroesophageal reflux owing to its viscosity (honey is 125.9 more viscous than distilled water at 37°C). Honey is found to improve nutrient uptake from gastro intestinal tract and to be useful for chronic and infective intestinal problems such as constipation, duodenal ulcers and liver disturbances. Honey can be used safely as a substitute for glucose in oral rehydration solution containing electrolytes. Also, honey shortens the duration of bacterial diarrhea in children. Diluted honey at 5% (v/v) concentration, decrease the duration of diarrhea in cases of bacterial gastroenteritis. However, in viral gastroenteritis no such effect could be observed. In rehydration fluid, honey adds potassium and water uptake without increasing sodium uptake and also acts as an anti-inflammatory agent on intestinal mucosa (Bansal *et al.*, 2005).

Treatment of allergies

There is some research to suggest that honey may be useful in minimizing seasonal allergies.

Treatment of common cold and benefits to the respiratory system

Honey has the ability to reduce night time coughing and to improve sleep quality in children with upper respiratory infection. World Health Organization (WHO) and American Academy of Pediatrics recommend honey as a natural cough remedy. In temperate regions and places with considerable temperature fluctuations, honey is a well known remedy for cold. It is also helpful for mouth, throat or bronchial irritations and infections. The benefits may be due to the soothing and relaxing effect of fructose present in honey.

Honey as food

Raw honey is most commonly consumed as extracted honey or comb honey or crystallized honey as food, medicine and as an ingredient in food recipes. Only a few developed countries, considered honey as food. In most of Asia, it is generally regarded as a medicine. Honey is used widely in baked products, confectionery, candy, marmalade, jam, spreads, breakfast cereals, beverages, milk products and many preserved food products.

Types of honey commercially available in the market

Raw honey: Raw honey is a minimally processed honey obtained by extraction, settling, or straining, without heating. Raw honey contains some pollen and may contain small particles of wax.

Comb honey: Comb honey is still in the honey bees' wax comb. Cut-comb honey is the simplest processing in which the honeycomb is removed from frame hives, top-bar hives or traditional hives and sell or consume it as such. Process involves collecting pieces of sealed and undamaged honeycomb, cutting them into uniform sized pieces and packaging them carefully in bags or cartons to avoid damaging the honeycomb. Since, the honeycomb is unopened, it is readily accepted to be pure, and it has a finer flavour than honey that is exposed to air or processed further. Cut-comb honey can therefore have a high local demand and fetch a higher price than processed honey. However, the honeycomb

is easily damaged by handling and transport, which makes distribution for retail sale more difficult. It requires protection by packaging materials that will absorb shocks or vibration (e.g. cushioning plastics such as 'bubble-wrap' and/or corrugated cardboard cartons) and packs should be carried carefully and not stacked, thrown or dropped to avoid damage to the honeycombs (http:// www.appropedia.org/Honey_Processing _(Practical_ Action_ Technical _Brief).

Crystallized honey: Crystallized honey occurs when some of the glucose molecules are spontaneously crystallized from solution as the monohydrate. The resultant honey is also called "granulated honey" or "candied honey". Depending on the source of the nectar collected by bees, some types of honey are more likely to granulate than others, but almost all honey will granulate if its temperature falls sufficiently. Granulation is a natural process and there is no difference in nutritional value between solid and liquid honey. Although, there is obviously a difference in the texture between liquid and granulated honey, there is no difference in the flavour or other quality characteristics. Honey that has granulated (or commercially purchased crystallized honey) can be returned to a liquid state by warming. Some customers prefer granulated honey, and if liquid honey is slow to granulate, the addition of 20 per cent finely granulated honey will cause it to granulate.

Pasteurized honey: Pasteurized honey has been heated in a pasteurization process which requires temperatures of 161 °F (72 °C) or higher. Pasteurization destroys yeast cells. It also liquefies any microcrystals in the honey, which delays the onset of visible crystallization.

Strained honey: Strained honey has been passed through a mesh material to remove particulate material (pieces of wax, propolis, etc.) without removing pollen, minerals, or enzymes.

Filtered honey: In this all or most of the fine particles, pollen grains, air bubbles, or other materials have been removed. In this process, honey is heated to 66–77°C and the filtered honey is very clear and will not crystallize quickly.

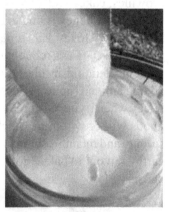

Ultrasonicated honey: Ultrasonicated honey has been processed by ultrasonication. Ultrasonication eliminates yeast cells and existing crystals in honey.

Creamed honey: Creamed honey is smooth with spreadable consistency produced by controlled crystallization. It is also called whipped honey, spun honey, churned honey, honey fondant, and set honey. Creamed honey contains a large number of small crystals, which prevent the formation of larger crystals that normally occur in unprocessed honey at lower temperature.

Powdered honey or dried honey: Water is extracted from liquid honey to produce completely solid, non-sticky powdered honey. Dried honey is used to garnish desserts and used in production of bakery goods.

Chunk honey: Chunk honey is simply a piece of honey comb immersed in a container that is filled with extracted liquid honey. It is the combination of comb and liquid honey.

Honey decoctions: Honey decoctions are made from honey or honey by-products which have been dissolved in water, then reduced (usually by means of boiling)

Baker's honey: Baker's honey is outside the normal specification for honey, due to a "foreign" taste or odour, or because it has begun to ferment or has been overheated. It is generally used as an ingredient in food processing.

Uses of honey

Honey in baked products

Honey in baked products confers uniform baking with more evenly browned crust at lower temperatures, improved aroma, spongy consistency and softness. Honey also keeps the product to have a lesser tendency to crack. Bread, cookies, brownies, cakes, muffins, etc. can be made with honey replacing white sugar. On average, honey is one to one and half times sweeter (on a dry-weight basis) than sugar. Honey can replace sugar in almost any recipe if few rules are followed as, always reducing the oven temperature by 25 degrees, reducing the amount of liquid in the recipe by ¼th cup for every cup of honey used and adding a pinch (up to ½ 2 teaspoon) of baking soda to neutralize the acidity of honey when trying custom recipes. With honey, the batter will be thinner, baked goods will brown more quickly in the oven and the finished products will be springy and retain freshness for a longer period of time. Honey imparts a unique, earthy flavour to baked goods that can be altered based on the variety of honey chosen.

Honey in milk products

Honey is the ideal sweetener in milk, because it provides desirable flavour. As the flavoured dairy product industry continues to grow, dairy manufacturers are looking for creative ways to reformulate and rebrand their products with many turning to honey due to its consumer appeal and versatility as a flavour and sweetener. Honey-milk, honey cream, honey cream cheese, honey butter and honey yoghurt are some of the popular recipes. Apart from energy recipes, honey milk  combination is ideal for skin care as the raw honey speed up the skins natural regeneration process. Propolis and bee pollen in raw honey stimulate growth of new skin. Amino acids in honey help skin retain moisture and sugar content makes honey excellent humectants (http://milkhoney.co.za/benefits.html).

Honey in confectionery

Consumers always recognized honey as a beneficial natural sweetener, but for some reasons honey rarely is considered as a prime ingredient in confections. This is especially true of baked sweets, which present formulation challenges due to honey being sweeter and, of course, wetter than sugar. Honey is used in few confectionery products and also used in very small quantities for the production of caramels.

Honey in breakfast cereal industry

Cereal is one breakfast options that *may* contribute to optimal health/body composition. Majority of cereals have added sweeteners/preservatives and are highly processed. Honey can be a better alternative to sugar in the breakfast cereal and can be used either mixed or applied as a component of the sweetening and flavouring agent.

Honey with dry fruits and nuts

Dry fruits/nuts can be placed directly into the honey, either whole or chopped. Pasteurization of both honey and fruits improves the hygiene and storability and reduces the risk of fermentation.

Honey in spreads

A variety of spreads with honey could be made either for flavour or as a substitute for sugar

Honey in nonalcoholic beverages

Honey is usually mixed with lemon juice and other fruit juices for flavour and as sweetener. Ice tea can be clarified and flavoured with the addition of lemon juice and honey.

Fermentation products of honey

Honey is used in making alcoholic drinks by fermentation. Those drinks are popularly known as honey beer, mead and wines. Mead is made by fermenting honey with water. Like beer, mead is sometimes flavoured with fruits, spices, grains or hops. But it is generally higher in alcohol than beer and more in line with grape wine—typically between eight and 20 percent ABV (alcohol by volume number).

Honey collection and its marketing in India are still not fully organised. The main uses of honey are in cooking, baking, as a spread on breads and as an addition to various beverages such as tea and as a sweetener in commercial beverages. Honey is also used in medicines. A number of small scale industries depend upon bees and bee products. Honey and bee products finds use in several industries which are under; pharmaceuticals, meat packing, beeswax in industries, bee venom, royal jelly, bee nurseries, bee equipment and hives etc. There is considerable demand for honey and other products. Outside thousands of homemade recipes in each cultural tradition, honey is largely used on a small scale as well as at an industrial level.

Quality and safety of honey products

Though, honey is preserved by itself because of its high sugar content, it must be handled hygienically and all tools used in the production of honey must be properly sanitized. The taste and flavour of honey are the important qualitative parameters. Higher moisture content (>20%), high HMF content (>80mg/kg of honey), higher proportion of pollen and contamination by pesticides, antibiotics and insects in honey lowers the quality of honey.

Honey should not be given to babies because of the risk of infant botulism. Honey consumption may interact adversely with existing allergies and high blood sugar levels.

Beekeepers extract honey in the open space and store mostly in containers made up of tin .This arrangement not only spoils the honey, but also turns it dark. To avoid such losses, the beekeepers should be encouraged to extract honey properly under roof or by putting tent so as to save it from direct sunlight and moisture. Metal react with honey and spoils it. Therefore, it should be stored in stainless steel or food grade plastic containers. As far as retail honey is concerned it should be made available in wide mouthed glass bottles well sealed and air light. Instances of bees' mortality through the use of pesticides/ insecticides are also reported. For this purpose, the farmers should be educated to use the chemicals in the evening in consultation with beekeepers in a periphery of 2-3 Km. The above policy measure shall go a long way in promoting and strengthening the beekeeping activities in the country. In the ultimate analysis, these measures shall help in making it effective to boost income and employment opportunities for the rural poor.

Beeswax

Beeswax is one of the main by-products of honey extraction. Bees built the comb with wax to hold greater amount of honey with least amount of wax. Beewax is secreted by 14-18 days old worker bees. Bees, after feeding on honey and pollen, hang themselves in festoons and secrete wax flakes/scales. These wax flakes are masticated by worker bees into pliable pieces with saliva and enzymes and then used for comb construction and capping brood and ripened honey combs. To produce wax, the bees have to consume about eight times as much of honey. Freshly produced beeswax is whitish to yellowish in colour with honey like odour. The colour may vary depending on the food of the honey bees. Beeswax is soft at warm temperature and becomes brittle at low temperature. Beeswax consists of at least 284 different compounds, mainly a variety of long-chain alkanes, acids, esters, polyesters, and hydroxy esters, but the exact composition of beeswax varies with location. Its density is 0.95 – 0.96 and melting point   is 62.5°C. The possibilities for using beeswax are nearly endless, including skin care, ornaments, church candles, cosmetics, shoe polish, car/mason polish, carbon paper, and in metal castings and moldings, certain adhesives and inks, waxed fabric, candy and chewing gums. It is also used for water proofing, scientific decorative models (batik), polishing optical lenses, bow strings in musical instruments as well as in electric and textile industries and crayon colour industry. In beekeeping industry, beeswax is mainly used to prepare comb foundation sheets (Ahnert, 2015). Beeswax can be made into an ideal moisturizing cream. Beeswax (2%) with rice bran oil protects fruits and vegetables over a month without refrigeration.

Institutions involved in the apiculture and honey processing

A host of institutions have been established in the country to contribute to the development of beekeeping and disseminate information in India (Agarwal, 2014).

i. The All India Beekeepers' Association has made laudable contributions to the development of beekeeping and has been disseminating information about honey trade through its informative publication 'Indian Bee Journal'.

ii. Host of institutions have been set up both by Government of India and State Governments to promote export of honey.

- Agricultural Products Export Development Authority (APEDA) under the aegis of the Ministry of Commerce and Industry, Government of India, is the nodal agency to promote export of honey.

- Tribal Cooperative Marketing Development Federation of India Ltd (TRIFED) has been playing an important role by providing training to tribals in the scientific cultivation and harvesting of wild honey.

- A lot of work has also been done on honey related issues like Indian bees and beekeeping by individuals, agricultural experts, agricultural colleges and institutions.

- The Central Bee Research and Training Institute, Khadi and Village Industries Commission (KVIC) not only contributed to the science of bees, bee plants and beekeeping but also developed several appropriate technologies suited to Indian beekeeping.

- The Export Inspection Council (EIC) under the Union Ministry of Commerce arranges for tests on residues, antibiotics, etc., in honey.

- ICAR through the AICRP on Honey bees and Pollinators has done exemplary work on honey bee rearing, pollination services, etc.

Success story of beekeeping in Kerala

Mr. Haridasan A, hailing from Muzhappilangad in Kannur District of Kerala, is a resource person in the field of apiculture. His services are utilized by various agencies for offering practical training in beekeeping. Mr. Haridasan took interest in beekeeping when he was working as a carpenter. His interest in beekeeping was triggered by a beekeeper who approached him to make beehives and who also gave him practical knowledge about the subject. Now, he is maintaining bee colonies and has been identified as a bee-breeder by Rubber Board and Horticorp. He also helps beekeepers in Kannur, Kasaragod, Kozhikode and other Districts. His tiny bee-box manufacturing unit supplies bee-boxes and other accessories to the beekeepers and other agencies. Mr. Haridasan's achievements as beekeeper earned him a national award instituted by Khadi and Village Industries Commission (KVIC) during 2010. Mr. Haridasan started his beekeeping vocation by initially capturing bee colonies of local bee species *Apis cerana indica* available in his locality and then setting them up in his own apiary at his native village and nearby areas. By that time, he also took training in beekeeping from KVIC and Khadi and Village Industries Board (KVIB) for learning scientific beekeeping and better bee management practices. He now runs a honey processing unit, a mini honey testing laboratory and bee nurseries at Muzhappilangad.

Conclusion

Beekeeping is promoted in India with the primary objective of honey production for supplementary farm income. There is a huge scope for value addition to honey and other bee hive products. Honey should be processed and standardized to meet the domestic and international standards. With the improvement of living standards, honey is finding a place in every house hold.

References

Agarwal, T. J. 2014. Beekeeping Industry in India: Future Potential. *Int. J. Res. Appl. Nat. Sco. Sci.* 2: 133-140.

Ahnert, P. 2015. *What Is Beeswax?* http://www.motherearthnews.com/natural-health/what-is-beeswax-ze0z1511zcwil?

Bansal, V., B. Medhi & P. Pandhi. 2005. Honey- A Remedy Rediscovered and its Therapeutic Utility. *Kathmandu Univ Med J.* 3: 305–309.

Eteraf-Oskouei, T. & M. Najafi. 2013. Traditional and Modern Uses of Natural Honey in Human Diseases: A Review. *Iranian J. Basic Med. Sci.* 16: 731-742.

Kejariwal, P. 2015. *An Overlook on the Indian Honey Industry.* http://www.cseindia.org/userfiles/food_safety-march12/Prakash_kejriwal.pdf

Krell, R. 1996. Value Added Products from Beekeeping. *FAO Agricultural Services Bulletin No.124*, Food and Agriculture Organization of the United Nations, Rome. http://www.fao.org/docrep/woo76Eoo.htm.

Olaitan, P. B., E. O. Adeleke & O. I. Ola. 2007. Honey: A Reservoir for Microorganisms and an Inhibitory Agent for Microbes. *Afr. Health Sci.* 7: 159–165.

Rakha, M. K., Z. I. Nabil & A. A. Hussein. 2008. Cardioactive and Vasoactive Effects of Natural Wild Honey against Cardiac Malperformance Induced by Hyperadrenergic Activity. *J Med Food* 11: 91–98.

16

Value Addition in Mushroom Processing- A Potential Sector for Budding Entrepreneurs

Lakshmy P S, Arun Prasath V, & Suman K T

Introduction

Mushrooms are fleshy spore-bearing fruiting bodies of fungi, produced above ground on soil or on their food sources. Total number of useful fungi having edible and medicinal values is over 2,300 species. Like green plants, they do not contain chlorophyll and as a result cannot manufacture their own food. In this respect, they are like animals because; they feed themselves by digesting other organic matter. They are a large heterogeneous group having various shapes, size, appearance and edibility.

The use of mushrooms as food is very old. People harvested mushrooms from the wild for thousands of years for food and medicine. The Egyptians regarded mushrooms as food for pharaohs. The Greeks and Romans described them as food for God and served only on festive occasions. The first professional growers of mushrooms were the Chinese. China is the major producer and consumer of both edible and medicinal mushrooms. As early as 1313, a document was published which described the cultivation method for Shiitake on wood logs. Even older is the cultivation of the wood ear mushroom (*Auricularia spp*) of which sources indicates that it was cultivated from the year 600 A.D. onwards. The white button mushroom, *Agaricus bisporus*, was domesticated in France.

Growing mushrooms under controlled condition is of recent origin. In India, its production earlier was limited to the winter season, but with technology development, these are produced almost throughout the year in small, medium, and large farms, adopting different levels of technology. Its popularity is growing and it has become an export-oriented business. Today, mushroom cultivation has been taken up in Kerala, Uttar Pradesh, Haryana, Rajasthan etc. while earlier it was confined only to Himachal Pradesh, J&K and Hilly areas.

Small scale mushroom production represents an opportunity for farmers interested in an additional enterprise, and is a special option for farmers without much land. Agricultural waste materials, such as sugarcane bagasse, wheat and rice straw, bean and soya bean stalks, maize stovers, banana pseudo stem and many others are excellent growing media for edible mushrooms.

Health benefits of mushrooms

Mushrooms are foods with high nutritional value, with high protein, essential amino acids, and unsaturated fatty acids. Moisture content of fresh mushrooms varies within the range of 70 - 95% depending upon the harvest time and environmental conditions, whereas it is about 10-13% in dried mushrooms. Protein content of the cultivated species ranges from 1.75 to 5.9 % of their fresh weight and on dry weight basis; it varies from 19 to 35%. Furthermore, mushroom protein contains all the nine essential amino acids required by man. Most abundant essential amino acid turns out to be lysine and the lowest levels among the essential amino acids are tryptophan and methionine. Amino Acid Score of the most nutritive mushrooms rank with those of meat and milk, and are significantly higher than those for most legumes and vegetables. Mushrooms are relatively a good source of vitamins including thiamine, riboflavin, ascorbic acid, biotin and niacin. They are low in calories, carbohydrates and calcium. Fat content in different species of mushrooms ranges from 1.1 to 8.3% on dry weight basis. At least, 72% of the total fatty acids are unsaturated, thus making mushrooms a health food. Mushrooms are a good source of minerals. It is calculated that the concentrations of K, P, Na, Ca, and Mg constitute about 56 to 70% of the total ash content. Potassium is particularly abundant and accounts for nearly 45% of the total ash. They are good in selenium, an antioxidant which helps to prevent cell damage, and copper, a mineral that aids in the production of red blood cells. In fact, mushrooms are the only produce that contains significant amounts of selenium.

Edible mushrooms also contain biologically active substances that contribute to the prevention and treatment of human diseases. In recent years much attention has been focused on various immunological and anti-cancer properties of certain mushrooms, apart from other potentially important benefits such as antioxidant property, anti-hypertensive and cholesterol lowering properties, liver protection, weight management etc. High fiber, vitamin C and potassium in the mushrooms promote cardiovascular health and alleviate the risk of heart stroke. Besides, sodium and potassium together work well to regularize the blood circulation and the habit of mushroom consumption reduce plasma viscosity and combat the problem of high BP. Regular consumption of mushrooms reduces LDL levels (bad cholesterol) also. These properties have attracted the interest of many pharmaceutical companies which are viewing the medicinal mushrooms as rich sources of innovative biomedical molecules.

Scope of entrepreneurship developments in mushroom processing sector

Mushroom farming business will be the perfect option for a person who has little knowledge in the science and technology of mushroom growing and who owns a building for having the farm. Mushroom cultivation is an art which requires both study and experience. Different types of mushrooms have different production cost and it is important to decide on a budget depending on amount of money available and the long term investment benefit. Kerala's hot humid climatic conditions aids in the cultivation of oyster mushroom and milky mushroom with minimal investment.

The best and the easiest way to market mushroom is in the fresh stage itself. But, if the entrepreneur is aiming for a bigger market, it is better to increase the value of fresh mushrooms. Adding value to fresh mushrooms usually means either developing processed products or drying surplus mushrooms for sale in the off-season, when prices are higher. Processed and preserved mushrooms will ensure the availability throughout the year. It is a good substitute for standard vegetables and a good replacement for meat or eggs. High quality commercial cultivation and processed products of mushrooms even on a small scale is a viable proposition as it is in good demand both in domestic and foreign markets.

The recent upsurge of interest in traditional remedies for various physiological disorders and the recognition of numerous biological response modifiers in mushrooms have led to the coining of the term "mushroom nutriceuticals". A mushroom nutriceutical is a refined/partially refined mushroom extractive which is consumed in the form of capsules or tablets as a dietary supplement and which have potential therapeutic applications. So, the entrepreneur who wants to start a good business can go for this tertiary level of mushroom processing.

Value added products of mushroom

Mushrooms are usually enjoyed fresh, but this can be problematic as most species should be consumed within three to four days of harvesting in order to avoid spoilage. Mushroom growers, large and small, always end up with harvested mushrooms that are not good enough to sell. Mushrooms may have cosmetic defects, be smaller than average, or have more mushrooms than the customers. One of the best ways to turn those extra mushrooms into profit is to make value added products. Value can be added to the mushrooms at various levels, right from grading to the readymade snacks or the main-course item. Real value added product in the Indian market is the mushroom soup powder.

Technologies for production of some other products like mushroom based biscuits, nuggets, preserve, noodles, papad, candies and readymade mushroom curry in retort pouches have been developed but are yet to be popularized (Wakchaure *et al.*,2010).

Minimally processed mushrooms

Minimal processing can provide fresh and quality products through a hurdle approach. Most of the mushrooms in urban areas are marketed like this (Fig.1). Processing steps included in minimal processing are grading and cutting and packaging. As mushrooms contain high amount of water, it should be dried before packaging.

Fig.1. Packaged fresh mushrooms

Dehydrated mushrooms

Most mushroom varieties can be dried whole or dried in slices. Conventional hot-air drying is considered as a comparatively simple, economical and efficient method to extend the shelf life of mushrooms. A properly dehydrated mushroom should snap apart like a cracker (Fig.2). After drying, they should be packed properly so that it remains airtight. Re-sealable glass jars work best, both for eye appeal at the farmer's market, and for ease of re-use.

Fig. 2. Dehydrated mushrooms

Freeze dried mushrooms

Freeze drying being a low temperature process causes less deterioration of aroma compounds in mushrooms. In this process water is eliminated by sublimation from a frozen state and the temperature during the process is very low. Freeze dried mushrooms have a very good shelf life and the reconstituted product is much superior in texture and flavour than that of oven dried mushrooms (Fig.3).

Fig. 3. Freeze dried mushrooms

Smoked dried oyster mushrooms

Smoked dried oyster mushrooms have a unique flavour, combining the subtle flavour of the mushrooms with the mild flavour of smoke. Smoked dried mushrooms can be rehydrated by keeping in boiling water for ten minutes and added to recipes for better taste.

Mushroom soup powder

High quality ready-to-make mushroom soup powder can be prepared (Fig.4). Mushroom soup powder using mushroom powder is produced from the dried button and oyster mushrooms. Dried button mushroom slices or whole oyster mushrooms are finely powdered in a pulveriser and passed through 0.5 mm sieve. Mushroom soup powder (Fig.4) is prepared by mixing this powder with milk powder, corn flour and other seasonings (Wakchaure et al., 2010).

Fig. 4: Mushroom soup powder

Mushroom biscuit

Delicious and crunchy mushroom biscuits can be prepared using button/oyster mushroom powder and ingredients like maida, sugar, ghee, coconut powder, baking soda, ammonium bicarbonate and milk powder. For making biscuits, mix ghee and powdered sugar for 5-7 minutes using dough kneader and make the mixture homogenous. Then, add all other ingredients and make dough. Water/milk can be added to make dough cohesive. Spread the dough in thin sheet, cut into different shapes and bake in an oven for about 15-20 minutes (Wakchaure et al., 2010).

Mushroom seasonings

A blend of finely ground dried mushrooms and coarsely-ground sea salt, or blend of mushroom powder and dried herbs, such as an Italian blend, masala blend etc. which complement chicken or a savory soup are popular items (Mahamud et al., 2012).(Fig.5).

Fig. 5: Mushroom seasoning

Pickled mushrooms

Mushrooms for pickling are either blanched or fried in oil till brown. Depending upon taste add various powdered condiments as per local preferences and practices. Condiments can also be fried in oil before adding to mushrooms (Fig.6).

Mushroom nuggets

To prepare mushroom nuggets, mix mushroom powder with pulse powder and prepare a paste by adding required quantity of water. Add spices to this paste and make round balls of 2-4 cm

Fig. 6. Mushroom pickle

diameter. Spread the balls over a steel tray and sundry. Nuggets can be relished either by deep frying as snack or can be used in vegetable curry along with other vegetables (Wakchaure *et al.*, 2010).

Mushroom ketch-up/sauce

Ketch-up is a common and popular product relished for its typical taste and texture as accompaniment with snacks. Freshly harvested button mushrooms after washing and slicing are cooked in 50% water for 20 minutes and prepare a paste using mixer grinder. Add spices, vinegar etc and cook to bring its TSS to 35°Brix. Fill the ketch-up in sterilized bottles or jars.

Mushroom preserve (Murabba)

Fresh button mushrooms after grading and washing are blanched in 0.05% potassium metabisulphite (KMS) solution for 10 min. Then, it is treated with 40% of its weight of sugar daily for 3 days. On the fourth day, take mushrooms from the syrup and add 0.1% citric acid and the remaining 40% of sugar in the syrup and bring its concentration to 65°Brix. Then, add mushrooms in the syrup to prepare good quality *murabba.*

Mushroom candy

The process for making candy is practically the same as that employed in the case of mushroom preserve, with the difference that the produce is impregnated with a higher concentration of sugar. Total sugar content of the impregnated produce is kept at about 75% to prevent fermentation. Fresh mushrooms after harvesting are washed and halved longitudinally into two pieces. Halves are blanched for 5 min in 0.05% KMS solution. After draining, treat mushrooms with sugar. Sugar treatment is given at the rate of 1.5 kg sugar per kg of blanched mushrooms. Initially, sugar has to be divided into three equal parts. On the first day, cover blanched mushrooms with one part of sugar and keep for 24 h. Next day, cover the same mushrooms with the second part of sugar and keep overnight. On the third day, remove mushrooms from syrup and boil with third part of sugar and 0.1% of citric acid and bring its concentration up to 70°Brix. Mix mushrooms with this syrup and again boil the contents for 5 min to bring its concentration to 72°Brix. After cooling, remove the mushrooms from the syrup and drain for half an hour. The drained mushrooms are placed on the sorting tables to reject defective and unwanted pieces. Finally, dry the mushroom pieces in a cabinet drier at 60°C for about 10 hours. As soon as these become crispy, pack in polypropylene bags and seal. The mushroom candy can be stored up to 8 months with excellent acceptability and good chewable taste (Wakchaure *et al.*, 2010).

Mushroom chips

Freshly harvested button mushrooms after washing and slicing (2 mm) are blanched in 2% brine solution. Keep the mushrooms in a solution containing 0.1% citric acid, 1.5% NaCl and 0.3% red chili powder overnight. Next day, drain off the solution, dry mushrooms in a cabinet dryer at 60°C for 8 hours. Then, fry in oil. Garam masala and other spices can be spread over the chips to enhance the taste.

Ready-to-serve mushroom curry

In view of the growing global market for the readymade/ready-to-eat food items and keeping in mind the popularity of Indian curry, a technology was developed at ICAR- Directorate of Mushroom Research (DMR), Solan to produce "Mushroom curry in flexible-retortable pouches" (Fig.7). The retort pouch of 105 um thick with polypropylene outer layer (80 μ), aluminium middle layer (12.5 μ) and polyester inner layer (12.5 μ) available in the market can be used for packing mushroom curry.

Canned mushrooms

Canning is a technique by which the mushrooms can be stored for longer periods up to a year and most of the international trade in mushrooms is done in this form (Fig.8). Canning process can be divided into various unit operations namely cleaning, blanching, filling, sterilization, cooling, labelling and packaging. In order to produce good quality canned mushrooms, process mushrooms as soon as possible after harvest. Mushrooms with a stem length of one cm are preferred and are canned whole, sliced as per demand. High loss in weight of mushrooms is the most serious problem in canning. This is also known as 'shrinkage', which

Fig. 7: Ready to eat curry   **Fig. 8:** Canned mushrooms

is caused by removal of water as well as solids from the mushrooms during processing operations. loss varies from 35-40 per cent and seriously affects the profitability of the cannery. Water binding additives viz., sodium polyphosphate, sodium alginate, agar-agar, methyl cellulose, carboxy methyl cellulose, pectin and pectin-calcium chloride have been used by various workers in increasing the drained weight.

Mushroom growing kits

Mushroom growing kits are simple to make and very profitable. These one-use kits allow customers to experience the joy of growing and harvesting their own mushrooms without the work involved in preparing and inoculating the substrate. It consists of just a plastic grow-bag filled with pre-inoculated substrate, such as sawdust inside a cardboard box.

Chitin and chitosan from mushrooms

Chitin is a ubiquitous biopolymer which occurs naturally as a major component in the skeletal or exoskeletal structures of lower animals. Chitin is also present in vast majority of fungi as the principal fibrillar polymer of the cell wall. The deacetylated form of chitin is chitosan. It has unique properties which make it useful for a variety of industrial applications such as viscosity control agent, adhesive, paper-strengthening agent and flocculating aid. Chitosan has recently been used in cosmetics, pharmaceuticals, food additives and agriculture. It is used as a component of toothpaste, hand and body creams, shampoo, lowering of serum cholesterol, cell and enzyme immobilizer, as a drug carrier, material for production of contact lenses, or eye bandages etc. On a commercial scale, chitosan is extracted from the exoskeleton of crustaceans employing harsh chemical treatments. However, this extraction process, together with the variability in source material leads to inconsistent physico-chemical characteristics of the chitosan produced. These characteristics make fungi a promising chitosan source as the physical properties of the extractable chitosan from fungi can be manipulated.

Mushroom nutraceuticals

The term "mushroom nutraceuticals" is used to describe those compounds that have considerable potential as dietary supplements, used for the enhancement of health and prevention of various human diseases. The important edible mushrooms with remarkable nutraceutical properties include the species *of Lentinus, Auricularia, Hericium, Grifola, Flammulina, Pleurotus, Lactarius, Pisolithus, Tremella, Russula, Agaricus, Cordyceps,* and so forth. A mushroom nutraceutical is refined/partially refined extractives from either mycelium or fruiting bodies, which is consumed in the form of capsule or tablet as dietary supplement (Fig.9).

Fig. 9: Nutraceuiticals

Apart from the above described value added products, mushrooms are used for making papad, chutney powder, jam, cake, dokhla, sev, idli, macroni , noodles, snacks etc.

New technologies in mushroom processing

Refrigeration/ Instant packing

Freshly harvested mushrooms are packed in 25 gauge polythene bags without making any holes. Immediately after packing, they are stored at 5°C in a refrigerator. This process extends the storage life for 3-5 days. This process helps in reducing the respiratory rate and minimizing the water loss. In addition, it reduces browning of mushrooms and off flavour development.

Dehydration

Clean mushrooms and blanch in boiling water for 2 minutes and immerse in cold water for 2 minutes. Dip mushrooms in water containing 0.2% potassium metabisulphite and 1% citric acid. Pretreated mushrooms are dried in open sunlight till it reaches 1/10th weight of the fresh product. After drying, it can be stored for 3 months, however colour may turn to brown and appearance of the final product may not be good. Mushrooms can be dried in a flow drier at 60°C with heated air for 6-8 hr after pretreatment. This process will reduce the final moisture level to 3-5%. Pretreated mushrooms can also be dried under vacuum at 40°C instantly. This process yield good quality mushrooms.

Canning

Canning is adopted on a very large scale, especially for preservation of button mushrooms. For canning purpose, mushrooms should be harvested at an early stage. Select mushrooms of uniform size, cut the stalks, wash in clean water to remove dirt and other foreign materials. Dip mushrooms in boiling water for 2 minutes, take it out and dip in cold water for 2 minutes. Fill mushrooms in specially made cans upto ¾ capacities. Prepare a solution consisting 2 per cent common salt, 2 per cent sugar and 0.3 per cent citric acid. Boil this solution and filter through muslin cloth. Pour this solution into the can up to the brim. Place the lid on the can and keep the cans in boiling water or steam till the temperature in the center of the cans reaches 80-85°C. Seal the can on a seamer to get an air tight seam. Sterilize the cans in an autoclave at 10 lb. pressure for 20-25 minutes; cool them by keeping in cold water. Wipe with a dry cloth and store in a cool dry place. This process extends the storage life up to 12 months (Lakshmipathy *et al.*, 2013).

Osmo-air drying/ Osmotic dehydration

It is a two-stage process, the first step consists of keeping the material in a concentrated salt solution called osmotic syrup and in the second stage, a stable dehydrated product is prepared after proper air drying. Main principle involved in the osmotic dehydration is removing moisture at lower temperatures avoiding thermal treatments to get a product with colour, flavour and textural qualities nearest to the natural one. Pretreatments of mushrooms in high concentrations of sucrose, followed by high salt concentration is the most effective method to remove water and to further lower the water activity in the mushrooms.

Freeze drying

Removal of water from a substance by sublimation from the frozen state to the vapour state is known as freeze drying. Mushrooms are sliced and immersed in 0.05% sodium metabisulphite and 2% common salt solution for 30 minutes. They are then blanched in boiling water for 2 minutes, followed by cooling. This is then frozen for one minute at −12°C and the moisture is removed by sublimation at a very low vacuum (0.012 mbar) for 12-16 h and then they are stored at −20°C. This process extends the storage life for 3-4 months. Freeze dried mushrooms have superior flavour and appearance but are brittle. The product can be stored up to 6 months without any change in its quality and appearance. However, this is a very costly and energy intensive process and the venture depend upon the demand and price for such products. In a freeze-drying system, original shape and size can be retained and the shrinkage, which is a problem with other drying methods, is almost negligible.

Fluidized-bed drying

Fluidized-bed drying is the process of removing moisture by exposing commodities into a high velocity of hot air. Fluidized-bed drying, besides providing high quality product, reduces drying time also. Quality of fluidized bed dried mushrooms is significantly influenced by the pretreatments as well as temperature. Optimum temperature for fluidized-bed drying of mushrooms is 50°Cwith an air flow rate of 35 m^3 / min.

Microwave drying

Increasing demand for foods that offer greater convenience in preparation and are time-saving has forced the food processors and consumers to go for microwave drying. Three stage drying is given to the mushrooms, where the initial power level as 100% for 20 min, followed by 60% for another 20 min and 50 min for the rest of the period of drying. Pretreated mushrooms take more drying time to reach 5-6 per cent moisture when compared to mushrooms without any pretreatments. Spices and condiments may be added prior to, during or after microwave drying for making ready-made snacks like chips from slices of button, paddy straw or whole oyster mushrooms. Mushrooms when dried with combined hot air-microwave further shorten the processing time and give good quality final product. Moreover, retention of the characteristic aroma compound (1- octen-3-ol) and its oxidation product (1-octen-3-one) are positively affected by microwave drying (Rai & Arumuganathan, 2008).

Success stories in mushroom processing sector

Mrs. Janaky, a Post Graduate in Literature, from a farming family of Trivandrum, started commercial mushroom cultivation during early 2001 under the brand name "Swadhishta Mushrooms". She underwent professional training on spawn production and cultivation of Oyster and Milky Mushrooms from Kerala Agricultural University. At present, her clientele for fresh mushrooms includes the leading hotels in Trivandrum District and HORTICORP a State Govt. enterprise. During 2003, she entered the market with value added mushroom products like cutlet, samosa and pickle. She is also focusing on value addition of mushrooms by starting a take home parcel service exclusively for mushrooms and equipped with a mobile unit for this purpose – the first of its kind in the country. During early 2003, she started sale of mushroom spawned beds, a novel method for popularizing mushrooms. Now, She has set up a large farm near Malayinkil in Trivandrum and expanded the cultivation of Oyster and Milky mushrooms and production of value added mushroom products.

Mrs. Meena from Ayroor, one of the trainees under the food security project of Government of Kerala implemented by KVK, Pathanamthitta attended several training programmes on food processing. She started her venture in mushroom production and value addition during 2007-2008. This unit linked by KVK Pathanamthitta and District Industries Centre gave her support to take SSI registration and to start a micro enterprise in processing. Mrs. Meena is now selling value added products in the name of 'Yummy' which includes processed mushroom products also.

Mr. Akbar, an auto rickshaw driver at the age of 38, decided to change his profession in view of the difficulties he was facing due to low returns. A platform was created by Krishi Vigyan Kendra, Villupuram District under Tamil Nadu Agricultural University to start the production of Oyster mushrooms. He saw better prospects in this venture and committed himself as a part-time cultivator of oyster mushroom in the early part of 2006. Initially, he used to get a net profit of Rs. 10,000/ month by directly selling mushrooms to retailers, other vendors and regular customers at a fixed margin. Through KVK intervention, he instituted "Bismi Milky Mushroom Growers Association" under the Societies Act, 1976 in 2009 with 22 members. The technical skill in the production of Milky and Oyster mushrooms improved slowly and he started marketing the fresh mushrooms in networking mode through five successful mushroom cultivating groups. Now, from Milky and Oyster mushroom cultivation, spawn production and value added mushroom products, he is earning Rs. 30,000-40,000 per month.

Effluent utilization

Mushroom growing is an eco-friendly activity as it utilizes the waste from agriculture, poultry, and brewery etc. and in turn produces fruit bodies with excellent and unique nutritional and medicinal attributes. Production of spent mushroom substrate after crop harvest is a matter of concern because it creates various environmental problems including groundwater contamination if not handled properly. Compost is considered "spent" when one full crop of mushroom, has been taken or when further extension of cropping becomes unremunerative. Diversified uses of spent mushroom substrate in managing agriculture, environment and recycling energy came in light recently and because of which its name has been changed from spent mushroom substrate to "used mushroom substrate". Material has been found to be good nutrient sources for agriculture. Addition of spent mushroom substrate in the nutrient poor soil leads to an improvement in soil texture, water holding capacity and nutrient status. Incorporation of spent mushroom substrate in soil does not have any adverse effect on its alkalinity while, its amendment in soil leads to an increase in both pH as well as the organic carbon content. Spent mushroom substrate (SMS) being rich in N, P and K acts as a good growing medium for vegetables. Uncontrolled release of industrial waste in the open and poor availability of pretreatment facilities contributes towards the increased levels of contaminants on the soil. Degradation of these contaminants mainly depends upon the physical and chemical conditions prevailing in the soil and the nature of microorganisms that thrive in the soil. SMS adsorbs the organic and inorganic pollutants and harbors diverse category of microbes having capabilities of biological break down of the organic xenobiotic compounds. The microbes, especially actinomycetes (*Streptomyces* sp. and *Thermomonospora* sp.) present in spent mushroom substrate also have strong pollutant catabolizing capabilities which results in decreased level of pollutants in contaminated soil after incubation with SMS. Spent substrate from paddy straw and oyster mushrooms was observed as better substrate in comparison to SMS of button mushroom for their use as substrate for vermicomposting (Ahlawat, 2011).

Edible fungi secrete many kinds of extra cellular enzymes that have strong capacities for decomposing cellulose, hemicellulose and lignin, and can therefore optimize the ratio of ingredients in an animal feed. SMS from mushroom cultivation contains about 14% protein and an abundance of vitamins and microelements such as Fe, Ca, Zn and Mg. Mycelia remaining in the SMS also contain essential amino acids which are absent in common feeds. Therefore, SMS represents a cheap and nutritious feed that can replace crude feeds such as grain and bran.

Spent mushroom substrate is an ideal material for producing marsh gas. There are many nutrient substances in SMS that provide the basis for the long-term propagation of methane-producing bacteria, and using SMS as a starting material for producing marsh gas shortens the fermentation time. Furthermore, since SMS is readily broken up into smaller pieces, there is less time and workload involved in the preparation and reloading of feedstock material.

Edible fungi use up approximately 70% of the available nutrients in the growth substrate and, if an appropriate amount of fresh substrate is added to the spent residue after harvesting, the latter can be used again for further mushroom cultivation.

Safety and quality aspects of mushroom products

Food safety has become a critical issue throughout the fresh produce industry as food service and retail buyers increasingly require growers and packers to develop and implement food safety plans. An increasing number of people are becoming fond of edible fungi because of the nutritional and health benefits, as well as delicious taste, but the quality and safety problems of edible fungi attract much attention. In the processed sector also, like any processed foods, industry wide food safety standards and procedures should be used to enhance safety of mushroom products. Producers and sellers should establish edible fungus quality safety records, strictly control the edible fungus base out and market access system, and carry out edible fungus self-check work earnestly. Relevant regulatory departments should intensify supervision at all levels and actively carry out routine monitoring, risk assessment, and sole rectification. The quality supervisions should investigate and take care of pesticides and additives illegally used in edible fungus production and also during the value addition and such actions will completely eradicate unqualified edible fungi and unsafe processed products in the market. The entrepreneurs should take FSSAI licence and should follow HACCP protocol for safe processing.

Mushrooms are at risk for contamination during and after harvesting. Growers should take action to prevent contamination of mushrooms during harvesting, when they are moved to staging areas, and when they are loaded onto trucks. Regular training and reinforcement is necessary to teach employees that they have an individual responsibility to assure the quality, safety, and wholesomeness of mushroom products. Entrepreneurs must make a long-term commitment to regularly communicate and train their work force on food safety principles and procedures. A regularly scheduled maintenance programme is an important part of a food safety plan because it ensures that equipment and instrumentation used to monitor safety related processes are in working order and that they do not become a source of contamination. If a food borne disease outbreak occurs, traceability procedures are useful for determining the source and distribution of suspected products. While making a value added mushroom product, quality of every raw material used for the production should be taken care of. Among the processed mushrooms, the canned mushrooms pose the highest safety issues. If under processed, it can pose serious food safety issues. Storage temperature, humidity and also the storage environment affect the processed mushrooms. In order to ensure the healthy and sustainable development of mushroom industry, strengthening the production of the main training, enhancing the quality and safety awareness, improving the safety technical standards, vigorously promoting the standardization of production, strengthening the monitoring team and improving the security service system to control the risk factors were put forward. Producers and administrators should take scientific and effective prevention and control measures by controlling every link of edible fungus production as effective means to ensure quality of value added edible fungus and safety to meet the demand of local and international markets, guaranteeing the health and sustainable development of the edible fungus industry.

Packaging of mushroom products

Packaging plays very important role in handling, marketing and consumption of the produce and products, protects the quality during storage and transport. Packaging of mushrooms from the production site up to the consumer including packaging for export market is an important aspect of post harvest handling. Generally, the see-through packaging increases the consumer confidence in the product. If the packaging and storage is not done properly, mushrooms not only deteriorate in their saleable quality but also in nutritional quality due to enzymatic changes.

Mushrooms after harvesting and cleaning are packed in polypropylene bags of about 100 gauge thickness with perforations having vent area of above 5 per cent. Though, the perforation causes slight reduction in weight during storage, it helps to maintain the freshness and firmness of the produce. The most common method of packing in developing countries like India is small polyethylene or polypropylene packets containing mostly 200 or 400 g of mushrooms and generally these small packets are stored for a few days by retailers or consumers. Other improved packaging systems especially in the developed countries are modified atmospheric packaging (MAP), controlled atmospheric packaging (CAP) and modified humidity packaging (MHP). Due considerations have to be given for other alternatives available like corrugated fibre board boxes, corrugated polypropylene bond boxes, plastic trays, crates, woven sacks, thermoformed plastic trays and stretch film and shrink wrapping.

Mushroom products like pickles and preserves are best packaged in glass jars. The principal limitation of glass is its susceptibility to breakage, which may be from internal pressure, impact, or thermal shock, all of which can be greatly minimised by proper matching of the container to its intended use and intelligent handling practices. Various flexible materials such as papers, plastic films, and thin metal foils have different properties with respect to water vapour transmission, oxygen permeability, light transmission, burst strength, pin holes and crease hole sensitivity, etc. and so multi-layers or laminates of these materials which combine the best features of each are used. Pickled mushrooms and mushroom soup powders are successfully marketed in laminates also.

Cellophane paper can be used for packing dried mushrooms. Polyethylene sheets are also used to pack dehydrated mushrooms and powders. They are flexible, transparent and have a perfect resistance to low temperatures and impermeability to water vapour. An important advantage is that these sheets can be easily heat-sealed. It is a good packing material for primary protection of dehydrated products. If a good protection is needed to prevent flavour and gas losses, it will be necessary to combine polyethylene with other materials. The appearance of freeze dried mushrooms is very similar to fresh mushrooms but as the product is brittle, it is packed in sturdy packing and cushion-packs flushed with nitrogen for better keeping quality.

Business plan for ready to serve mushroom curry/canned mushroom

Canned ready to serve mushroom curry are very popular abroad particularly among the Indians. These are consumed directly with chapathi or rice after heating. Growth rate of ready to serve mushroom curry is very good in international market due to increasing number of Indians residing in UK, USA, Middle East, and other countries. The products being for direct consumption can be distributed locally also at hotels, restaurants, super markets, etc.

Basis and presumptions
- The project is based on single shift basis and 300 working days per annum.
- The rate of interest has been taken @15% on an average.
- Labour wages have been taken as per market rates.
- To run the unit viably throughout the year, other fruits and vegetables can be canned with the same machinery and equipment whenever mushroom is not available.
- Yield of canned mushrooms has been considered as 60% based on fresh mushrooms. Drain weight of canned mushrooms has been taken as 440 g, in each A 2½ can.
- Cost of machinery and equipment/materials indicated refers to a particular make and the prices are approximate to those prevailing at the time of preparation of this profile.
- It is presumed that the unit will manufacture mushroom curry, mixed vegetable mushroom curry, potato mushroom etc. The above products may change according to the markets as well as customers demand and orders in this regard.

Implementation schedule

Approximate time required for various activities is given below; However, it may vary from place to place depending upon the local circumstances and on the enthusiasm of the entrepreneur:

S.No.	Activities	Time required
1.	Selection of the site	1 Month
2.	Registration as Small Scale Industry	1 Week
3.	Project report preparation	1 Month
4.	Availability of finance	
5.	Machinery procurement,	6 Months
6.	Erection, commissioning and trial run, etc.	3 Months

Technical aspects

Items to manufacture mushroom curry, vegetable curry, etc. are first washed thoroughly to remove dust, dirt, stones, etc. Mushrooms and vegetables are sliced/ cut manually or mechanically. Potatoes/carrots are peeled in peelers to remove the outer skin. Vegetables and mushrooms are cooked together in a steam jacketted pan on one side and on the other side gravy for curried vegetables is prepared in a separate pan from chopped tomato, onion, garlic, butter, etc. After mixing, mushroom curry is filled into pre sterilized cans. Cans are then exhausted, sealed, and processed. After this, cans are cooled, wiped and labelled, etc.

- Production capacity: 600 MT canned mushroom curry per annum.
- Value: Rs. 300 Lakhs.
- Motive power: 50 HP

Pollution control

The solid waste of the plant such as vegetable skin and pieces of mushrooms should be collected separately and put in the pits which may be used as manure later on. Water can be used for irrigation purposes after storage and sedimentation. No objection certificate may be obtained from the concerned State Pollution Control Board.

Quality control and standards

In general, canned products should conform to the specifications of FSSAI. ISO 9000/ISO 14000 certificate is necessary for European Markets. As the products are mainly for export, strict hygienic conditions at each stage of operation are essential. As steam generation and distribution are involved, proper care should be taken to save heat loss due to leakage in the distribution pipe lines, etc.

Financial aspects

A. *Fixed capital*

i. *Land and building*

Land and building	Amount (Rs.)
Land 2000 sq. mtr. @ Rs 350/sq. mtr.Built up area which includes main processing building 500 sq. mtr., raw material storage 300 sq. mtr., finished products storage 300 sq. mtr., laboratory office 150 sq. mtr., and boiler/workshop 150 sq. mtr.	7,00,000
Total 1400 Sq. mtr. @ Rs 1500 per sq. mtr.	21,00,000
Bore-well and fencing	5,00,000
Total cost of land and building	33,00,000

ii) *Plant and machinery*

Sl.	Description	Quantity (Nos.)	Rate (Rs.)	Total (Rs.)
1	Preparation and filling tables 235 × 135 × 82 cm with Al Top	12	11,000	1,32,000
2.	Flattened can reforming unit consisting of (a) Can reformer with rubber rolled pulley (b) Hand flanger table model with dies for 301, 401cm size	1	41,000	41,000
3	Hand flange rectifier with dies for 301 to 401 size cans	2	7,400	14,800
4	Can sealer table model automatic roller operation complete with chuck for A 2½ size	2	44,000	88,000
5	Set of can tester with pressure gauge and hand pump	2	3,500	7,000
6	Lid embossing machine with coding arrangement with one set of numeral from 1 to 9.	1	15,500	15,500
7	Canning retorts for processing size 810 mm x 915 mm with dial thermometers, etc.	3	50,000	1,50,000
8	Crates for canning retort	8	7,000	56,000
9	Blancher	1	60,000	60,000
10	Exhaust box with tunnel of 3962 mm with 2 HP motor	2	75,000	1,50,000
11	S.S. Steam jacketted kettles 50 gallons capacity complete	4	32,000	1,28,000
12	Semi automatic can seamer	1	80,000	80,000
13	Mini pulveriser with 5 HP motor	1	35,000	35,000
14	Potato peeler	1	30,000	30,000
15	Power slicer with 1 HP motor to slice mushroom, carrots potatoes, onion capacity 50 to 70 kg/hour	1	20,000	20,000
16	SS tank on trolley for transporting peeled/crushed material with push handles	2	13,000	26,000
17	Trays of aluminum size 16" × 14" × 4" with flaps for lifting	50	600	30,000
18	Steam generator with motor, and water feeding pump	1	1,50,000	1,50,000
19	Washing machine rotary rod washer equipped with spray	1	60,000	60,000

Contd.

	arrangement, collection tank, etc.			
20	Pulper capacity 1.5 to 2 tons/hour	1	44,000	44,000
21	S.S. storage tanks	4	20,000	80,000
	Total			13,97,300
	Electrification and installation charges @ 10% of cost of plant and machinery			1,39,730
	Total			15,37,030
	– Misc. equipment such as plastic bucket, weighing balance, concrete tank for cooling cans, strapping machinery and sorting conveyors, etc.		L.S.	3,50,000
	– Laboratory equipment		L.S.	50,000
	– Office furniture and equipment		L.S.	1,00,000
	Total cost of plant and machinery			20,37,030
iii)	Pre-operative expenses			1,50,000
	Total fixed capital (i+ ii+ iii)			54,87,030

B. *Working capital (per month)*

(i) *Personnel*

Sl.No.	Designation	Nos.	Salary (Rs.)	Total (Rs.)
1.	Production manager	1	20,000	20,000
2.	Production supervisor	2	12,000	24,000
3.	Export manager	1	10,000	10,000
4.	Accountant	1	8,000	8,000
5.	Office clerk	2	8,000	16,000
6.	Foreman	1	7,000	7,000
7.	Skilled workers	4	5,000	20,000
8.	Un-skilled workers	10	4000	40,000
9.	Peon	2	3000	6,000
	Total			1,51,000
	Perquisites @ 15%			22650
	Total			1,73,650

(ii) *Raw material*

Sl.No.	Raw material	Total (Rs.)
1.	Button mushroom, different vegetables and pulses (onions, garlic, green chillies, spinach, rajmah, white chhole, cream, salt, yogurt, spices, butter, vegetable oil, paneer and other misc. ingredients	5,50,000
2.	Cans A 2½ size 59000 @ Rs 14/can	8,26,000
3.	Labelling and packing cases for cans 4900@Rs 20/each	98,000
	Total	14,74,000

(iii) Utilities

Sl.No.	Utilities total	Total (Rs.)
1.	Power 50 HP 80% of 40×8×25 = 6400 KWH @ Rs 3/unit	19,200
2.	Coal 50 M.T. water from own bore well	1,00,000
	Total	1,19,200

(iv) Other contingent expenses

Sl.No.	Other contingent expenses	Amount(Rs.)
1.	Postage and stationery	2,000
2.	Telephone	2,000
3.	Consumables	4,000
4.	Repair and maintenance	2,000
	Advertisement and publicity	1,00,000
	Sales expenses	1,00,000
	Miscellaneous expenses including transport	90,000
	Total	3,00,000

(v) Working capital (per month) (i+ii+iii+iv) = 20,66,850

(vi) Working capital (for 2 months) = 41,33,700

C. Total capital investment

Sl.No.	Capital investment	Amount (Rs.)
1.	Fixed Capital	54,87,030
2.	Working Capital for 2 months	41,33,700
	Total	96,20,730

Financial analysis

Cost of production

Sl.No.	Cost of production	Amount (Rs.) (per annum)
1.	Total recurring cost	2,38,84,500
2.	Depreciation on building @8%	2,08,000
3.	Depreciation on machinery @ 10%	1,93,703
4.	Depreciation on furniture @ 20%	20,000
5.	Interest on total investment @ 15%/annum	14,20,167
	Total	2,57,26,370

Turnover (per annum)

Turnover (per annum)	Amount (Rs.)
By sale of 600 MT of canned curried vegetables at an average rate Rs. 50/kg	3,00,00,000
Net profit (per annum) Rs.	42,73,630
Net profit ratio= (42,73,630 × 100)/ 3,00,00,000	= 14%
Rate of return= (42,73,630 × 100)/96,20,730	= 45%

Break-even point

i) Fixed cost (per annum)	Amount (Rs.)
Depreciation on machinery @ 10%	1,93,703
Depreciation on furniture @ 20%	20,000
Interest on total investment	14,20,167
40 per cent of salary and wages	4,66,440
40 per cent of other expenses	14,40,000
Total	**35,40,310**
B.E.P. = (35,40,310 × 100)/ (35,40,310 + 42,73,630)	= 45.3%
= (35,40,310 × 100)/ 78,13,940	

This project profile is prepared for canning of mushrooms mostly for Export-Oriented Unit. The *Agaricus bisporus* (white button) type of mushroom is suitable and preferred for commercial canning.

Some of the benefits for Export Oriented Units are mentioned below

- The unit may import free of duty capital goods, raw materials, components prototypes, office equipment and consumables for office equipment, material handling equipment, etc.

- An Export Oriented Unit/Export Processing Zone (EOU/EPZ) Unit may export goods manufactured through an Export House/Trading House/ Star Trading House.

- Foreign equity up to 100% is permissible in case of EOUs and EPZ units.

- EOU/EPZ units will be eligible for concessional rent for lease of industrial plot and standard design factory building sheds.

- EOUs/EPZ units will be exempted from payment of corporate income tax for a block of five years during the first eight years of operations.

- Net foreign exchange earned by an EOU/EPZ unit can be clubbed with the net foreign exchange of its parent/associate company in the domestic tariff area for the purpose of according Export House, Trading House or Star Trading House status later.

Addresses of Machinery Suppliers

- M/s. Gardners Corporation, 158, Golf Links, New Delhi-110003.
- M/s. Techno Equipment, 31 Parekh Street, Girgaon, Mumbai-400 004.
- M/s. T. Alimohammed and Co., 144/45, Sarang Street, Near M.J. Phule Market, Mumbai-400003.
- M/s. S.S. Enterprises, 299, Katra Pera, Tilak Nagar, Delhi-110006.
- M/s. Devendra Cottage Industries, Sector-22-G,Chandigarh.

Conclusion

Mushroom cultivation has proven to be a profitable business either as a household industry or as a large-scale enterprise with the produce marketable as fresh, canned, dried or as powder. The focus of Indian mushroom industry is predominantly on trade of fresh produce while most of the export is in the preserved form. Mushrooms are not only a nutritious protein-rich food, but also provide nutriceutical and pharmaceutical products. Moreover, it will reduce environmental pollution as well as heal the soil through mushroom mycelial activities. Mushroom cultivation and value addition provides livelihood to many poor families in one way or other because of low capital investment. Mushroom farming if done in proper manner and with care, can become an established and profitable enterprise. India has tremendous potential for mushroom production and to produce different edible and medicinal mushrooms commercially. There is increasing demand for quality products at competitive rate both in domestic and export markets. Interaction of government, entrepreneurs, farmers, marketing agencies and processing industries will help to realise the potential of this venture to handle the problems of poverty, unemployment and malnutrition prevalent in the country

References

Ahlawat, O. P. 2011. Recycling of Spent Mushroom Compost. In: *Mushrooms: Cultivation, Marketing and Consumption* (Singh, M., B. Vijay, S. Kamal and G.C. Wakchaure eds.). Directorate of Mushroom Research, Solan, India. pp. 189-196.

Lakshmipathy, G., A. Jayakumar & S. P. Raj. 2013. Studies on Different Drying, Canning and Value Addition Techniques for Mushrooms (*Calocybe Indica*). *African J. Food Sci.* 7(10): 361-367.

Mahamud, M. M., M.R.I. Shirshir & M. R. Hasan. 2012. Fortification of Wheat Bread using Mushroom Powder. *Bangladesh Res. Publications J.* 7(1): 60-68.

Rai, R. D. & T. Arumuganathan. 2008. Post Harvest Technology of Mushrooms, *Technical Bulletin*- NRCM, ICAR, Chambaghat, Solan.

Wakchaure, G. C., M. Shirur., K. Manikandan & L. Rana. 2010. Development and Evaluation of Oyster Mushroom Value Added Products. *Mushroom Research* 19(1): 40-44.

17
Importance of Quality and Safety of Processed Products

Sudheer K P & Indira V

Introduction

Food safety and quality is the most important issue in the global food supply chain. Quality and safety of fresh as well as processed foods have an important role in the export of these commodities. Increasingly stringent food safety and agricultural health standards in industrialized countries pose major challenges for developing countries to establish in the international markets for high-value food products, such as fruits, vegetables, fish, meat, nuts and spices. Yet, in many cases, such standards have played a positive role, providing the catalyst and incentives for the modernization of export supply and regulatory systems and the adoption of safer and more sustainable production and processing practices. This has created an increased need for updated research and development to demonstrate and provide adequate evidences for their ability to identify and control food safety hazards, and deal specifically with food quality and safety.

Food quality is an increasingly important factor in the production and marketing of biological products. Consumers are becoming more discerning as their affluence increases, and hence the food industry/suppliers of products must meet these demands if they are to maintain or increase market share. The term 'quality' is one of the most defined terms used in food industry today. Quality may be defined as 'the totality of features and characteristics of a product that bear on its ability to satisfy a given need'. The first part, 'the totality of features and characteristics of a product.....' concerns objective factors related to the product. The second part, '...... to satisfy a given need', concerns subjective factors related to user or consumer of goods.

Food safety implies absence or acceptable and safe levels of contaminants, adulterants, naturally occurring toxins or any other substance that may make food injurious to health on an acute or chronic basis. "Food Safety Management

System" means the adoption of good manufacturing practices (GMP), good hygienic practices (GHP), hazard analysis and critical control point (HACCP) and such other practices as may be notified by the Food Authority, by the food business engaged in the manufacture, processing, sale, storage and distribution of food. There are several possible hazards due to natural contaminants, synthetic toxicants, microbial contamination etc., which affects the safety of food. A hazard can be a biological, chemical or physical agent in food or condition of food with the potential to cause an adverse effect on the health of consumer of such food. Details of these factors are briefed under the following sessions.

Importance of quality

Quality may be equated to meeting standards required by a selective customer. In this context, customer is the person or organization receiving the product of each point in the production chain. This is important, because quality is perceived differently depending on needs of particular customer. Customer purchasing fruit for consumption usually judges product on the basis of its appearance, including its shape, firmness, and colour, as well as freedom from defects such as spots, marks, or rots. Consumer will judge fruit on its eating quality as well as its keeping qualities in home. Packing shed operator may be much more concerned about percentage of good fruit in a batch, and how easy it is to handle and grade. There are many different factors that can be included in any discussion of quality.

Concept of global village and advances in technology enabled exporters to supply markets around the world with high quality product, and in some cases to introduce and develop new markets. This has included introducing new crops and often defining quality standards for these crops. The post harvest technologist has an important role to play in enabling producers to define and meet quality requirements.

Factors influencing product quality

To produce high-quality products, processor needs to be aware of quality attributes which consumer discerns as most important and which are most relevant in determining acceptability. Most consumers would initially judge acceptability of products on their appearance, flavour, texture and perceived nutritional benefits. Each of these attributes is a function of biochemical and physico-chemical composition of fruit or vegetable, which is influenced by quality and composition of raw materials, effects of processing and effects of environmental factors, such as temperature, oxygen, light and moisture encountered during storage and distribution, customer handling and use and barriers to these factors provided by packaging.

Food safety is an important factor in quality of processed foods. There are several possible problems due to natural contaminants, microbial contamination, synthetic toxicants etc. Details of these factors are briefed under the following session.

Natural contaminants

These include cyanogenic compounds in lima beans and cassava, nitrates and nitrites in leaf vegetables, oxalates in rhubarb and spinach, thioglucosides in cruciferous vegetables, and glycoalkaloids (solanine) in potatoes. Traditional processing and cooking practices usually remove bulk of these contaminants, but there could be problems if new consumers are introduced to new crops. Mycotoxins (fungal toxins), bacterial toxins, and heavy metals occur naturally and can be present in some crops.

Microbial contamination

This is considered as the most serious health concern by health professionals. Consumers are giving less priority to it compared to chemical toxicants. Bacteria can be introduced onto fresh fruits and vegetables through use of untreated organic fertilizers (e.g., manure), or through insufficiently treated wastewater. Inadequate hygienic standards in packing sheds and anywhere else in food chain can also cause problems. Problem is exacerbated because fruits and vegetables often are eaten fresh. Washing fresh produce is a help, but water used should be clean and free of contaminants.

Synthetic toxicants

Agricultural chemical residues and introduced environment pollutants such as lead, can be a problem unless production systems are controlled and monitored carefully. In developed countries, levels of these are usually very low. In many countries, consumers are very much concerned about levels of chemical-spray deposits found in and on fresh products. Evidence of pesticide residues above acceptable levels, or presence of unapproved or banned spray deposits, renders entire shipments of otherwise sound crops unsaleable. Growers therefore must be especially concerned with ensuring that no illegal sprays are used on crops, and that withholding periods are strictly enforced.

Measurement of product quality

The only test of quality is the response of consumer. Eating quality can be assessed most accurately by using taste panels. These consist of a selection of customers from market who are trained to assess quality attributes being examined. This is a very expensive and time consuming task. Furthermore, it can only be attempted when the product reached eating stage of its lifetime, by which time it is too late to do much about quality. Hence, taste panels cannot be used easily as part of the production process. As an alternative, it is necessary to resort to readily usable fast test methods using equipment. Reliability of these methods depends on how well they correlate with the views of consumers, instead of scientific objectivity of test.

Another way of assessing product quality is to monitor sales and customer complaints. A high volume of sales and less complaints about the product indicates that the product meets the requirements of consumer. However, food manufacturers may not completely depend on this statistics alone for quality control. In general, quality of products is measured using instrumental, immunoassay and sensory evaluation techniques.

Instrumental

Quality assessments carried out using technical experts or trained sensory panels are disadvantages by their requirement for man power, and perceived subjective nature of results. One ideal instrumental method for measuring quality would be cheap, non destructive to sample, easy to use and not subject to variation or fatigue with a rapid response and wide application. In addition, instrumental methods must be linked to assessments made by a trained sensory panel.

Quality is assessed with the help of measuring parameters like colour, flavour, texture etc. Colour may be measured by using colour meters such as Gardner or Hunterlab colour meters. This type of meters evaluates colour by means of three photocells, which measure the hue, saturation (intensity of colour) and luminosity of a sample viewed in the reflected light from a standard source. Other instruments for colour measurement are Munsell discs and Lovibond Tintometer.

Flavour of food is caused by chemical substances present in food and consists of a combination of taste, odour and other sensations. This is measured using a gas chromatograph coupled to a mass spectrophotometer by identifying flavour constituting compounds present.

Textural characteristics which consumers appear to value in fruits and vegetables are crispness and firmness. Factors which contribute to texture are turgor pressure within cells which is important to crispness, and strength of cell walls which, in combination with adhesive properties hold cells together, provide rigidity, firmness and resistance to shear. Different texture measuring devices are available which comprise penetrometers, texture analyser and instron universal testing machines.

These instrumental methods are further classified broadly into destructive and non destructive methods. Destructive indices are again classified into mechanical destructive tests, visual destructive tests, chemical tests and biological destructive tests. Non-destructive measurements are sub divided into mechanical non-destructive tests and visual non-destructive tests.

Destructive tests

i). Mechanical destructive measurements

Firmness measurements in fruits and vegetables are used as a guide to assess the quality of product. Firmness meters attempt to record a value that represents how easily the product can be deformed under a pressure applied to a limited area of its surface. They range from laboratory systems costing several lakhs of rupees to much cheaper and simpler devices, which can be used in field. Devices have been applied to a wide range of fruits and are often the main test specified to establish acceptability of product for a particular market or storage condition.

a. *Penetrometer*: Most common device used to assess firmness is the penetrometer. This has a cylindrical probe, end of which is pushed into the object to be measured. The force required to give a preset penetration is noted.

b. *Universal testing machine*: This is another modified version of penetrometer and texture analyser. Here also probe diameter is changed according to requirements of fruit or vegetable. Unlike penetrometer, in the case of universal testing machine, the plunger is moved automatically and it removes the operator bias altogether. However, this device is costlier and lacks portability due to its heavy weight.

c. *Texture analyser*: It is an instrument that measures the response of a sample to tension, compression, penetration and bend. Texture analyser is a highly versatile and useful instrument, which can replace many other existing texture measuring instruments. Research laboratories commonly use texture analyser to determine textural properties of foods.

d. *Tenderometer*: This device is generally used to measure pea tenderness. It consists of a grid assembly which simulates jaw action in eating of peas. Upper and lower sets of grids are hinged together and sample is first compressed and then sheared and extruded. It is operated hydraulically and force is shown by an indicating hand.

e. *Denture tenderometer*: It is designed to simulate denture surfaces and motions of mastication in the mouth. A complete set of human dentures is used for mechanical chewing. Forces are measured by strain gauges fitted on arm connecting to upper jaw. An electronic modification of this device is adopted for the measurement of tenderness of peas.

f. *Warner bratzler*: This instrument is primarily designed for meat texture. This can be also used to assess texture of some stem vegetables too. Device consists of a blade and two shear bars. Sample is placed through a hole in a plate, and plate slides between two other metal plates, shearing through specimen at edge of hole.

g. *Succolumeter*: It measures volume of extractable juice under controlled conditions of time and pressure. Its measurements are correlated with maturity of set corn, storage quality of apple, and oil and water content of products.

ii). *Visual destructive tests*

a *Physiological changes*: Some indices require visual assessment of cut sections of fruits and vegetables. Some fruits (e.g. banana) change their cross-sectional shape noticeably as they approach maturity, and cross-sectional area therefore can be used as an index. Other fruits split open as they mature, to give a simple criteria for maturity.

b *Colour*: Colour can be defined by three parameters. Humans see colour differently from electronic equipment, and use different scales. People can distinguish level of lightness or colour intensity of an object, its hue (i.e., its colour name such as red, blue, or green), and its chroma (degree of colour purity, saturation, brightness, or grayness). Colour meters give an absolute determination of colour using a standard three-component specification, known as the Hunter Lab scale. This uses lightness (L), red-green character in absence of yellow or blue (a), and yellow-blue character in absence of red or green (b).

c *Refractometer*: A measurement of chemical composition, level of soluble solids in fruits and vegetable juices can be determined by measuring refractive index of juices. Laboratory and field devices require a small sample of juice placed on a glass cover. Refraction of light produces an indication on a scale that gives a measure of soluble solids directly.

iii). *Chemical tests*

a *General chemical assessment*: Relative proportions of chemicals in juices of fruits and vegetables often are a good indicator of quality. During maturation, there are significant changes in chemical composition, and these can be measured by appropriate analytical means. In avocado, oil content is a key maturity indicator (the level depending on variety).

b   *Starch test*: A common maturity test is the starch level in a fruit, found by cutting the fruit in half along equator and placing cut surface in an iodine solution (4% potassium iodide and 1% iodine). Iodine stains the section in regions of high starch but does not affect sugars. The resulting stain pattern can be compared against photographs of standard pattern, and a starch index is determined.

c   *Chemical residues*: Normal analytical chemical tests can be used to test for spray residues. Gas chromatographs also are used extensively. These detect characteristic distribution of molecular weights associated with targeted organic substances.

d   *Moisture content measurement*: In addition to electrical measurement, moisture content can be determined accurately by gravimetric method (oven drying), to establish dry weight after removal of moisture.

iv). Biological destructive tests

Presence of pathogens and bacteria can be determined by a range of cell-culture techniques. Samples of skin or tissue are placed in a suitable growth media and incubated for several days. Pathogens reproduce to levels at which they can be detected and counted.

Non-destructive tests

Non-destructive methods offer significant advantages over destructive methods. Obviously there is a saving in quantity of food wasted, but there are other advantages: First, the same sample can be retested several times throughout its lifetime, giving a reduction in variability as a result of random sampling of food and storage trials (each sample becomes its own control). This means that test procedures should become more reliable and rigorous, because measurements can be correlated better with performance of food sample by tracking development and storage behaviour of individual sample. This improves predictions of storage life. Other advantages are that samples taken from packets do not need to be replaced. This has major advantages for quality-control inspection procedures. There is no mess or problem of disposal of samples as they can be repacked or returned to packing line. On-line assessment of every sample is a possibility. In scientific experiments, because the same sample can be used many times without interrupting its normal life cycle, number of samples required can be reduced.

i). Mechanical non-destructive tests

a   *Mass and bulk density*: Mass is one of the most obvious quality indices and is easy to measure. Density can be measured using Archimedes' principle.

b   *Quasistatic low-pressure indenters*: A large number of different techniques involve applying low-pressure compression to intact fruits and vegetables. Some are quasistatic, and in general, these can be set up using a modern texture analyser.

c   *Dynamic low pressure indenters*: Other indenters involve impact devices using small accelerometers to measure impact parameters. In soft-sense system, fruits are dropped from a small height onto a force transducer. Fruit applies a small impact load to sample without producing apparent damage. Typical analyses involve measurement of dwell time (the period that the impact force remains above 50% or 80% of the peak value), or area under force-time curve.

d   *Non-destructive use of texture analysers*: Some researchers use modern texture analysers capable of following complex loading patterns. Equipment must be sufficiently sensitive and is capable of measuring force and displacement continuously. A wide range of non-destructive measurements can be taken. Measurements made include compressibility, energy absorption during a loading cycle, and whole-fruit modulus of elasticity determined by compression between flat surface and the upper fixture.

e   *Acoustic methods*: Acoustic-response method is used for estimation of overall or global texture of fruits that are approximately spherical. Acoustic impulse response method makes use of sound signal emitted by fruit as it vibrates in response to a gentle short-duration shock, produced by tapping with a small rod or pendulum. A microphone captures the signal, and the principle frequency of vibration is then calculated by means of a fast Fourier transforms. Technique has been applied successfully for the determination of melon ripeness.

ii). Visual non-destructive measurements

a   *Size and shape*: As fruits grow, their shape and size changes. Assessment of size is one of the simplest methods for assessing maturity prior to and at harvest for many vegetables. In some cases, size determines the product. Mango maturity can be assessed in some cultivars by examining position and angle, the shoulder makes with stalk and its point of attachment to the fruit.

b   *Colour*: Surface colour is used widely for maturity and quality assessment and is probably the most common characteristic used in selection and harvesting of many fruits. Generally, assessments are still performed manually, particularly in field, although inside packing sheds, colour sorters are now available commercially for many crops. Although, manual measurement may be subjected to operator fatigue, human error, and variability, automatic systems are still often considerably inferior, because humans provide more subjective assessments than machines.

c   *Near infrared spectroscopy*: Researchers studied use of electromagnetic radiation from ultraviolet to infrared. Near Infrared Spectroscopy (NIRS) analysis uses absorptions in electromagnetic spectrum from 750 to 2500 nm which have previously been correlated to an analytical or quality factor of food sample.

Near infrared is one of the most practicable and exciting modern analytical techniques. Ability to scan a spectrum rapidly, coupled with ability to measure extremely low light levels, by means of high circulating or microprocessor is quantified into different parameters along with standard calibration. It is absolutely necessary to calibrate the instrument with range of values obtained from classical methods and calibration to be realised on each type of product.

NIRS technique is used to determine various parameters in food and agriculture products, like moisture, oil or fat, nitrogen/protein, crude fibre, carbohydrates, active principles in essential oils, menthol, vitamins, amino acids, cholesterol etc. Near-infrared reflectance is also used to detect bruising in apples.

Gamma- and X-rays are used to detect internal disorders in fruits and vegetables. Radiographic systems have detected water core in apples, hollow heart in potatoes, split pits in peaches, and maturity of lettuce heads at harvest.

d   *Delayed light emission and transmittance*: Some fruits re-emit radiation for a short time after exposure to a bright light. Amount of delayed radiation is a measure of chlorophyll present, and this is inversely dependent on maturity. The method has been used with tomatoes and papaya. Other studies used transmittance and reflectance of light for papaya and citrus. Reradiated energy is affected by wavelength and intensity of exciting radiation, sample thickness, area of excitation and ambient temperature and there is considerable variation among similar samples.

Immunoassay

Original food assays were developed for quality control purposes, specifically for testing quality of bulk raw materials. Two specific areas of interest in food industry were species of origin and drug contamination. These assays represent beginning of the immunoassay in the food industry.

Enzyme-linked immune-sorbent assays (ELISA) can be simple and inexpensive to perform requiring a little expertise or equipment so they can be used away from specialised laboratory, even in field. These assays utilize enzyme labelled antibodies that are specific to analyte in question. Detection is usually by production of a coloured end product from a near colourless substrate. Increasing interest in quality assurance in food industry resulted in development assays for a plethora of analytes. These include allergens, hormones and drug residues. Other analytes in food for which immunoassays are developed include pesticides, plant gums, vitamins, contaminants, natural toxicants, fungal toxins and pathogens. Tests for food processing such as irradiation are also reported.

Sensory evaluation

Using instruments to predict sensory quality has long been the goal of many research workers, with the belief that such techniques will be without the human failings of subjective nature and of fatigue.

Total sensory quality of a product, however, can only be predicted to a certain extent by instrumental, chemical and physical methods and accuracy of these predictions varies according to product under test. As a consequence, Martens summarised the following points in favour of using sensory analysis as measure of product quality.

- Sensory analysis cannot easily be replaced by other methods for quality evaluation.

- Sensory analysis uses human senses as an instrument and can therefore measure many variables at one time.

- Sensory analysis is more rapid and cheaper than many of the traditional chemical methods because, colour, flavour and texture may be measured on same sample.

- Sensory analysis, used in correct way, has no measurement noise compared to many traditional chemical analyses. In fact, since human beings are involved, there is an intelligent self correction in a sensory panel that is often lacking in other instruments.

Clearly, the correct way to proceed is to take advantage of sensory evaluation as a measurement technique and to integrate these with advantages of appropriate instrumental techniques into a coordinated method for measuring and controlling quality.

Selecting assessors for sensory evaluation must be carried out with as much care as is necessary in selecting any instrument used to measure food quality. For all types of tests, however, it is important that assessor is well motivated and has correct attitude towards the test. Poorly motivated assessors quickly lose interest in their work, become hasty, careless and apparently poor at discrimination. They can also start to influence others leading to a general despondency in the panel. It is important, therefore, to ensure that assessors are aware of their contribution to the success of maintaining and improving quality of products, and to be aware that their employers have faith to act on their recommendations and observations.

Approach used by assessors is of paramount importance. All observations should be independent, objective and be made by individuals who are confident in their own judgment. Domineering assessors who seek to impose their views on others should not be selected. For the same reason, it is always wise to avoid a mix of senior and junior staff, and to exclude assessors with any particular bias.

Quality control measures

Some of the important points to be considered for maintaining good quality products are as follows:

- Only sound fruits or vegetables of sufficient maturity are to be used for processing.

- Adequate hygienic practices should be followed during processing of product.

- The inspector must be aware of pesticides and other chemicals used in production of the raw materials. Necessary laboratory analyses can then be arranged to ensure residue levels in final product.

- At commencement of and during processing, inspector should pay attention to the state of raw materials, preparation of raw materials for processing (peeling, slicing, dicing, blanching, etc.), preparation and density of packing medium (sugar syrup, salt, brine, etc.), state of containers to be used (cleanliness and strength), pasteurisation or freezing process (time/ temperature relationship), bottle filling and capping and bottle/container storage.

- People who work in processing plant must maintain a high degree of personal cleanliness and conform to hygienic practices while on duty.

- Persons who are monitoring sanitation programmes must have education and/or experience to demonstrate that they are qualified.

- Plant construction and design shall provide enough space for sanitary arrangement of equipment. Equipment must be self-cleanable as far as possible. Cleaning operations must be conducted in a manner that will minimise possibility of contaminating foods or equipment surfaces that contact food.

- Check final product to ensure vacuum and headspace, packing medium strength and container conditions. Statistically based sampling plans should be adopted for examination of final product to ensure that it meets requirements of export regulations.

- Each processing unit should have its own sufficiently equipped laboratory and staff to carry out physical, chemical and microbiological quality examinations of goods.

- Practice proper sanitary handling procedures. Cleaning operations must be conducted in a manner that will minimise possibility of contaminating foods or equipment surfaces that contact food

Good Manufacturing Practices - requirements

Important requirements for Good Manufacturing Practices (GMP) are personal hygiene, plant hygiene, regular sanitary operations, good sanitary facilities and controls, proper designing of equipment and utensils, and an overall control on material and process. Each component is explained in subsequent sections.

Personal hygiene

Persons who have an illness, infected wounds, or any other abnormal source of microbial contamination must not work in a food-processing centre. They must wear clean outer garments, maintain a high degree of personal cleanliness and conform to hygienic practices while on duty. People, who are actually handling food, should remove any jewellery that cannot be properly sanitized from their hands; it is necessary to wear effective hair restraints, such as hairnets, caps, headbands or beard covers. Also, eating food, drinking beverage or using tobacco must not be allowed in food processing area; all necessary steps have to be taken by supervisors to prevent operators from contaminating foods with microorganisms or foreign substances. Persons who are monitoring sanitation programmes must have education/training about personal hygiene.

Plant hygiene

Premises of a food processing centre must be free from conditions such as improperly stored equipment; litter, waste or refuse; uncut weeds or grass close to buildings; excessively dusty roads, yards or parking lots; inadequately drained areas; inadequately operated systems for waste treatment and disposal.

Plant construction and design shall provide enough space for sanitary arrangement of equipment and storage of materials; floors, walls and ceilings must be constructed so that they are cleanable and must be kept clean and in good repair. Provide effective screening or other protection to keep out birds, animals and vermin such as insects and rodents. Also adequate ventilation may be provided to prevent contamination of foods with odours, noxious fumes or vapours.

Sanitary operations

Entire plant must be kept in good repair and maintained in a sanitary condition. Cleaning operations must be conducted in a manner that will minimise possibility of contaminating foods or equipment surfaces that contact food. Insecticides and rodenticides may be used as long as they are used properly. These pesticides must not contaminate food or packaging materials with illegal residues.

Utensils and equipment surfaces that are in contact with food must be cleaned as often as necessary to prevent food contamination. Where there is the possibility of introducing undesirable microorganisms into food, all utensils and equipment surfaces that contact food must be cleaned and sanitized before use. Equipment used in a continuous production operation must be cleaned and sanitized on a predetermined schedule.

Sanitary facilities and controls

Water used in the processing plant or processing equipment must be safe and of adequate sanitary quality. Proper sewage disposal should be practised through other adequate means. Plumbing must be of adequate size and design so that it supplies enough water to areas in plant where it is needed and dispose sewage liquid waste from plant. Adequate floor drainage may be provided.

Toilets and hand-washing facilities must be provided inside fruit and vegetable processing centres. Toilets must be kept sanitary and their doors must not open directly into areas where food is exposed. Hand-washing facilities must provide running water at a suitable temperature and clean towel service or suitable drying devices.

Equipment and utensils

Equipment and utensils must be designed and constructed so that they are adequately cleanable and will not adulterate food with lubricants, fuel, metal fragments, contaminated water, etc. Contact surfaces shall be smoothly bonded and must be made of non-toxic materials and must be corrosion-resistant.

Holding, conveying and manufacturing systems, including gravimetric, pneumatic, and automated systems, shall be maintained in a sanitary condition. Instruments and controls used for measuring, regulating or recording temperatures, pH, water activity, etc. shall be adequate in number, accurate and maintained properly.

Materials and process controls

There must be individuals, responsible for supervising overall sanitation of plant. This includes a control over raw materials, ingredients and manufacturing operations. Raw materials must be inspected, washed and sorted to ensure that they are clean, wholesome and fit for processing into human food. It shall not contain levels of microorganisms that may produce food poisoning or other diseases. Raw materials shall be stored in containers, and under conditions, which protect against contamination.

Food processing equipment must be kept in a sanitary condition through frequent cleaning and sanitizing. It is necessary to process, package and store food under conditions that will minimise undesirable microbial growth, toxin formation, deterioration or contamination. Food shall be held under conditions that prevent growth of undesirable microorganisms. Any processing operations shall be performed in such a way that food is protected against contamination by use of a quality control operation in which Critical Control Points (CCP) are identified and controlled during manufacturing. Areas and equipment that are used to process human food should not be used to process non-human food-grade animal feed, or inedible products unless there is no possibility of contaminating human food.

Hazard Analysis Critical Control Point (HACCP)

Safety of food supply is the key to consumer confidence. In the past, periodic plant inspections and sample testing were used to ensure quality and safety of food products. HACCP was introduced as a system to control safety as product is manufactured, rather than trying to detect problems by testing the finished product. This system is based on assessing inherent hazards or risks in a particular product or process and designing a system to control them. This is a tool to assess hazards and establish control systems that focus on preventive measures rather than relying mainly on end product testing. It is now widely embraced by food industries and by the government regularity agencies around the world as a most cost-effective means of minimising occurrence of identifiable food borne biological, chemical and physical hazards and maximising product safety. It is a system, which targets critical areas of processing and in doing so, reduces risk of manufacturing and selling unsafe products.

HACCP can be applied throughout food chain from primary producer to final consumer. It enhances food safety. Other benefits include better use of resources and more timely response to problems. In addition, application of HACCP systems aids inspection by regulatory authorities and promotes international trade by increasing confidence in food safety. Successful application of HACCP requires full commitment and involvement of management.

Principles of HACCP

The HACCP system consists of seven principles. It is a system, which identifies specific hazard(s) and preventive measures for their control. These principles outline how to establish, implement and maintain a HACCP system.

First Principle (Hazard Analysis): Identify potential hazard(s) associated with food production at all stages, from growth, processing, manufacture and distribution, occurrence of hazard(s) and identify preventive measures for their control.

Second Principle (Identify Critical Control Points): Determine points/ procedures/ operational steps that can be controlled to eliminate hazard(s) or minimise its likelihood of occurrence (Critical Control Point [CCP]). A "step" means any stage in food production and/or manufacture including raw materials, their receipt, harvesting, transport, formulation, processing, storage, etc.

Third Principle (Establish Critical Limits): All CCP's must have preventive measures which are measurable. Critical limits are the operational boundaries of CCP's which control food safety hazard(s).

Fourth Principle (Monitor CCP's): Establish a system to monitor control of CCP by scheduled testing or observations.

Fifth Principle: (Establish Corrective Action): HACCP is intended to prevent product or process deviation. However, should loss of control occur, there must be definite steps in place for disposition of product and for correction of process.

Sixth Principle (Verification): Establish procedures for verification, including supplementary tests and procedures to conform that HACCP system is working effectively.

Seventh Principle (Record keeping): Establish documentation concerning all procedures and records appropriate to these and their application.

Application of principles of HACCP

In the process of hazard analysis, consideration must be given to impact of raw materials, ingredients, manufacturing practices, role of manufacturing processes to control hazards, likely end-use of product, and consumer populations at risk and food safety. Aim of HACCP system is to focus control at CCPs. HACCP should be applied to each specific operation independently. For identification of CCPs, all aspects of food chain need to be examined. The HACCP application should be reviewed and necessary changes has to be made, when any variation is made in product, process or any step. A flow diagram for the typical application of HACCP is shown in Figure 1.

Implementation of HACCP

Implementation of HACCP is made through following tasks. HACCP is implemented by people. If they are not properly trained and experienced, then the HACCP system is likely to be ineffective. There should be a multidisciplinary team that has specific knowledge and expertise appropriate to product, for example; food engineers, production personnel, biochemists, microbiologists, medical experts, food technologists, chemical engineers, agronomists and veterinarians according to particular study. If the expertise is not available in process plant, expert advice should be obtained from other sources.

Description of the product and its use

It should be noted what the product actually is and how/by whom it will be used. A full description of product should be drawn up, including information on composition and method of distribution. Intended use should be based on expected uses of product by end user or consumer. In specific cases, vulnerable groups of population, for example; institutional feeding may have to be considered.

Preparation of process flow diagram

Process flow diagram is used as basis of hazard analysis and should, therefore, contain sufficient technical details for study to progress. It should be constructed by the HACCP team. Each step within specified area of operation should be analysed for the particular part of operation under consideration to produce flow diagram. When applying HACCP to a given operation, consideration should be given to steps preceding and following the specified operations. When the process flow diagram is complete, it should be verified by HACCP team. It should confirm processing operation against flow diagram during all stages and hours of operation and amend flow diagram where appropriate.

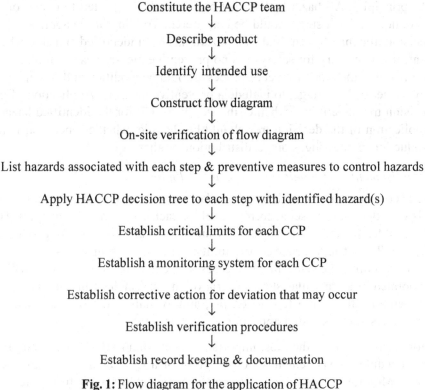

Constitute the HACCP team
↓
Describe product
↓
Identify intended use
↓
Construct flow diagram
↓
On-site verification of flow diagram
↓
List hazards associated with each step & preventive measures to control hazards
↓
Apply HACCP decision tree to each step with identified hazard(s)
↓
Establish critical limits for each CCP
↓
Establish a monitoring system for each CCP
↓
Establish corrective action for deviation that may occur
↓
Establish verification procedures
↓
Establish record keeping & documentation

Fig. 1: Flow diagram for the application of HACCP

Identification of all hazards

When process flow diagram is completed and verified, HACCP team should list all biological, chemical and physical hazards (First principle) that may be reasonably expected to occur at each step and describe preventive measures that can be used to control these hazards. For inclusion in list, hazards must be of a nature such that their elimination or reduction to acceptable levels is essential to production of a safe food. The team must then consider what preventive measures, if any, exist that can be applied for each hazard. Preventive measures are those actions and activities that are required to eliminate hazards or reduce their impact or occurrence to acceptable levels. More than one preventive measure may be required to control a specific hazard(s) and more than one hazard may be controlled by a specified preventive measure.

Identification of the CCP's

A critical control point is a point/step/procedure where a food safety hazard can be prevented, eliminated or reduced to acceptable levels. Identification of a CCP in the HACCP system is facilitated by application of a decision tree

(2nd principle). All hazards that may be reasonably expected to occur, or be introduced at each step, should be considered. Training in the application of decision tree may be required. If a hazard has been identified at a step where control is necessary for safety, and no preventive measure exists at that step, or any other, then the product or process should be modified at that step, or at any earlier or later stage, to include a preventive measure. Application of the decision tree determines whether the step is a CCP for the identified hazard. Application of the decision tree should be flexible, whether operation is for production, processing, storage, distribution or other.

Establishment and monitoring of critical limits

Since critical control point defines boundaries between safe and unsafe products, it is vital that they are set at correct level for each criterion (3rd principle). The HACCP team should, therefore, fully understand criteria governing safety at each CCP to set appropriate critical limits. Critical limits must be specified for each preventive measure. In some cases, more than one critical limit will be elaborated at a particular step. Criteria often used include measurements of temperature, time, moisture level, pH, and available chlorine, and sensory parameters such as visual appearance and texture.

Monitoring is one of the most important aspects of any HACCP system. It is the scheduled measurement or observation of a CCP relative to its critical limits. Monitoring procedures must be able to detect loss of control at the CCP (4th principle). Further, monitoring should ideally provide this information in item for corrective action to be taken to regain control of process, if there is a need to reject product. Data derived from monitoring must be evaluated by a designated person with knowledge and authority to carry out corrective actions when indicated. If monitoring is not continuous, then amount or frequency of monitoring must be sufficient to guarantee that CCP is in control. Most monitoring procedure for CCPs has to be done rapidly (online processes), as there will not be time for lengthy analytical testing. Physical and chemical measurements are often preferred to microbiological testing because they may be done rapidly and can often indicate microbiological control of product. All records and documents associated with monitoring CCPs must be signed by person(s) doing monitoring and by a responsible reviewing official(s) of company.

Corrective actions and verification procedures

Explicit corrective actions (5th principle) must be developed for each CCP in HACCP system to deal with deviations, when they occur. Actions must ensure that CCP has been brought under control. Actions taken must also include proper disposition of affected product. Deviation and product disposition procedures must be documented in the HACCP record keeping. Corrective action should also be taken, when monitoring results indicate a trend towards loss of control at a CCP. Action should be taken to bring process back into control before deviation leads to a safety hazard.

The HACCP system should include verification procedures (6th principle) to provide assurance that HACCP system is being complied with on day-today basis. This can be done most effectively by using audit method. Establish procedures for verification that the HACCP system is working correctly. Monitoring and auditing methods, procedures and tests, including random sampling and analysis can be used to determine if the HACCP system is working correctly. Frequency of verification should be efficient to validate HACCP system. Examples of verification activities include; review of the HACCP system and its records, review of deviations and product dispositions, and validation of established critical limits.

Record keeping and documentation

Well-organized and precise record keeping (7th principle) is essential to application of HACCP system. Records need to be kept of all areas, which are critical to product safety as written evidence that HACCP system is in compliance with documented system. Documentation of HACCP procedures at all steps should be included and assembled in a manual. Records are useful in providing a basis for analysis of trends as well as for internal investigation of any food safety incidents, which may occur. It is extremely useful to allocate a unique reference number to each HACCP record.

Labelling

Labelling packages helps handlers to keep track of produce as it moves through post harvest system, and assists wholesalers and retailers in using proper practices. Customers and consumers expect the labelling on food to be a true description of what they are buying. Labels can be preprinted on fiberboard boxes, or glued, stamped or stenciled on to containers. Branded labelling packages can aid in advertising for the products, producer, packer and/or shippers. Some shippers also provide brochures detailing storage methods or recipes for consumers.

Misleading labelling is an unfair trade practice that cannot be entertained. Most of the countries now have labelling laws stipulating how foods are to be labelled and what information labels must contain. The most important requirements that the label should bear:

- Common name of the product.
- A declaration of net contents (weight or number of pieces).

- Brand name.
- Size and grade.
- Recommended storage temperature.
- Names of approved waxes and/or pesticides used on the product.
- A statement of identity and a true, as distinct from misleading, description of the product.
- The name and address of the manufacturer, packer, distributor or consignee, and
- A list of ingredients (in descending order of volume or weight).

In addition, labels may also be required to include, the country of origin, date of manufacture or packing, a use-by or expiry date, nutritional qualities or values of the food, storage directions, a quality grade and directions for preparing the food.

Improper labelling sometimes results in consignments being rejected, but more often in them being withheld from entry until the labelling is corrected or new labelling applied. In either case, trade is interrupted and the cost involved may make sales unprofitable. It is essential therefore, that exporters be familiar with the food labelling requirements of importing countries.

Rejections and Recalls

Over an extended period, the Indian food trade earned a reputation for product quality and marketing service. India has been encountering intensified competition in the world market for horticultural produce. Recent regulatory changes in selected destination markets, together with evolving requirements among major commercial buyers have triggered a variety of responses by Indian producers and processors/exporters and other governmental agencies. Its exporters have faced increased scrutiny by buyers and regulators for product quality and microbiological or chemical contamination. To increase competitiveness in these areas, effective use of the installed technological capacities that have been put in place over the past decade will have to be made. In addition there is a need to intensify efforts to promote 'good agricultural practices' and improved post-harvest practices among crop growers. Furthermore, measures need to be taken to apply and enforce regulations dealing with pesticides and domestic food safety.

In the last few years, India has shown a tremendous increase in export of bulk and processed products. India, however, has encountered a number of food safety problems in its food exports including high pesticide residues, aflatoxin contamination and the use of prohibited food colourants. In the mid-nineties, Indian dry chilly exports faced several rejections including rejections in Spain due to pesticide residue in excess of permissible MRL's, and in the United States because of residues of quinalphos, a pesticide not registered in the United States. In February 2005, a massive recall of some 600 food products took place in UK because of the detection of Sudan 1 in Worcester sauce. The source of the Sudan 1 dye in the sauce was traced to chilly powder imported from India in 2002. This was the largest ever food recall in U.K. and it affected all major retailers as well as large numbers of food manufacturers and food service companies, as the Worcester sauces had been used in the preparation of a large number of different products. It is estimated that this recall, and associated expenses, cost the U.K. and other European food manufacturers some 200 million Euros (Jaffee, 2005).

Food safety management system

Internationally, there are many Food Safety Certification systems available to check the quality of food products. The Key elements of any FSMS are Good Practices/ PRPs, Hazard Analysis /HACCP, Management Element/System, Statutory and regulatory requirements and Communication. In India, there are many systems which meet these requirements. These are HACCP, ISO 22000, and many more. These are voluntary certifications to strengthen the food safety system. However, under current Indian regulation defined by the FSS Act 2006, Food Safety Management System (FSMS) means the adoption of Good Manufacturing Practices, Good Hygienic Practices, Hazard Analysis and Critical Control Point and such other practices as may be specified by regulation for the food business.

ISO 22000 specifies requirements for a food safety management system where an organization in the food chain needs to demonstrate its ability to control food safety hazards in order to ensure that food is safe at the time of human consumption. It is applicable to all organizations, regardless of size, which are involved in any aspect of the food chain and want to implement systems that consistently provide safe products. The means of meeting any requirements of ISO 22000 can be accomplished through the use of internal and/or external resources.

ISO 22000 enables an organization to plan, implement, operate, maintain and update a food safety management system aimed at providing products that, according to their intended use, are safe for the consumer. It also complies the following points.

- Demonstrate compliance with applicable statutory and regulatory food safety requirements
- Evaluate and assess customer requirements
- Demonstrate conformity with those mutually agreed customer requirements that relate to food safety
- Enhance customer satisfaction
- Communicate food safety issues and conformity to various stakeholders in the food supply chain
- Ensure that the organization conforms to its stated food safety policy, and
- Seek certification or registration of its food safety management system by an external organization

It specifies requirements for establishing, implementing and maintaining prerequisite programmes (PRP) to assist in controlling food safety hazards. Food manufacturing operations are diverse in nature and not all of the requirements specified in ISO 22000 apply to an individual establishment or process.

Prerequisite programmes (PRPs)

The organization ensures that PRPs are established, implemented, maintained, reviewed, improved and updated to assist in: controlling or preventing the introduction of food safety hazards through the work environment, to eliminate, prevent or reduce the biological, chemical and physical contamination of the product(s) including cross contaminations between products to an acceptable level and to control, minimise and/or prevent food safety hazard levels in the ingredients, finished product, and product processing environment.

ISO 22000 specifies detailed requirements to be specifically considered under PRP's. It includes; a) cleaning and sanitizing, b) supplies of air, water, energy, and other utilities, c) pest control, d) temperature control, e) personnel hygiene, f) structure and infra structure, construction and layout of buildings and associated utilities, layout of premises including workspace and employee facilities, g) suitability of equipment and its accessibility for cleaning, maintenance and preventive maintenance, h) supporting services, including waste and sewage disposal, i) management of purchased materials, j) measures for the prevention of cross-contamination, k) physical and chemical contamination, and l) work methodology. In addition, ISO 22000 adds other aspects which are considered relevant to manufacturing operations viz: i) rework, ii) product recall procedures, iii) warehousing, iv) product information and consumer awareness, and v) food defence, biovigilance, and bioterrorism.

All PRP's are approved by the food safety team, their relevance and the reason for their inclusion is documented in the hazard analysis including details of why the PRP is appropriate to the organisation and the control of food safety hazards. PRP's are categorised into two types; Infrastructure and Maintenance Programmes and Operational Prerequisite Programmes

Infrastructure and Maintenance Programmes includes; local environment layout, construction and design of buildings including temporary buildings, construction and layout utilities, supplies of air, water, energy and other supporting services, including waste and sewage disposal layout, construction and design of workspace layout, construction and design of staff facilities, equipment appropriateness including: preventative maintenance, sanitary design, ease of maintenance, ease of cleaning, etc.

Operational prerequisite measures includes; the controlling measures of food hazard controlled by the monitoring procedures, including parameters viz; frequencies and records that demonstrate the operational PRP is working, the corrections and corrective action to be taken, and responsibility and authority for each operational PRP. Operational prerequisite measures are determined during hazard analysis to control chemical, microbiological and physical hazards and are described in the operational prerequisites programmes manual.

Hazards involved in food supply chain

Various hazards are classified as biological hazards (bacterial infections, virus infections, hazards caused by parasites), chemical contaminants (herbicides, pest control substances and other chemicals such as mercury) and physical hazards (ground glass, metal or plastic fragments). The concept "hazard" in the HACCP terminology is expressed in terms of a danger to food safety from a biological, chemical or physical point of view. The term "hazard" refers to any part of a production chain or a product that has the potential to cause a safety problem.

Biological hazards

Biological hazards can be divided into three types: bacterial, viral, and parasitic (protozoa and worms). Many HACCP programmes are designed specifically around the microbiological hazards. HACCP programmes address this food safety problem by assisting in the production of safe wholesome foods.

The various biological hazards identified for crop production and its supply chain are *Clostridium perfringens, E. coli, Bacillus cereus, Salmonella, S. aureus* and moulds (*Aspergillus flavus*). These microorganisms are entering into the processing due to the lack of GAP, GMP and GHP. Poor storage facilities at high temperature and relative humidity provide congenial atmosphere for the mould growth. Control mechanism will be based on the external factors under storage or handling chain along with the various stress factors applied on the food products. Various growth conditions /stress limits viz; pH, water activity of these biological hazards is shown in Table -1.

Table 1: Growth conditions for biological hazards

Sl.No	Hazard	Water Activity	pH	Temperature
1.	*Clostridium perfringens*	0.95	5.0- 8.3 (7*)	12-50(43-47)*
2.	*Bacillus cereus*	0.92	4.9- 9.3 (7*)	4-5to 50 (30-37)*
3.	*Salmonella*	0.95	4-4.5 to 8-9.6 (7*)	6- 49.5 (35-37)*
4.	*Aspergillus flavus*	0.8	2-11.2 (4-6*)	12-48(25-37*-42)
5.	*Staphylococcus aureus*	0.87 (0.83-0.86)	4.5 (7*)	6-48 (37)*
6.	*Escherichia coli*	0.95	4.4- 9.0 (7*)	7-50(37)*

Source (www.icd-online.org/) *-Optimum value/condition

Chemical hazards

Webster defines a hazard chemical as any substance used in or obtained by a chemical hazard process or processes. All food products are made up of chemicals, and all chemicals can be toxic at some dosage level. However, certain hazardous chemicals are not allowed in food and others have allowable limits. Two types of chemical hazards in food are naturally occurring ones and added chemicals. Both may potentially cause chemical intoxications if excessive levels are present in food. Various chemical hazards like aflatoxins, artificial colours viz; sudan I, pesticide residues, heavy metal contamination, etc, in foods are reported by many research groups.

Physical hazards

It is often described as extraneous matter or foreign objects and includes any physical matter not normally found in food, which may cause illness (including psychological trauma) or injury to an individual. Most common physical hazards encountered are metal parts, wooden pieces, glass splinters and stones, which may enter during primary processing and further handling processes. Regulatory action may be initiated when agencies find adulterated foods or foods that are manufactured, packed or held under conditions whereby they may have become contaminated and may be injurious to health.

Reasons for poor safety and quality assurance

- Weak regulatory systems relating to the import, production, and sale of pesticides.

- Limited farmer knowledge of alternative pest management approaches and appropriate use of pesticides

- Limited application of HACCP principles by packers/exporters (especially SME's)

- Improper storage and packaging facilities

- Limited/lack of systems for traceability (especially from small holders)

- Poor waste management

- Unhygienic handling and transport

- Use of banned food additives (colours /chemicals) as preservatives

- Cross contamination due to unhygienic processing environment

To meet these challenges, India needs to strengthen its regulatory framework. This process would include upgrading testing facilities to meet international as well as importing-country requirements; upgrading human capabilities or empowering personnel in the areas of testing, risk analysis, and development and auditing of HACCP plans; developing GMP/GHP/HACCP modules for implementation at both domestic and export levels; and establishing databases on requirements of importing countries. India is either funding these upgrades itself or seeking assistance under programmes funded by the Food and Agriculture Organization of the United Nations (FAO).

Strategies to enhance safety and quality

The scenario of present post harvest processing of crops is not encouraging in India. It leads to export rejection. Due to unscientific processing methods, large quantities of processed products produced are also spoiled every year. So, it is very urgent to implement developmental programmes, especially in the processing sector, to enhance the export quality of food products.

The Country could not take full advantage of the emerging opportunities in processing industry mainly for want of domestic supply of raw materials of desired quality. This is due to unscientific application of basic inputs and poor post harvest handling practices. Over dosage of fertilizer and chemicals, improper storage practices leading to increased fungal/mould growth, etc. are some of the examples for these unscientific practices. Unhygienic processing methods aggravate the situation. A thorough understanding and application of good production and manufacturing practices can improve the safety and quality of produce and their value added products. This will ensure higher returns and improve the economic security of growers and processors of the Country.

Quality of food product is assessed by its intrinsic as well as extrinsic characters. The former consists of chemical quality, i.e. the retention of chemical principles while the latter emphasizes physical quality. These include appearance, texture, shape, presence or absence of unwanted things, colour, etc. In addition, certain health requirements are also implemented as export quality standards viz. pesticide residue, aflatoxin, heavy metals, sulphur dioxide, solvent residues and microbiological quality. However, physico-chemical quality remains the ultimate attribute, while considering export requirement of food products as these properties delineate its grade in the market. These qualities vary unpredictably. The quality of the food product can be maintained with the help of proper food safety and management systems like Good Manufacturing Practices (GMP), Good Agricultural Practices (GAP), and Hazard Analysis and Critical Control Points (HACCP).

Upgrading of laboratories

Laboratories are being strengthened in terms of equipment, manpower, and systems. A significant amount of investment has been made over the past decade in laboratory facilities and equipment, with individual labs being extended beyond their initial focus on physical and chemical parameters to include testing for pesticide residues, aflatoxin, and, in a few cases, heavy metals. To meet the requirements for testing, specifically for testing for chloramphenicol, nitrofurans, and other antibiotics, the Export Inspection Council (EIC) labs and other government labs now have the capability to test at 0.02 parts per billion.

Training and technical assistance

Training efforts in India focus on developing and upgrading skills of industry and government personnel. A Human Resource and Quality Development Centre have been established under the EIC. It offers certification personnel a chance to keep abreast of the latest developments and take training programmes for implementing and monitoring food-product certification. Similar training and awareness programmes are being organized for industry on various issues, including HACCP, testing of food quality parameters.

Establishing a database on importing-country requirements

Information on regulations and specifications regarding methods of sampling, inspection, and testing in various countries is often unavailable or available only in the language of the importing country. This lack of clarity about specific requirements can sometimes lead to rejection at the point of import. Now, India is building a database of requirements of major import partners that can be accessed by exporters. Technical assistance in this area has been sought from the EIC. Some importing countries are imposing unjust measures that conflict with Codex and impede trade. Some of these measures include applying standards more stringent than Codex without carrying out a risk analysis, destroying nonconforming consignments, imposing new requirements without notification or information, and applying test methods that may be different from internationally specified ones. To work out solutions for such issues, India is entering into dialogue with importing governments.

Role of public and private sectors in enhancing food safety

Addressing food safety issues in India will require the adoption of more appropriate legislation and their enforcement. Joint efforts by the government and the private sector will be needed in a number of areas. These include better risk management, the promotion and adoption of good agricultural, manufacturing and hygienic practices, greater collective action and some targeted public investments. Responsibilities for these functions need to be shared between the private and public sectors. While, there are many critical regulatory, research and management functions that are normally carried out by governments, the private sector also has an important role in the actual compliance with food safety requirements as detailed below.

Role of public sector	Role of private sector
1. Policy and regulatory environment	1. Implementation of GAP, GMP, HACCP, etc.
2. Risk assessment and management	2. Development of traceability system
3. Awareness building and promoting good practices	3. Develop training, advisory, and conformity assessment services
4. Public expenditures in common facilities	
5. International trade diplomacy	

Conclusion

Though, India is self-sufficient in the production of most of the agricultural produces our contribution is negligible in the world market. To enhance our world share and to become, competitive in export and processing sector, India must make effective use of the installed technological and testing capacities put in place over the past decade. Furthermore, there need to be an intensification of efforts to promote "Good Agricultural Practices" among the crop growers. The lack of harmonization of international standards is a cause for some uncertainty within the trade and added costs for exporters, since they must use different technologies and employ different types of tests to satisfy different markets. The harmonization of international standards would reduce this uncertainty and enable more uniform procedures. Adoption of new and hygienic production and post production methods coupled with infrastructure development for testing methods and sampling procedures will enhance the return of crop growers, processors and exporters through the increased domestic and international market.

The challenges for ensuring food safety in the domestic market and in its food exports remain large. Improving the food safety and quality management information is essential to make the best use of available knowledge and resources to increase the export potential and to prevent food-borne illnesses. India has made some progress in the last decade to strengthen food safety measures at home and in meeting food safety and sanitary and phytosanitary standards abroad. The challenge for the future will be to adopt a more strategic, rather than crisis management approach. This will be essential to ensure the sustainability and cost effectiveness of these efforts.

References

Aksoy, A. & J. Beghin. 2005. Global Agricultural Trade and Developing Countries. *Trade and Development Series*. Washington D.C. World Bank.

Alerts. 2004. *Rapid Alert System for Food and Feed. European Commission*, http://europa.eu.int/comm/ food/food/rapidalert/index_en.htm.

Aydin, A., E.M. Erkan., R. Baskaya & G. Ciftcioglu. 2007. Determination of Aflatoxin B_1 Levels in Powdered Red Pepper, *Food Control* 18: 1015–1019.

Codex Alimentarius: http://www.codexalimentarius.net/ mrls/pestdes/ pest_ref /MRLs.pdf.

Diop, N. & S. Jaffee. 2005. Fruits and Vegetables: Global Trade and Competition in Fresh and Processed Product Markets, in Global Agricultural Trade and Developing Countries, (eds.) M. A. Aksoy & J. Beghin. Washington D.C. World Bank.

European Union. 2003. *Emergency Measures Regarding Hot Chilly and Hot Chilly Products*. Commission Decision 2003/460/EC.

Jaffee, S. 2005. Food Safety and Agricultural Health Standards: Challenges and Opportunities for Developing Country Exports, *Research Report* No. 31207, Poverty Reduction & Economic Management Trade Unit and Agriculture and Rural Development Department.

Jaffee, S. & S. Henson. 2004. Standards and Agri-Food Exports from Developing Countries: Rebalancing the Debate. *Policy Research Working Paper 3348*. Washington D.C. World Bank.

Pritty, S. B., K. P. Sudheer., M. C. Sarathjith & A. A. Tina. 2012. Development of GMP and HACCP Protocol for Pepper Industry *Int. J. Agril. Fd. Sci .Technol.* 3(1): 22-39.

Sudheer, K. P. & V. Indira. 2007. *Post Harvest Technology of Horticultural Crops*. New India Publishing House, Pitampura, New Delhi, India.

WHO (http://www.who.int/mediacentre/factsheets/fs125/en/).

18

Marketing Strategies and Supply Chain of Horticultural Products

Ranjit Kumar E G

Introduction

Traditionally, Indian economy was characterised as an agrarian economy for the reasons that majority of the population were dependant on agriculture and allied activities and more than fifty per cent of the Gross Domestic Product (GDP) was derived from agricultural sector. However, paradoxically, despite a drastic decline in the share of agricultural sector to GDP wherein it declined from 56.5 per cent in 1950-51 to 15.11 per cent in 2016-17 (Statistics times.com) Indian economy is still characterised as an agrarian economy. But, it would not be inappropriate to state that Indian agricultural sector is at crossroads, or it would be rather safe to comment that for any further advancement of the agricultural sector, value addition is imperative. During the post independence period, the sector had witnessed commendable progress attributing to a shift from an undesirable situation of 'ship to mouth' to 'farm to mouth'. It is obvious that as a result of various programmes, projects and schemes tailored for development of agricultural sector, India had attained self sufficiency in food crops and other agricultural/horticultural products. However, the darker side is that the Indian peasants are still deprived of a remunerative price for their products as well as a respected status in the society which in turn is arresting further advancement and progress of the sector. Thus, it could be concluded that future of agricultural sector in India vests on the adoption of appropriate marketing strategies, supply chain management and value addition initiatives for the prosperity of our farming community in specific and agricultural sector in general. It is in this backdrop, the NITI Aayog has proposed various reforms in India's agriculture sector, including liberal contract farming, direct purchase from farmers by private players, direct sale by farmers to consumers, and single trader license, among other measures, in order to double rural income in the next five years. Ministry of Agriculture, Government of India, has been conducting various consultations and seeking suggestions from numerous stakeholders in agriculture sector, in order to devise a strategy to double the income of farmers by 2022.

India is the largest producer, consumer and exporter of spices and spice products. India's fruit production has grown faster than vegetables, making it the second largest fruit producer in the world. India's horticulture output is estimated to be 287.3 million tons (MT) in 2016-17 after the first advance estimate. It ranks third in farm and agriculture outputs. Agricultural export constitutes 10 per cent of the country's exports and is the fourth-largest exported principal commodity. Agro industry in India is divided into several sub segments such as canned, dairy, processed, frozen food to fisheries, meat, poultry, and food grains.

Department of Agriculture and Co operation under the Ministry of Agriculture is responsible for the development of agriculture sector in India. Given the importance of agriculture sector, Government of India, in its Budget 2017–18, planned several steps for the sustainable development of agriculture. Total allocation for rural, agricultural and allied sectors for FY 2017-18 has been increased by 24 per cent year-on-year to Rs 1,87,223 crores.

Agriculture sector in India is expected to generate better momentum in the next few years due to increased investments in agricultural infrastructure such as irrigation facilities, warehousing and cold storage. Factors such as reduced transaction costs and time, improved port gate management and better fiscal incentives would contribute to the sector's growth. Furthermore, the growing use of genetically modified crops will likely improve the yield for Indian farmers. All these efforts shall not reap the designated advantage unless an efficient post harvest product handling system is developed and standardised. It includes diverse activities like procurement, processing, grading, and marketing. Considering the peculiarities of Indian agricultural sector, all these distinct activities should be brought under an integrated institutional network with active participation of farmers. In this context, the successful existence of the integrated approach in the dairy sector viz., the renowned "Anand Pattern" is to be remembered. Thus, the co-existence of an efficient and effective post harvest marketing system along with the concerted efforts in the pre-production and production scenario is the ultimate solution to solve the problems and hardships of Indian farmers.

Marketing- A general perspective

According to Kotler and Armstrong "Marketing is the social process by which individuals and organizations obtain what they need and want through creating and exchanging value with others". The Chartered Institute of Marketing states that "Marketing is the management process for identifying, anticipating and satisfying customer requirements profitably. However, the American Marketing Association has defined marketing more comprehensively that reads that "Marketing is the activity, set of institutions, and processes for creating, communicating, delivering, and exchanging offerings that have value for customers, clients, partners, and society at large". Thus, marketing is an economic activity for all the players involved in the chain of activities from production to the delivery of goods or services to the ultimate customers. It is obvious that at each point of activity value is being added to the product physically or otherwise which ultimately decides the price of the product (www.marketingteacher.com).

Marketing is a crucial component to the success of any venture especially in a predominantly agrarian economy. Peculiar characteristics of agricultural commodities/products pose a challenge for itself, the challenges being multi-dimensional and multifaceted. Moreover, agricultural marketing is totally distinct from the marketing of industrial commodities due to these inherent peculiarities which makes it a complex task. Besides the peculiarities of the agricultural products, the pattern of production process also adds complexity to the system. With very large rural population engaged in cultivation spread over geographically (scattered production units) and low marketable surplus (trivial output), the necessity for an organised and orderly marketing system is the need of the hour. An organised marketing helps not only in integrating production and reducing post harvest losses but also helps in ensuring a remunerative price to the producers and supply goods at a reasonable cost to the consumers.

Marketing of horticultural products includes planning production, growing and harvesting, assembling, processing, grading and packing, transportation, storage, distribution and sales, feedback etc. All of these activities are links in the production-marketing chain. Those who carry out marketing have a strong incentive to increase the value of rural trade, because increased sales should lead to higher profits. Rural businesses include suppliers of inputs, buyers of produce, transporters, storage companies, processors and wholesalers. They can range in size from individual entrepreneurs to large-scale agribusinesses, but whatever their size, all stand to gain from improvements in the marketing process.

At present, sustainability of agriculture is determined by the quantum of production and productivity but equally important is the existence of an efficient and effective marketing system. Farmers by and large are generally highly skilled in agricultural techniques but marketing requires learning new skills, new techniques and new sources of information. Armed with business and marketing skills farmers will be better able to run their farms profitably. Indian agriculture is dominated by small and marginal farmers. As such, the small-scale farmers face the biggest marketing problems. Thus, all efforts should be prioritised in favour of small farmers so that they get maximum support and their success depends on getting the best prices possible. This can be done by obtaining better information about marketing and the different marketing options available to them.

Present marketing systems existing in the agricultural sector are dynamic and volatile. Moreover, they are competitive and involve continuous change and improvement to suit to the changing business, legal and technological environment. In fact, the suppliers who command control over overheads and maintains lower costs are the more efficient players. This in turn helps them to deliver quality products to the customers and finally they are the ones who survive and prosper. In contrast, those who have high costs do not adapt to changes in market demand and provide poorer quality goods and services are often forced out of business. Private businesses are often said to be exploiting farmers and making unfair profits. They certainly try to maximize their profits, but without such businesses farmers would not be linked to markets and would not be able to sell all their produce. Thus, it is obvious that there need to be intermediary for linking the farmers to the markets, but the present chain of intermediaries is exploiting the farmers. Thus, a new system should be evolved to help farmers to find new markets and lower their costs. All of this would lead to improved production opportunities and higher income for farmers.

Rural marketing businesses are often small, have limited resources and are traditional in outlook. Identifying new markets, advising on technologies and improving understanding of markets are some of the ways by which farmers could be supported to increase production, productivity and marketable surplus. Governments can also extent helping hands to farmers without actually working with them directly. Promotion of competition, provision of market information and improvement of market infrastructure are powerful ways to ensure good returns for farmers.

Marketing strategy

A marketing strategy is a coherent and agreed upon process formed with the aim of increased revenue and market share. Marketing strategies can help you to maximize the resources at your disposal. Marketing strategies refer to the set of actions designed to meet your business goals and should address the following issues: Who your target customers are? How to reach your target customers? How to retain your customers?

However, the following aspects should also be considered while designing our strategy.

Unique Selling Point **(USP):** USP is vital as it will help to differentiate our products and services from that of our competitors and will also help to convey the value of our products and services to our customers.

Pricing: Pricing is a critical element of marketing strategy. Pricing of our products should be based on the expected market share, the market positioning, the proposed promotion strategy to be adopted, the brand image to be created, and also the unique features we offer. However, before deciding on the price, we should also consider the price of rival products, our target customer's purchasing power and our own institutional objectives.

Positioning: Although positioning is related to pricing it has relevance of its own. While deciding our positioning strategy the key point to be decided is whether we aim at serving an established segment of the market or we go head to head with established rivals or whether we should carve out a niche market.

Offers: Offers are special promotional measures designed to win new customers or to attract existing customers to repeat their demand. However, a cost benefit analysis should be administered before an offer is launched so as to ensure the economic viability of the organisation as well as to finalise the type of offer to be placed to customers.

Promotion: Success of the marketing efforts and initiatives of an organisation is directly linked to the promotional strategies adopted. Choosing among the different promotional means is a vital component for management. Should we adopt the traditional promotional means or should we go for the modern cost effective promotional means? The choice could be whether to go down the route of paid advertising or handle the promotion in-house with a series of blog posts and social media activities?

Referral strategy: One of the best and cost effective ways to win new business is a referral by an existing customer and as such firms should probe ways to stimulate referrals. This is a strategy that could be adopted by a newly organised firm.

Challenges in marketing of horticultural products

Horticultural commodities are characterised by high perishability. Their markets are thin, sparsely spread and inefficient which is evidenced by price gap between producer's price and consumer rupee. In India, the marketing of horticultural commodities are mainly chanelised through the mandis established under Agricultural Produce Marketing Committee (APMC). These mandis are by and large controlled by a few big local traders which in turn make the operations non-transparent and finally results in escalation of various marketing costs.

The net result of all these is that it all works at the cost of the poor peasants which alarmingly lowers the realisation of income by the poor and unorganised farmers.

It is obvious that in India, most of the horticultural commodity markets generally operate under the normal forces of demand and supply. The trade or exchange process of buying and selling of horticulture produce happens in the market yards where a large number of market functionaries/intermediaries are involved. A study on price spread by the Government of India (2001) concluded that the farmer fetches only Rs. 1.00 out of every Rs 3.50 paid by the consumer, the retailers earn Rs 0.75, the wholesaler bags Rs 0.50 and the remaining Rs 1.25 flows to other market intermediaries. Thus, it shows that the farmers are not the price getters but they are the price takers in the sense that the farmers cannot decide the price for their produce but it is being fixed by the market intermediaries. Some other studies indicate that the share of producers varies from 56 to 83 per cent in food grains and 79 to 95 per cent in pulses, 65- 96 per cent in oilseeds and 33 to 75 per cent in the case of fruits and vegetables (Dastagiri. *et al.*, 2009). However, it should be remembered that the upper limit in the price spread cannot be generalised because those are not being enjoyed by small and marginal farmers but by a few large farmers. The share of producer in consumer price depends upon the types of marketing channels followed in sales transactions by farmers. In order to provide remunerative prices to farmers, there is a need to eliminate the chain of middlemen by introducing innovative marketing channels like direct marketing, contract farming etc.

Marketing initiatives/strategies for fruits & vegetables

India amended APMC Act and allowed cooperatives and private entrepreneurs to set up special markets for fruits, vegetables and flowers. Economic reforms led India to open up a number of innovative liberalized markets in the WTO regime to eliminate middlemen and facilitate direct contact between producer and consumers as well as a new private retail liberalized markets for fruits and vegetables. Some of the new initiatives in agricultural marketing for fruits and vegetables in India are discussed in the following paragraphs:

- The Rythu Bazaar of Andhra Pradesh was established to plug exploitation of farmers and consumers by middlemen. Main objective of Rythu Bazar is to ensure remunerative price to farmers and provide quality fresh vegetables to consumers at reasonable rates. Rythu Bazars provide direct interface between farmers and consumers eliminating intermediaries in trade. Farmers are greatly benefited by this kind of business since they sell directly to the customers and do not pay any commission to the agents. Customers are also getting good quality produce. Thus, they have become popular, creating a demand for the produce of small farmers.

- Apni Mandis were set up in Punjab which is an adapted format of the Saturday markets existing in United Kingdom and United States which aims at eliminating market middlemen.

- Uzhavar Sandhai (Farmers' Market) was setup in Tamil Nadu in 1999 to establish direct negotiation between farmers and consumers. Main aim of Uzhavar Sandhai includes to provide fresh vegetables and fruits at reasonable price daily without any interference of middlemen, to provide correct measurement to consumers, to give full satisfaction to farmers and public, to provide higher price than that of wholesale price to farmers for their vegetables and fruits, to provide fresh fruits and vegetables at a lesser price than that of retail price to consumers, to function as a Technical Information Centre and to act as a Technical Training Centre to farmers and to provide seeds and other inputs to farmers. Under these markets, farmers are issued cards authorizing them to sell their products in the markets.

- The Hardaspar Vegetable Market is a successful model market for direct marketing of vegetables established in Pune city.

- Shetkari Bazars were established in Maharashtra for marketing fruits and vegetables by eliminating the market intermediaries and establishing direct links between producers and consumers thereby reducing the price spread and enhancing producer share's in consumer rupee.

- Krushak Bazars were established in Orissa during 2000-01 to empower farmer-producers to compete effectively in the open market to get a remunerative price and also to ensure products at reasonable price to consumer.

- Mother Dairy Booths are the retail outlets of milk marketing in Delhi organised under the world renowned Operation Flood programme. It was also entrusted the task of handling retail vegetable marketing to supply vegetables to consumers at reasonable prices.

- Marketing Co-operatives were established throughout our nation as a solution to various defects and exploitation that existed in the agricultural marketing system. Co-operatives being a nationally well knit organisational network were expected to eliminate some or all intermediaries to the advantage of both producers and consumers. Successful integrated model for milk marketing was the guiding force behind the organisation of marketing cooperatives. A few successful co-operative marketing societies for fruits and vegetables are the Co-operatives for marketing banana in Jalgaon District, vegetable Co-operatives in Thane District, HOPCOMS, Bangalore and Gujarat Co-operative cotton marketing society.

- Contract farming/Contract marketing is a kind of guaranteed marketing wherein an agreement is arrived between farmer producers and an agribusiness firm. As per the agreement/contract, the farmers are to produce commodities confirming to prefixed quality standards and supply the designated quantity of produce at the price, time and other terms of contract agreed upon. This initiative by and large helps in reducing transaction costs by establishing a genuine farmer- processer linkage.

- The Mother Dairy Fruit & Vegetable Pvt Ltd (SAFAL) is a novel initiative by National Dairy Development Board (NDDB) started for marketing fruits and vegetables by establishing a retail chain in Delhi. This market is a move to introduce a transparent and efficient platform for sale and purchase of fruits and vegetables by connecting growers through Grower's associations.

- Forward and future markets are another form of marketing initiative regulated through the Forward Contracts (Regulation) Act, 1952. Price stabilization and risk management with respect to physical delivery are the key issues addressed by this system. Forward markets are used to contract for the physical delivery of a commodity. By contrast, future markets are 'paper' markets used for hedging price risks or for speculation rather than for negotiating the actual delivery of goods. Prices in both physical and future markets tend to move together because traders in future contracts are entitled to demand or make delivery of physical product against their future contracts.

- Commodity exchanges: A commodity exchanges/market is a market that trades in primary economic sector rather than manufactured products. Commodity exchanges are divided roughly into three types: Hard commodities such as metal exchanges, fuel exchanges and soft (agricultural) commodity exchanges. Other exchanges deal in currencies and commodity indices. These exchanges help to narrow marketing, storage and processing margins. National Multi-Commodity Exchange of India Ltd (MCX) and National Commodity and Derivatives Exchange of India Ltd. (NCDEX) are the two commodity exchanges in India.

- E-trading: National Agriculture Market (NAM) is a pan-India electronic trading portal which networks the existing APMC mandis to create a unified national market for agricultural commodities. The NAM Portal provides a single window service for all APMC related information and services. This includes commodity arrivals & prices, buy & sell trade offers, provision to respond to trade offers, among other services. While material flow (agriculture produce) continues to happen through mandis, an online market reduces transaction costs and information asymmetry.

NAM addresses various challenges by creating a unified market through online trading platform, both, at State and National level and promotes uniformity, streamlining of procedures across the integrated markets, removes information asymmetry between buyers and sellers and promote real time price discovery, based on actual demand and supply, promotes transparency in auction process, and access to a nationwide market for the farmer, with prices commensurate with quality of his produce and online payment and availability of better quality produce and at more reasonable prices to the consumer. At present there are 455 enrolled Mandis in the NAM network.

- Food retail super markets: The traditional retailing channels existing in India are highly unorganized, fragmented and dominated predominantly by small and family owned business with poor access to capital, technology and regulations. The emergence of new retail initiative, the Food Retail Chains (FRC) witnessed after liberalization facilitated the entry of corporate into retailing business. The changing political, demographic and economic features like high economic growth, increase in disposable income, proliferation of brands, consumer awareness, advancement in technology etc., accelerated and favoured the emergence of FRCs. Although the food retail chains are yet to be accessed in full potential by the general consumers, presently, the retail chains are viewed as niche markets for rich consumers in capital cities which has got transformed themselves into shopping destinations.

These new marketing initiatives by and large has contributed substantially to enhance and augment the farm income, improve the well being of the farmers and bring in stability in prices of horticultural crops, but still we have to go a long way to elevate the farmers to the status of price setters rather than price getters. In India, during the post liberalisation period, a number of new liberalized public markets for fruits & vegetables were opened in the WTO regime. Main objectives and functions of these farmer markets are to empower the farmers to participate effectively in the open market to fetch a remunerative price for their produce, to avoid exploitation of both the farmers and the consumers by the unscruplous middlemen, to enhance the distributional efficiency of the marketing system, to bring in price stabilization and finally for efficacious risk management. This post liberalisation strategy adopted by India paved way for attracting huge investments by leading Indian corporates and presently the traditional markets are making way for the new retail formats like departmental stores, hypermarkets, supermarkets and specialty stores and western-style malls. Besides, their number, scale of operation and turnover in retail business has been undergoing phenomenal growth in the past few years.

In the marketing of horticultural products too there exists a wide gap between prices received by the farmers and those paid by urban consumers, reflecting inefficient marketing arrangements. Generally, the marketing model adopted by private business houses functions more efficient than government markets for various reasons. Therefore, there is an immediate need to replicate such models in a much larger scale to cover not only the cities but also the interior villages in the country. Both public and private retail markets have to adopt the new marketing models to enhance the distributional efficiency of the marketing system. Further, necessary steps should be initiated to amend outdated laws if any, which restricts the establishment of markets and also allow cooperatives and private entrepreneurs to set up modern markets.

Farmer Producer Organisation (FPO)

In the earlier sections we had discussed the various initiatives/strategies adopted for marketing of agricultural commodities in India. However, it may be noted that most of these are micro strategies confined to a crop/location/organisation or group and lacks a macro approach. Cooperatives may be an exception. Though, the cooperatives has a national label, their success as a strategy for marketing of agricultural commodities is confined to milk and milk products and are yet to succeed in the marketing of agricultural commodities in general. It was in this backdrop the new initiative of people's organisation, the Farmers Producers Organisation (FPO) was mooted by Late Dr. V. Kurien, the chief architect of the world renowned "Operation Flood Programme" and unrelenting believer of people's organisation in bringing prosperity to the poor peasants. However, the new concept of producer companies is based on the recommendations of an expert committee led by noted economist, Y. K. Alagh.

It is a fact that majority of the farmers in India and especially in Kerala are marginal and small farmers with limited means producing small quantities for sale with little or no bargaining power. The organised traders and middlemen take advantage of this situation, the farmers being at the mercy of these unscrupulous trade intermediaries are forced to dispose their produce at a price offered by these intermediaries. However, the large-scale farmers, who are small in number produce large quantities of a consistent quality standard and have no difficulty in attracting buyers and receive the true market price for their output. The only way small-scale farmers can compete with these large farmers and the intermediaries is to co-operate with each other to form producer's organisation. This ultimately helps the farmers in increasing their share in the consumer price, resolving issues of trader exploitation, exploring new markets, accessing timely credit and quality inputs thereby strengthening their sustainable agriculture based livelihoods.

Producer Organisation (PO) is a generic name for an organization of producer's agricultural products, non-farm products, artisan products, etc. A Producer Organisation is a legal entity formed by primary producers, viz. farmers, milk producers, fishermen, weavers, rural artisans and craftsmen. A PO can be a producer company, a cooperative society or any other legal form which provides for sharing of profits/benefits among the members. In some forms like producer companies, institutions of primary producers can also become member of PO. Producers Organisation could be registered under various legislations, however, majority of them are registered under the Companies Act. Subsequently, the Indian Companies Act, 1956 was amended in the year 2013 to facilitate the registration of FPOs'. (NABARD, 2015)

The relevance and prospective future of FPO is perspicuous from the words of Shri. Arun Jaitley, Hon'ble Union Finance Minister, Govt. of India which reads as"The issue of profitability of small holding based agriculture has assumed importance in view of increasing proportion of small and marginal farmers in the country. I propose to supplement NABARD's Producers' organization development fund for Producer's development and upliftment called PRODUCE with a sum of Rs 200 crore which will be utilized for building 2,000 producers organizations across the country over the next two years."

NABARD, Small Farmers' Agribusiness Consortium (SFAC), Government departments, Corporates and Non-governmental organisations (NGO) are some of the Producer Organisation Promoting Institution (POPI) providing support for promotion and handholding FPOs. NABARD is spearheading in organising farmer producer organisations across the nation and has evolved a basket of initiatives for supporting the producer organisations with a flexible approach to fulfill farmer's needs and aspirations. The "Producers Organization Development Fund" (PODF) set up in April 2011 is a credential to this effort. The Fund is utilised to support Producers Organizations in three dimensions viz. financial support, capacity building & market linkage with an objective to ensure sustainability & economic viability of these organisations. Besides the above three categories, support is also extended for preparation of detailed project report (DPR) and grant support for converting SHG/farmers clubs/producer groups to producer organisation.

Branding- concept and need

Branding is a concept which is difficult to define or the definition is short lived. This conveys the volatility of the term brand. Decades ago, branding was defined as "a name, slogan, sign, symbol or design, or a combination of these elements that identify products or services of a company" (www.businessdictionary.com). The brand was identified by the elements that differentiated the goods and or service from that of the competitor. Today, brand is a more complex phenomenon with lot of intrinsic and extrinsic connotations, but is inevitable in the present world of marketing. Product positioning is an important challenge for any organisation destined to deliver goods and services. Branding is an important component that helps an organisation in positioning their product and or services.

Brand is the perception that a consumer has when they see, hear or think of the organisation name, service or product. Branding in the present context is so powerful that a colour or a chord of music reminds us of a product or service. It may be well said the word "brand" or "branding" is a moving target and evolves with the behaviour of consumers, or the mental picture of the organisation that represents to consumers and is influenced by the elements, words, and creativity that surround it.

Branding is not only about getting the target market to select our product over the competitor but about getting our prospects to visualise the organisation as the sole provider of a solution to the customer needs, aspirations and problem. Therefore, a successful brand should:

- Clearly deliver the proposed message.
- Confirm credibility of the organisation.
- Emotionally connect the target prospects with the product and or service.
- Motivate the buyer to buy and instigate repeated buying.
- Convert latent demand into demand.
- Convert consumer into customer.
- Help retention of customers.
- Create User Loyalty.

A strong brand is invaluable for an organisation as the battle for retaining customers intensifies day by day between organisations. Branding is an important component of marketing and sales promotion irrespective of the type of product. However, it is more common for industrial commodities and trivial for agricultural commodities. Business houses spends lot of time investing in researching, defining, and building brands because brands are the source of unconditional promise to the customers and the platform for marketing communication. Brand not only creates loyal customers, but also creates loyal employees. It is an undisputed fact that the customers and employees are the two vital components that decides the destiny of an organisation. Brand gives them something to believe in and something to stand behind. Thus, this strength of a brand is the foundation and the driving force which helps any organisation to sustain in the present competitive world and to face the business onslaughts with courage, confidence and determination.

HOW BRANDS INFLUENCE CONSUMERS

The figure tries to summarises how a brand influences the customers (Chris, 2012). A product gets registered in the minds of the customers based on various components and it also aids the customers in distinguishing between different brands. It is a complex process and is the outcome of differential influences of the various components on the customers and it varies among customers. The circle at left shows the six main components that distinguish a brand in consumers' minds. The success of the marketing strategy of an organisation depends how best it identfies the priorities of these components and workout appropriate strategies to gain command over these components. This ultimately helps to minimise the intermediary interventions and influence consumers directly (dotted arrow) - resulting in direct purchase by consumers, leverage with buyers and added value for producers.

Relevance of branding for horticultural commodities

Agricultural commodities play an important role in development. But traditional commodity trading, based on exporting produce in bulk at low prices, limits how much of the profits from these products flows to producers in developing countries. Branding adds value to agricultural commodities. Thus, the focus should be on product brands rather than service brands, aimed at consumers rather than businesses. Product brand can be broadly classified into four: Producer, Varietal, Geographical and Certification brands (Chris, 2012). These branding types vary in ownership and also in the way it is used to create an advantage for the brand owner.

The producer brands are those commodities which are specific to a particular producer which helps in distinguishing the different products and their producers. In these kinds of commodities the producer over years of experience has succeeded in developing and standardizing the production/farming practices which assigns a unique quality to the products of producer farmer concerned. This kind of branding is commonly applicable for fruits and also to value addition in food industry; however, it is also applicable to certain agricultural commodities. Internationally, Chiquita bananas are a good example of how the brand of a traditional private-sector producer has evolved the agricultural commodity sector. The company Chiquita Inc. owns the character "Miss Chiquita" as a trademark which distinguishes its bananas from those of other suppliers and adds value both at the retail level and as a business-to-business device. The branded Chiquita bananas are sold at a price higher than those of their competitors in most retail markets even though there is little physical difference between their products and those from other producers. A quotable example for the state is that Kerala's indigineous medicinal paddy, Njavara (*Oryza sativa*) is being cultivated organically at an eco-farm in Chittur taluk of Palakkad District by a committed farmer, Mr. Narayanan Unni for whom cultivation of njavara rice is a life pursuit.

Varietal branding is a kind of branding which closely resembles the producer brands but tries to establish a distinction between products in agricultural commodity categories. For example, the 'Pink Lady' is the brand name for a patented apple variety, Cripps Pink, established through natural breeding techniques in Australia. In varietal branding, neither the variety nor the brand is owned by the producers. The brand and variety owner allows production and branding under license (producers pay a levy to the owner) and imposes strict quality standards. The benefit to producers is an established market for their products at a premium price, with the brand owner ensuring that Pink Lady is marketed effectively and supported by strong product and supply chain management. 'Sundowner' branded apples, 'Zespri' kiwi fruit and 'Tenderstem' broccoli follow a similarly successful supply chain and marketing model. This is a kind of branding which has been successfully tested and implemented for industrial products which is now expanded to the agricultural commodities too.

Geographical branding is a kind of branding based on the specific location of the origin of a commodity. Geographical branding is an assurance of distinctiveness of a product/commodity attributable to its origin in a defined geographical area. It is also termed as Geographical Indication. With reference to the earlier examples cited Chiquita bananas are sourced from multiple countries (Producer brand) and Pink Lady apples grown under license around the world (Varietal brand) whereas, geographical indicators are defined by a single specific location. The brand ownership is registered to the location of the product or the country of origin. Darjeeling tea is an example for geographical branding. The Chegalikodan banana variety is a GI brand of banana of Kerala. Chengalikodan banana is known for its unique shape, size, colour and taste. Vazhakkulam pineapple, Pokkali rice, Palakkadan matta rice are some of the other agricultural products with GI from Kerala. The uniqueness of the product emanates from number of factors like geographical and climatic factors, humidity, rainfall, soil characteristics and above all the cultivation practices and processing methods. GI confers legal protection to commodities and promotes economic benefit to farmers and helps consumers to get quality products. Geographical Indications of Goods are defined as "that aspect of industrial property which refers to the geographical indication referring to a country or to a place situated therein as being the country or place of origin of that product. Typically, such a name conveys an assurance of quality and distinctiveness which is essentially attributable to the fact of its origin in that defined geographical locality, region or country". Under Articles 1 (2) and 10 of the Paris Convention for the Protection of Industrial Property, geographical indications are covered as an element of IPRs. They are also covered under Articles 22 to 24 of the Trade Related Aspects of Intellectual Property Rights (TRIPS) Agreement, which was part of the Agreements concluding the Uruguay Round of GATT negotiations. (Geographical Indications Registry, GOI)

India, as a member of the World Trade Organization (WTO), enacted the Geographical Indications of Goods (Registration & Protection) Act 1999, has come into force with effect from 15th September, 2003. (Geographical Indicators Registry).

Certification brand is a distinct kind of branding which may not fit into the traditional outfit of branding. Certification brands distinguish products through ethical or social certification marks. It behaves in the same way as that of producers brand and further adds a social or ethical dimension to the product. Certification aims at developing and promoting environment friendly agriculture and to offer recognition to farmers committed to this method of production. Ecocert is an inspection and certification body established in France in 1991 by agronomists. Ecocert operates in India from its offices at Aurangabad and Gurgaon. It has largely contributed to the expansion of organic farming by persistently introducing much affordable schemes of certification for small and marginal farmers. "Agmark" is a certification for quality and purity of agricultural commodities governed by the Agricultural Produce (Grading & Marking) Act, 1937 for export and domestic trade. Certification process is advantageous both for the consumers and the producers (Directorate of Marketing & Inspection, Govt. of India).

Supply chain management and marketing of horticultural produce

Supply chains are principally concerned with the flow of products and information between supply chain member organizations-procurement of materials, transformation of materials into finished products, and distribution of those products to end customers (MANAGE). India is the world's largest producer of many vegetables but there still exists huge gap between per capita demand and supply due to enormous waste during post-harvest handling & marketing. These losses are a missed opportunity to recover value for the benefit of farmers. The deploying of appropriate strategic and operating models, will allow the efficient closure of gaps between demand and supply so as to contribute to doubling farmers' income. The gap between demand and supply is due to ineffective market links, poor handling and lack of consolidation on both the demand-side and supply-side. On the supply side, the government has agenda to promote modern cultivation practices and collaborative farming. On the demand side, the government has to reinforce its effort to handle the excess supply of produce resulting from the various government initiatives. This could be possible only if the government succeeds in developing a fool proof replicable model for marketing of fruits and vegetables.

Post-harvest supply chain systems

Supply chain management represents the management of the entire set of production, manufacturing/transformation, distribution and marketing activities by which a consumer is supplied with a desired product in desired quantity at desired time. Post-harvest supply chain encompasses the planning and managing of all activities involved in procurement, preconditioning and delivery system of farm produce (Rais & Sheoran, 2015).

Marketing of horticultural crops is quite complex and risky due to the perishable nature of the produce, seasonal production and bulkiness. The marketing arrangements at different stages play an important role in price levels at various stages viz. from farm gate to the ultimate user. These features make the marketing system of vegetables to differ from other agricultural commodities, particularly in providing time, form and space utilities. While the market infrastructure is better developed for food grains, for vegetables, markets are not that well developed and markets are congested and unhygienic.

Generally, the middlemen and wholesale businessmen purchase the agricultural products from the farmers at a lower price. In turn, fresh vegetables and fruits purchased at the lower price from the farmers are sold out to retail businessmen at higher price and the retail businessmen sell those products further at higher price to the consumers thus widening the price spread. Vegetable farmers are the most vulnerable category because even if prices soar to one of the highest levels, they get only a third or fourth of the prices in retail markets. The situation is worsened further for the farmers because the input costs and other production overheads are rocketing beyond their control. Besides, the high inflation rates which has a direct impact on trade and pricing along with the fall in other economic development indicators has generated cost pressures resulting in an adverse effect on the returns to the farmers.

Supply chain development not only benefits the agricultural sector but also creates spin- offs that stimulate development of social, economic and environmental sustainability in the region viz., employment generation, value addition, minimization of product and post harvest losses, price stabilisation, market guarantee etc. which ultimately elevates the confidence and morale of the farmers. The specific gains of an efficient and effective supply chain system can be summarised as:

- Reduction of product losses in transportation and storage
- Increasing of selling radius and revenue from sales
- Productivity improvement
- High customer satisfaction
- Increased profit
- On time delivery
- Tracking and tracing to the source
- Better control of product safety and quality
- Better information about the flow of products, markets and technologies
- Transparency of the supply chain
- Dissemination of technology, capital and knowledge among the chain partners
- Large investments and risks are shared among partners in the chain
- Gross Capital Formation at back-end and in agriculture allied business.

Value addition - Panacea to minimise post-harvest losses

The economic intensification and shift in dietary patterns in favour of vegetarian diet has redefined the paradigm of production and consumption of fruits and vegetables in our country. Various initiatives and efforts of the government and other institutional support system have helped to increase the production and productivity. However, despite the positives on the production scenario, the sector suffers significantly from post-harvest losses both in terms of quantity and quality of the produce which ultimately diminishes the returns of the producers and leads to farmer discontent and causalities. Major factor contributing to this phenomenon is the non-existence of a foolproof system for value addition. Various plans and programmes for promoting agriculture and allied activities designed by the government since independence was skewed towards pre-production and production sector and the post-production or the post-harvest sector was either neglected or poorly addressed. Of late, it was realised and recognised that for agriculture to be sustainable, an equal balance should be drawn between the input sector (pre-production), farm sector (production) and product sector (post-production). This realisation was an eye opener to the policy makers to think differently and to draw appropriate strategies for each of these sectors, value addition and efforts to plug post harvest losses gained prominence.

It is apparent that tremendous opportunities exist for vertical diversification within the fruit and vegetable sector in terms of increase in farm income, poverty alleviation, food security, and sustainable agriculture. Value addition is not a process to be confined to post harvest sector. Value addition could be adopted in various forms in the input sector and the farm sector. For example, selection of quality inputs like superior quality planting materials, scientific selection of land, selecting appropriate timing to start agricultural operations and identifying an appropriate institution for availing financial assistance are kinds of value addition in input sector. Whereas adopting Good Agricultural Practices (GAP) and undertaking other farm related activities is an example of value addition at farm sector. Thus, value addition is a chain of activities which should start from the stage of decision making to produce a commodity and continue till the product is delivered to the ultimate consumer.

The vibrant agro-food industry have unlimited potential in the form of processed or value added products and consumers all over the country will get an opportunity to enjoy them throughout the year. In this transaction, processing of food crops into a variety of products with extended shelf life is achieved. Adding value to the original crop helps the farmer not only to overcome the spoilage and losses, but also fetches high returns due to the newly added technology. Value addition enterprises aims at assigning value to the raw commodities into multiple final products and semi-processed food which will fetch remunerative prices to farmers. Besides, it also provides convenience and safe food to consumers and promotes diversification and commercialization of agriculture by effectively establishing linkage between consumers and farmers. Moreover, it will also expand the horizons of markets beyond national boundaries. In this context, an important aspect to be borne in mind is that there exists a need for technology generation and commercialization at small scale which are of critical importance for growth and diversification. The development of a dynamic agro-food industry will depend on innovative research and the deliberate engagement of the national inventive system comprising academia, industry, and the government sector.

Conclusion

Marketing of horticultural products is a complex process especially due to its characteristics such as perishability, seasonality, bulkiness and delicate nature of the products coupled with colossal post-harvest losses in storage and handling due to improper bagging without crating, lack of temperature controlled vehicles, absence of cold chain facilities and inadequate transportation facilities. Both supply chain and value chain management has to be improved by adopting global best practices in post harvest activities and also by disintermediation and participation of organized players with a view to benefit both farmers as well as ultimate consumers. Though, in India we have multiple organisational initiatives for marketing of horticultural products, none of them could address the core issue of the farming community or in other words the hardships of the poor peasant community persists. This may be attributed to the reason that these are all isolated successful micro models that lacks a macro outlook and approach. Further, other factors contributing to this outcome is the scattered and unorganised nature of the Indian farmers coupled with other socio-economic intricacies. Thus, the need of the hour is to have a model which could address the present situation. An integrated approach similar to that of the world renowned 'Anand Model' successfully tested and proven for procurement and marketing of milk, a highly perishable commodity will have to be thought of for the marketing of a comparatively less perishable horticultural produces. Ensuring farmer participation through a farmer promoted and farmer oriented organisation is the foundation of the Anand model integrated approach. The new concept of Producer Organisation harmoniously blends the advantages of both a cooperative organisation and that of a limited company. Besides, the Government of India is committed to extent all support and patronage for the promotion of FPOs' along with NABARD, SFAC and other POPIs destined to re-engineer the future of marketing of horticultural products. The support of institutional arrangements, practicing of global GAP and HACCP has also helped a long way to improve and maintain quality of horticultural products. Thus, the years to come will bring in hopeful dawns and dusks for the farming community and the consumers of our nation.

References

Chris, D. 2012. Branding Agricultural Commodities: The Development case for Adding Value through Branding, International Institute for Environment and Development/Sustainable Food Lab.

Dastagiri, M.B., B. G. Kumar & S. Diana. 2009. Innovative Models in Horticulture Marketing in India, *Indian J. Agricultural Marketing* 23(3): 83-94.

Directorate of Marketing & Inspection, Government of India.

Geographical Indicators Registry, Government of India.

Government of India. 2013. Policy & Process Guidelines for Farmer Producer Organisation, Department of Agriculture and Cooperation. GOI.

Grahame, D. 2005. *Horticultural Marketing*, FAO, Rome.

MANAGE, Training Programme on Supply Chain Management in Agriculture, National Institute of Agricultural Extension Management, Hyderabad.

NABARD. 2015. Farmer Producers Organisation, National Bank for Agriculture and Rural Development, Mumbai.

Parveen, S., I. Bushra., K. Humaira., S. Shazia & M.A. Azhar. 2014. Value Addition: A Tool to Minimize the Post-harvest Losses in Horticultural Crops. *Greener J. Agricultural Sciences*. 4(5): 195-198.

Pradeep, K. 2013. *Rural Marketing*, Dorling Kindersley (India) Pvt. Ltd., Delhi.

Rais, M. & A. Sheoran. 2015. Scope of Supply Chain Management in Fruits and Vegetables in India. *J Food Process Technol*. 6: 427.

Vasant, P., N. V. Gandhi & Namboodiri. 2017. Marketing of Fruits and Vegetables in India: A Study Covering the Ahmedabad, Chennai and Kolkata Markets., https://core.ac.uk/download/files/ 153/6443656.pdf (Accessed on 13.07.2017).

www.dmi.gov.in/.../Final%20Text%20Matter%20related%20to%20QC%20Section.pdf (Accessed on 25- 07- 2017)

www.statisticstimes.com/economy/sectorwise-gdp-contribution-of-india.php (Accessed on 25-07-2017).

www.businessdictionary.com/definition/branding.html (Accessed on 25-07-2017)

www.marketingteacher.com/what-is-marketing-2/ (Accessed on 25-07-2017)

www.nccd.gov.in/PDF/Analysis_NDDB_veg_model.pdf, Analysing NDDB Cluster Model for Marketing of Vegetables, (Accessed on 15.07. 2017).

19

Model Business Plan for Dried Jackfruit Flakes Production

Sangeetha K Prathap & Sudheer K P

Introduction

Jackfruit (*Artocarpus heterophyllus*) is the largest edible fruit in the world. It is a seasonal fruit found in almost all the humid tropical regions of the world. It is very popular all over India and is believed to be of Indian origin. Flesh of jackfruit is starchy and fibrous and is a source of dietary fiber. Varieties are distinguished according to characteristics of the fruit's flesh. In Kerala, two varieties of jackfruit predominate: *varikka* and *koozha*. *Varikka* has a slightly hard inner flesh when ripe, while the inner flesh of ripe *koozha* fruit is very soft and almost dissolving. Strong aroma of the fruit accounts popularity of its processed products. Each year, approximately 30-50% of the total harvested fruit is spoiled because of lack of post-harvest processing in India

Jackfruit has more protein, calcium, thiamine, riboflavin and carotene than banana. Edible bulbs of ripe jackfruit are consumed for their fine taste and pleasant aroma. The fibre content helps to protect the colon mucous membrane by decreasing exposure time as well as binding to cancer-causing chemicals in the colon. Fresh fruit has small amounts of carotenoid and flavonoid pigments such as ß carotene, xanthin, lutein and cryptoxanthin-ß. Together, these compounds play vital roles in antioxidant and vision functions.

However, the fruit is perishable and cannot be stored for long time because of its inherent compositional and textural characteristics. Every year, a considerable amount of jackfruit obtained in the glut season (June-July) goes waste during harvesting, transporting and storing. Proper post-harvest technology for prolonging shelf life is therefore necessary. Besides, alternate ways of using jackfruits in on-season plays significant roles in reducing post-harvest losses. In order to process the fruit, the skin has to be removed first. Sticky latex from the fruit will ooze out when the tough scaly skin is cut causing some difficulties. Therefore, it is usual for workers to smear their hands and knives with a film of edible oil before cutting the fruit. Preservation of fruits by processing has been the research pursuits of many developed and developing countries and has yielded quite a number of technologies.

Drying is a process used widely to preserve fruits and vegetables during off season by removing water content by evaporation. This process reduces weight of the fruit or vegetable and can be preserved in room temperature in an air tight packing. Drying process reduces usage of preservatives and also occupies less space to store. It adds diversified and attractive food items in dietary menu and contributes to income generation and employment.

Technical feasibility

Objectives

- To produce and sale dried jackfruit in Kerala.
- To generate employment opportunities in food processing sector.
- To utilize the peel for animal feed.

Uniqueness

Jackfruit, an organic fruit, available in plenty at affordable price is not fully utilized for commercial purposes. At present, a huge proportion of fruit is wasted causing enormous loss of this resource of the State. Many varieties are available in the State which has tremendous potential to be processed into specific products. And one such product is the dried jackfruit. Its unique features includes

- The raw material is cheap and easily available
- It is natural
- Field problem like pest and diseases are absent
- They are devoid of harmful chemicals
- They survive the heat of summer and have better longevity
- Shelf life is much high compared to fresh fruit
- Uniformity in quality is better compared to fresh fruit
- Minimum loss of quality (risk in production is less)
- Effective waste management is possible
- They are eco-friendly and biodegradable
- Low investment
- Can be used for any occasion and require little care
- They offer wide range of products like sweets, chips etc.

Raw material

Mature unripe jackfruit is the main raw material for this unit. Jackfruit is grown in large quantities in Kerala. Actual recovery of bulbs or edible portion is in the range of 75% to 85%. However, transportation of jackfruit is expensive and hence the factory should be as far as possible, located in the jackfruit growing area. It should also be procured from farmers, local agents, farmers clubs etc on contractual agreement.

Process description

Drying of jackfruit involve various unit operations *viz.,* cleaning, grading, cutting, removal of bulb, deseeding, slicing, blanching, drying and packaging.

Flow chart for dried jackfruit

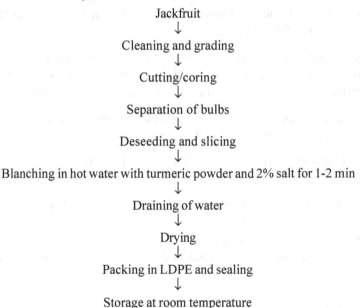

Jackfruit
↓
Cleaning and grading
↓
Cutting/coring
↓
Separation of bulbs
↓
Deseeding and slicing
↓
Blanching in hot water with turmeric powder and 2% salt for 1-2 min
↓
Draining of water
↓
Drying
↓
Packing in LDPE and sealing
↓
Storage at room temperature

Cleaning and grading: Cleaning and grading are the primary and most vital operations for processing jackfruit. It is carried out to remove undesirable foreign materials and grading shall be based on the maturity and ripening stage of jackfruit. Jackfruit is cleaned and graded manually. Remove immature, over-ripe, damaged and misshapen fruits. Wash fruits using chlorinated water (100 ppm) to remove dirt, latex stains and any field contamination. Drain fruits properly to remove excess moisture from the surface for further processing or storing.

Cutting and coring: Jackfruit is cut in to desirable pieces and stalk is cored out in order to remove bulbs from the fruit. The cutting knife may be greased with coconut oil to prevent sticking of latex.

Separation of bulbs: Bulb of the jackfruit is removed manually. Hands are greased with coconut oil before removing the bulbs in order to prevent sticking of latex.

Deseeding and slicing: Deseeding may be done manually. Deseeded jackfruit flesh may be sliced into strips using knife and cutting board into desired size of 2 - 3 mm width. The increased surface area of sliced jackfruit will shorten the drying time there by increase the tough-put of dryer.

Blanching: Blanching is carried out to retain colour of jackfruit after drying and to prevent enzymatic activity. Blanching may be done in boiling hot water with 2% salt and 1% turmeric powder at 95 - 100°C for 1- 2 min. A hot water blancher/steam blancher can be used for this process.

Draining of water: Excess water may be removed immediately after blanching process using a strainer. Retention of water may increase drying time and affect the quality of dried product.

Drying: Drying could be done in a cabinet dryer by placing slices in a single layer and care should be taken to avoid overlapping of slices in order to have uniform dried product. Drying process will take about 3–3.5 hrs at 65°C.

Packaging: Dried jackfruit slices are cooled to room temperature and packed in 400 gauge LDPE film/ laminated pouches and sealed.

Storage: Dried jackfruit slices are shelf stable products with shelf life of 12 months when stored under room temperature/ambient condition.

Cleaning, grading, cutting and separation of bulbs are carried out at five different locations. So, the outer pericarp is utilised in those locations for compost. Separated bulbs are only brought to the processing site for further processing. Deseeding, slicing, blanching, drying, and packaging operations are done at the factory.

By-products

After cutting the fruits, seeds are removed from the perianths and the outer layer of the seeds is removed manually. The seeds are rich in carbohydrates and can be used for culinary purposes. Processed seed flour can be used to fortify bakery products like cookies and cakes. Cabinet dryer which is already available for drying of jackfruit can be used for seed drying. Moreover, utilization of Jackfruit seed flour is an environment friendly technology since it could solve the waste disposal problem of residues.

Other auxiliary materials like weighing balance, working tables, stainless steel utensils, apron, hand gloves, packaging films, storage racks etc. are also required.

Scope for commercialization

Processed jackfruit products have excellent market in the domestic and export front. Traditional products of jackfruit are relished by Keralites at home and abroad. Seed flour serve as a raw material for bakery products and health drinks. Waste arising from jackfruit could be converted into good animal feed and is a valuable resource for biofuel.

Commercial Jackfruit processing venture should be attempted on a multiline approach starting from the mature unripe jackfruit dried products to seed flour and waste utilization, if it is to be economically viable and successful.

Processing layout for Jackfruit drying plant

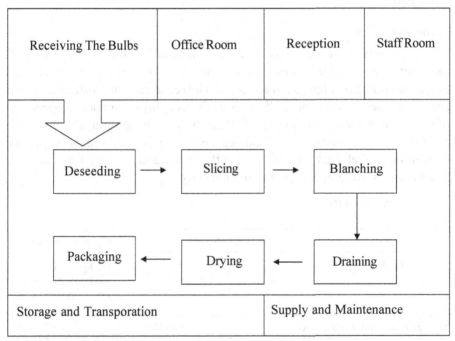

Benefits of processing

- Converts raw food into edible, usable and palatable form.
- Helps to store the seasonal agricultural commodities to be available throughout the year even in off season, avoid glut in the market, check post-harvest losses.
- Minimally processed product of jackfruit saves time for cooking.
- Helps in value addition of the fruit and make marketing and distribution tasks easy.
- Increases round the year availability of many foods.

Location of plant

The processing unit will be located at a place where adequate raw material, access to cheap labour, power, water, transportation facility of input and output, storage facility, suitable waste management system are available.

Infrastructural facilities

The area of proposed site for jackfruit drying unit is about 2500 sq. ft, which will be sufficient for establishing a unit with a floor area of 2000 square feet structure. This structure includes receiving area for raw jackfruit and storage space for finished products, processing area, office buildings, and rest room. The entire building may be diverted into zones for utilising the above said facilities efficiently.

Plant & machinery

It is suggested to have annual production capacity of 100 tons of dried jackfruit. Jackfruits are available around 7- 8 months. Production operations envisaged in the present plan is limited to 8 months. However, sale and distribution runs round the year. To overcome the limitation of seasonality of production operations of the unit, either storage of raw jackfruit can be introduced during the expansion stage or alternate seasonal fruits and vegetables available during the respective seasons can be selected for drying using the existing machinery. Currently, the plant is expected to operate with 80% capacity.

Production capacity

S. No	Product	Production capacity	Quantity produced per day
1	Dried jackfruit	250 kg	175 kg (70% capacity) 200 kg (80% capacity) 225 kg (90% capacity)

Machinery & equipment capacity	Quantity	Capacity
Blancher	1	100 kg
Dryer	1	250 kg
Sealing machine	1	150/hr

Miscellaneous assets

Other items like office furniture, working tables with aluminum tops in the factory, exhaust fans, storage racks and bins etc.

Utilities

Total power requirement shall be 8 HP and water requirement for production process, potable and sanitation purposes shall be around 1000 liters every day.

Technical parameters

Parameters	Remarks
Plant capacity	250 kg/day
Production per day	175 kg (70% capacity),200 kg (80% capacity)
	225 kg (90% capacity)
Number of working days in a month	25 days
Number of production months	8 months
Production per month	4375 kg, 5000 kg, 5625 kg
Number of Labourers	21
Procurement source of jackfruit	Jackfruit cultivators
Cost of 1 jackfruit (purchased from farmers) (Rs.)	50
Selling price of dried jackfruit (Rs./kg)	500 (in first two years), 600(in third & fourth year)
	700(in fifth year)
Selling quantity	100 g, 250 g, 500 g, 1 kg and 2 kg packet
Electricity	210.4 units per day
Unit cost for electricity (Rs.)	4.75
Monthly use of electricity	5260 units
Depreciation	10%

Marketing plan on dried jackfruit

Jackfruit is a widely accepted fruit. Like any other fruit, it is perishable in nature. It is grown in very limited parts of India and hence is not much popular in other parts. It is heavy and bulky and hence transportation is not very easy and is costly as well. Therefore, its down the stream products with longer shelf-life can be easily transported and shall also have value addition. Consumption of natural value depleted and at the same time highly noxious food items, poisoned by pesticide residues is causing serious health hazards, prompting increasing number of people to turn to organically grown and processed ethnic food, noted for nutritional value, taste and safety. The situation world over is that supply is not adequate for meeting the growing demand for this type of food.

There is a prospective market for these products in India as well as outside the country. It has a good export market potential especially in Middle East countries. There are millions of Keralites in gulf countries who knows the nutritional value of jackfruit. In view of the above, it is seen that there is good scope for setting up jackfruit processing units in jackfruit growing areas. This will not only help the farmers to utilize the perishable raw material, but also generate more employment opportunities in rural areas. Dried jackfruit can be produced in the unit with unripe fruits. Cleaned and processed fruits in hygienic atmosphere will be prospective if marketed in attractive 250 g, 500 g, 1 kg, 2 kg packs at premium prices in India and abroad. The processing unit can be started as a sole proprietorship, partnership, company or as co-operative.

Current market potential

- Gradual conversion of **ready to cook** products to **ready to eat** products
- comparatively lesser competition
- Satisfies the off season demand for jackfruit
- High export potential
- Wide varieties of dishes can be prepared
- High shelf life (1 year)
- Even diabetic patients can consume
- Anti cancerous property attracts health conscious consumers.

Competition

At present, market of dried jackfruit faces lesser competition. Very few numbers of local companies are operating in this market. Agrahar Food Products, Palakkad and People's Service Society, Palakkad (PSSP) are the major competitors in the local market. Jackfruit 365 is the competitor in the national market. Foreign players already exist in the market which holds the majority market share. Woodland Foods and Newark Nuts Company are global players. Major players are also trying to expand their product range in this market to tap different market segments.

SWOT Analysis

Strength

- Unique product
- High quality
- Availability of raw materials
- Low price of raw materials
- High shelf life (1 year)
- Natural.

Weakness
- Cannot be consumed as it is
- High fixed cost
- Bulk quantity of raw material is required for small amount of produce.

Opportunity
- Huge potential market
- Government support
- Favourable public attitude
- Opportunity to export.

Threats
- Government regulation
- Public unawareness about the food value of dried jackfruit
- Political instability
- Lack of subsidies
- Seasonality.

Opportunity analysis

Since, in Kerala, Jackfruit is grown in about 89701.52('000) hectares of land and the annual production of jackfruit is about 299.689('000) million, the company can easily collect raw materials at low price. The production unit can opt for large scale production that will give an opportunity to export. Besides this, the product has a large potential market as most customers want **dried jackfruit** with good and hygienic quality.

Marketing strategy

Marketing strategies of **"Dried jackfruit"** are primarily aimed at increasing market share and establishing brand more dynamically. A unique product strategy can be pursued with proper packaging, positioning, pricing, distribution and extensive promotional strategy.

Positioning strategy

Jackfruit contains high amount of carbohydrates and calorie that provides energy instantly. It is rich in antioxidants which protect from cancer, ageing and degenerative diseases. Due to rich antioxidants it is good for eye sight and protect from conditions like cataract and muscular degeneration. It is a good source of potassium which maintains fluid and electrolyte balance. It also improves bone and skin health. Hence, it is beneficial to all age groups.

- It offers fully natural ingredients.
- Price is relatively low when compared to quality and competitors prices.
- It is rich in antioxidants which protect from cancer, ageing and degenerative diseases.

Potential market for dried jackfruit in Kerala

State	Kerala.
Population group	Urban, Suburban, Rural.
Social class	Upper/Middle income group.
Occasion	Regular, Occasional.
Perceived benefits	Health conscious, natural products loving
User status	Potential.

Marketing mix

Product strategy

The product strategy includes development of product and proper packaging. In Kerala, Jackfruit is grown in about 89701.52('000) hectares of land from which it is possible to ensure channelization of raw materials.

Product development

Drying of jackfruit involves various unit operations *viz.,* cleaning, grading, cutting, removal of bulb, deseeding, slicing, blanching, drying and packaging.

Package development & branding

- It is eye-catching with suitable appealing brand name
- Protection and preservation of the product
- Temperature and moisture proof
- More convenient to use

The package sizes of dried jackfruit are the following:

100 g (small), 250 g (small), 500 g (medium), 1 kg (big), 2 kg (family size).

Pricing strategy

Products focused on global market usually adopt premium pricing. One kilo gram of dried jackfruit cost Rs 150 and is being packed in 100 g, 250 g, 500 g, 1 kg and 2 kg packs. As the local markets are targeted, price is being fixed at an amount that covers cost of production and certain profit margin.

Pack size(Packet)	MRP (Rs)					
	Our price	Pssp	Agrahar	Jack 365	Newark nuts company	Wood land foods company
1 kg	500	-	-	-	-	-
500 g	325	-	-	-	-	-
250 g	150	125	150	-	-	2375
180 g	-	-	-	360	-	-
100 g	75	-	-	-	-	-
50 g	-	-	-	-	450	950

Promotion strategy

Advertising and promotions will support the introduction of exploration of the product into the market. Since it is a new product, considerable efforts need to be given for educating the customers of the product. Key programmes include new arrival stand, merchandise, free samples and press advertising.

- *New arrival stand:* The product can be displayed on the new arrival stand in stores to attract the customers which induces them to seek information about the product.

- *Merchandise:* A person with enough information about the product will be assigned to promote the sale of the product in the stores.

- *Free samples:* providing free sample of product to consumer.

- *Print media advertising:* Considerable print media advertising in popular dailies and popular magazines to enhance brand image.

Distribution strategy

The target and potential consumers of **"Dried jackfruit"** live in all parts of the state. So to ensure availability of the product to each and every consumer, following activities are ensured.

- The product will be supplied to the stockists who will in turn supply to wholesale and retail outlets.

- A web store displaying jackfruit products is also proposed along with physical store. Provision for catering to online orders will also form part of the business.

Financial feasibility

Capital budgeting is concerned with the allocation of firm's scarce financial resources among the available market opportunities. The consideration of investment opportunities involves the comparison of the expected future streams of earnings from project with the immediate and subsequent streams of expenditures for it. This part of the report is organized under the following heads.

Techno-economic parameters

Investment in business ventures requires commitments on fixed basis during the initial phase. Working capital is needed during the period of operation. The cash flow from operations is calculated on the basis of the expected expenditure and revenue out of the project.

Fixed capital requirement: In order to establish processing facility for dried jackfruit, following fixed capital requirements are necessary. It is assumed that the land is owned and no cost towards acquiring land is provided for. Building/ modifying the facility/rooms required for the establishment of the unit is provided for. Purchase of truck worth Rs. 15 lakh is provided for (Instead of owning the truck, it is possible to make alternate arrangements with carriage and freight agents which may be opted suitably). The expenses incurred for installation of the plant and machinery and other necessities for establishment of the processing facility can be accounted under capital expenditure. Total fixed capital required for the proposed processing unit for dried jackfruit is estimated as Rs. 45,20,881/-, of which, detailed expenditure heads are furnished below.

Particulars	Specification	Unit Cost (Rs.)	Amount (Rs.)
Technological cost			50000
Construction of building			2000000
Truck	1		1500000
Gate	4	7500	30000
Total			**3580000**
Water supply			
Well	1		25000
PVC pipe(30m@100m	30	100	3000
Foot valve	1		200
Pump set 2HP	1		10000
Water purifier	1		50000
Total			**88200**
Electrical installation			
Wiring charges			55000
Fan	1		1200
Exhausting fan	1		1500
Misc.exp(switch board, CFL, tube light)			3800
Total			**61500**
Machinery			
Weighing machine	1		6000
Cutting knives	20	500	75000
Blancher –cum-drier	1		300000
Total			**381000**

Contd.

Equipment

Trolley(3@2500	3	2500	7500
Lab equipments (moisture meter etc.)			10000
Firefighting equipments	1		4500
Knives	12	100	1200
Vessels(pan, spoons)			18500
Misc. (gloves, aprons, caps			25000
Total			66700

Furniture& fixtures

Chair(15@300)	15	300	4500
Working table(5@5000)	5	5000	25000
Cupboard	1		3500
Computer	1		25000
LPG stove	1		5000
Total			63000

Preliminary & Pre-operative expenses

Incorporation of company	
VAT registration	10000
FSSAI registration	5000
License fee	250
Project preparation	5000
Market survey	5000
Travelling &consultancy	5000
Administrative expenses	2500
Total	**27750**
Interest during project implementation period	125322
(construction and installation of machinery)	
Working capital margin	127409
Grand total	45,20,881

***Estimation of working capital*:** The project is expected to operate at 70% capacity in the first year. Detailed estimate of raw materials and manpower along with consolidated statement of working capital for the proposed dried jackfruit is given below.

Raw material and utilities

Particulars	Qty /month	Unit cost (Rs.)	Cost (Rs.)
Jackfruit	10000 kg	50	500000
Ingredients (Salt, turmeric)			25000
Water	25000	1	25000
Electricity	5260	4.75	24985
Packaging material	20000	2	40000
Total			614985

Man power

Sl.No	Designation	No. of persons required	Salary per month / person (Rs.)	Total salary (Rs.)
1.	Manager cum Supervisor	1	20000	20000
2.	Accountant	1	12000	12000
3.	Marketing manager	1	15000	15000
4.	Machine operator*	1	8000	8000
5.	Sales man	2	6000	12000
6.	Unskilled worker*	15	5000	75000
	Total			142000

*Machine operator and unskilled workers are employed for 10 months in a year due to seasonality in operations assumed. Other staff is maintained round the year for sale and distribution purposes.

Working capital requirements: The working capital requirement of the proposed project is summarized as follows. The working capital requirement of the project is estimated to be Rs. 11,66,985/-per annum.

Particulars	Cost/month (Rs.)
Raw material and ingredients	550000
Packaging	40000
Transportation	50000
Power and fuel	25000
Electricity charges	24985
Wages	142000
Repairs and maintenance	10000
Administrative expenses	10000
Sales and distribution expenses (12% of the sales value)	175000
Promotional expense	140000
Total	1166985

Sources of finance

The proposed project comes under the classification micro enterprises. For the enterprises engaged in the manufacture or production, processing or preservation of goods, where investment in plant and machinery does not exceed Rs. 25 lakh is termed as micro enterprise.

The capital requirement for the project may be financed out of owned funds and bank loans. At present, banks are providing requisite funding for the MSMEs at concessional terms. Schemes to promote industries are in vogue. The Ministry of MSME, Government of India and SIDBI set up the Credit Guarantee Fund Trust for Micro and Small Enterprises (CGTMSE) with a view to facilitate flow of credit to the MSE sector without the need for collaterals/ third party guarantees. Main objective of the scheme is that the lender should give importance to project viability and secure the credit facility purely on the primary security of the assets financed.

Banks have however been advised to sanction limits after proper appraisal of the genuine working capital requirements of the borrowers keeping in mind their business cycle and short term credit requirement. As per Nayak Committee Report, working capital limits to SSI units is computed on the basis of minimum 20% of their estimated turnover up to credit limit of Rs.5 crore.

In terms of RBI circular RPCD. MSME & NFS.BC.No.5/06.02.31/2013-14 banks are mandated not to accept collateral security in the case of loans up to Rs 10 lakh extended to units in the MSE sector. Further, banks may, on the basis of good track record and financial position of MSE units, increase the limit of dispensation of collateral requirement for loans up to Rs.25 lakh with the approval of the appropriate authority.

Total capital requirement for the establishment of the unit is Rs.45.20 lakh. It is proposed to mobilize the required capital from the following sources: (i) Bank loan and (ii) Own contribution.

It is proposed to mobilise Rs. 31.64 lakh as term loan from the commercial bank. Expected rate of interest is 12%. Repayment will be made in 5 year equal installments.

Indicative proposal for term loan from commercial bank

Particulars	Amount (Lakhs)
Project cost for loan purpose	Rs. 45.20 lakh
Margin money @ 25%	Rs13.56 lakh
Loan amount	Rs. 31.64 lakh
Interest rate	15%
Repayment period	5 years

The repayment schedule of the loan for a period of five years shall be as follows

Indicative loan repayment schedule: Term Loan

Year	Principal (Rs.)	Interest @ 12% (Rs.)	Total repayment/ Annum (Rs.)	Balance (Rs.)
1.	632923	379754	1012677	2531693
2.	632923	303803	936726	1898770
3.	632923	227852	860776	1265847
4.	632923	151902	784825	632923
5.	632923	75951	708874	0

Working capital loan: Based on the working capital estimates, a working capital loan of Rs. 16.98 lakhs is proposed to be taken. The working capital margin (owner's stake) has formed a part of the term loan. Working capital loan can be obtained for maximum 30 month period after which it can be renewed or applied fresh. A conservative estimate of usage of entire cash credit being sanctioned is done in the business plan. Interest expenses of Rs. 2.54 lakh per annum are chargeable under this head.

Cash flow statement

Cash flow includes cash inflows and out flows-cash receipts and cash payments-during a period. Movements of cash are of vital importance to the management. The short term liquidity and short term solvency positions of a firm are dependent on its cash flows.

S.N.	Particulars	0th year	1st year	2nd year	3rd year	4th year	5th year
1.	Initial investment	4520881					
2.	Cash inflow	-	14000000	14000000	19200000	19200000	25200000
3.	Cash outflow	-	9506984	9597108	11037230	11156421	12816828
4.	Net cash inflows	-	4493016	4402892	8162770	8043579	12383172

For the present calculation commercial taxes are not included; taxes apply as per taxation policy of the Government as and when applicable. At present all food processing industries are exempted from taxes during initial years of establishment.

Capital budgeting techniques

The financial feasibility of the project is analyzed by the application of the following capital budgeting techniques:

Payback period

The payback period is one of the popular methods useful to ascertain the number of years required to recover the initial investment of the project. Thus, this method reveals the length of time required for the inflow of cash proceeds generated from the investment.

Payback period of the current project is estimated to be almost one year (plus three days to be exact).Thus, the present proposal may be accepted.

Net Present Value

NPV may be defined as the excess of present value of project cash inflows over that of cash outflows. The discount rate is 15%.

Year	Discounting factor @15%	Cash inflows in the project (Rs.)	Cash inflows (Rs.)
1	0.870	44,93,016	39,06,970
2	0.756	44,02,892	33,29,219
3	0.658	81,62,770	53,67,154
4	0.572	80,43,579	45,98,943
5	0.497	123,83,172	61,56,625
a.	Total present value of cash flows (Rs.)	2,33,58,911	
b.	Initial investment (Rs.)	45,20,881	
c.	Net Present Value [a-b] (Rs.)	1,88,38,031	

The present value of cash inflow is greater than zero. Hence, the project satisfies the criteria of having positive value for NPV and hence viable.

Internal rate of return

Internal rate of return (IRR) is the interest rate at which the net present value of all the cash flows from a project or investment equal zero. Internal rate of return is used to evaluate the attractiveness of a project or investment. If the IRR of a new project exceeds a company's required rate of return, that project is desirable. If IRR falls below the required rate of return, the project should be rejected and the investment with the highest IRR is usually preferred.

Particulars	Amount Rs.
Initial Investment	(4,520,881)
Cash Inflow for year 1	4,493,016
Cash Inflow for year 2	4,402,892
Cash Inflow for year 3	8,162,770
Cash Inflow for year 4	8,043,579
Cash Inflow for year 5	12,383,172
Internal Rate of Return	115%

IRR of the present project is high, hence, the project is feasible.

Profitability index (PI) or Benefit Cost Ratio (BCR)

Profitability index is the ratio of the present value of cash inflows, at the required rate of return, to the initial cash outflow of the investment. It is the ratio of payoff to investment of a proposed project. It is a useful tool for ranking projects because it helps to quantify the amount of value created per unit of investment. A profitability index of 1 indicates breakeven. Value of lower than one indicates that the project's present value of cash inflows is less than the initial investment. As the value of the profitability index increases, the financial attractiveness of the proposed project also tends to increase.

Sl. No	Particulars	Amount(Rs.)
a	Present value of cash inflows	2,33,58,911
b	Initial Investment	(45,20,881)
	Profitability Index [a/b]	5.17

Profitability index obtained is 5.17 which is greater than the optimum criteria of 1. So, the project can be accepted.

Conclusion

The financial, technical and economic analysis of the proposed Jackfruit processing unit is viable and has good market potential. The tools used for the financial analysis including Pay Back Period, Net Present Value, Internal Rate of Return and Profitability Index show the worthiness and bankability. Investment in the proposed project will be worthwhile. The model business plan depicted here has not taken into account any subsidy component that is available under present schemes offered by Government. In the instance of availing subsidies meant for the purpose, it is definite that profitability of the business will improve.

Printed in the United States
by Baker & Taylor Publisher Services